Gopnik, A. (2009) *The Philosophical Baby: What Children's Minds Tell Us about Truth, Love and the Meaning of Life*. London: Bodley Head.

Gopnik, A., Meltzoff, A. N. and Kuhl, P. K. (1999) *The Scientist in the Crib: Minds, Brains, and How Children Learn*. 1st edn. New York: William Morrow & Co.

Gray, J. (2002) *Straw Dogs Thoughts on Humans and Other Animals*. London: Granta Books.

Grazzini, C. (2013) 'Maria Montessori's Cosmic Vision, Cosmic Plan, and Cosmic Education', *The NAMTA Journal*, 38(1), pp. 107–116.

Gregg, M. and Seigworth, G. J. (eds) (2010) *The Affect Theory Reader*. Durham, NC: Duke University Press.

Grosz, E. (2005) 'Chaos, Territory, Art, and the Framing of the Earth', *IDEA Journal*, 6(1), pp. 15–29.

Guattari, F. (2000) *The Three Ecologies*. London: The Athlone Press.

Hackett, A. (2017) 'Parents as Researchers: Collaborative Ethnography with Parents', *Qualitative Research*, 17(5), pp. 481–497.

Hackett, A. (2018) 'Inchoate Literacies; Leaking, Entwining, Messyways of Parents and Children Being in the World', in *Mattering, Knowing, Ethics and Care; Post-human Approaches to Parenting in Neo-liberal Times: RECE Reconceptualising Early Childhood Education Conference*. Copenhagen, Denmark.

Hackett, A., MacLure, M. and Pahl, K. (2020) 'Literacy and Language as Material Practices: Re-thinking Social Inequality in Young Children's Literacies', *Journal of Early Childhood Literacy*, 20(1), pp. 3–12.

Hackett, A., Pahl, K. and Pool, S. (2017) 'In Amongst the Glitter and the Squashed Blueberries: Crafting a Collaborative Lens for Children's Literacy Pedagogy in a Community Setting', *Pedagogies: An International Journal*, 12(1), pp. 58–73.

Hackett, A. and Rautio, P. (2019) 'Answering the World: Young Children's Running and Rolling as More-than-Human Multimodal Meaning Making', *International Journal of Qualitative Studies in Education*, 32(8), pp. 1019–1031.

Hackett, A. and Somerville, M. (2017) 'Posthuman Literacies: Young Children Moving in Time, Place and More-than-Human Worlds', *Journal of Early Childhood Literacy*, 17(3), pp. 374–391.

Halberstam, J. (2013) 'The Wild Beyond: With and for the Undercommons', in Harney, S. and Moten, F. (eds) *The Undercommons: Fugitive Planning & Black Study*. Wivenhoe: Minor Compositions.

Haraway, D. J. (1992) 'The Promises of Monsters: A Regenerative Politics for Inappropriate/d Others', in Grossberg, L., Nelson, C. and Treichler, P. A. (eds) *Cultural Studies*. New York: Routledge, pp. 295–337.

Haraway, D. J. (1997) *Modest_Witness@Second_Millennium.FemaleMan_Meets_OncoMouse: Feminism and Technoscience*. New York: Routledge.

Haraway, D. J. (2015) 'Anthropocene, Capitalocene, Plantationocene, Chthulucene: Making Kin', *Environmental Humanities*, 6(1), pp. 159–165.

Haraway, D. J. (2016) *Staying with the Trouble: Making Kin in the Chthulucene*. Durham: Duke University Press (Experimental Futures: Technological Lives, Scientific Arts, Anthropological Voices).

Harman, G. (2007) 'Aesthetics as First Philosophy: Levinas and the Non-human', *Naked Punch*, 9, pp. 21–30.

Harper, H. (1997) 'Disturbing Identity and Desire Adolescent Girls and Wild Words', in Todd, S. (ed.) *Learning Desire: Perspectives on Pedagogy, Culture, and the Unsaid*. New York: Routledge, pp. 141–161.

Hayles, N. K. (1999) *How We Became Posthuman: Virtual Bodies in Cybernetics, Literature, and Informatics*. London: University of Chicago Press.

Hayles, N. K. (2017) 'Inside Out, Outside In: Recursive Dynamics in Posthumanism and in Nam June Paik's Artwork', in International Symposium Publication, presented by Nam June Paik Art Centre *Coevolution: Cybernetics to Posthuman*. Gyeonggi-do, Korea: Nam June Paik Art Center (Nam June Paik Reader #7).

Health Service Executive. (2011) *The Child, Health, Welfare & Development Assessment Guide*. Dubin: Health Service Executive.

Hickey-Moody, A. (2016) 'A Femifesta for Posthuman Art Education: Visions and Becomings', in Taylor, C. A. and Hughes, C. (eds) *Posthuman Research Practices in Education*. New York, NY: Palgrave Macmillan.

Hickey-Moody, A. and Page, Tara (eds) (2015) *Arts, Pedagogy and Cultural Resistance: New Materialisms*. London: Rowman & Littlefield International.

Hicky-Moody, A. (2016) 'A Femifesta for Posthuman Art Education: Visions and Becomings', in Taylor, C. A. and Hughes, C. (eds) *Posthuman Research Practices in Education*. Basingstoke: Palgrave Macmillan, pp. 258–266.

Hickey-Moody, A. C. (2013) 'Deleuze's Children', *Educational Philosophy and Theory*, 45(3), pp. 272–286.

Higgins, M. (2017) 'Post-qualitative Mo(ve)ments: Concluding Remarks on Methodological Response-Abilities and Being Wounded by Thought', *Reconceptualizing Educational Research Methodology*, 8(3), pp. 89–101.

Highmore, B. (2010) 'Bitter After Taste: Affect, Food and Social Aesthetics', in Gregg, M. and Seigworth, G. J. (eds) *The Affect Theory Reader*. Durham: Duke University Press, pp. 118–137.

Hill, L. and Lyons, A. (2020) 'Strawing: Perpetual Line', in Trafí-Prats, L. and Schulte, C. M. (eds) *Language, Sense and the Event of Drawing in Childhood*. Switzerland: Springer.

Hill, L. and O'Gorman, A. (2020) 'For the Love of Small Stuff: Materialising Theory in an Early Years Artist Residency', in Hackett, A., Holmes, R. and MacRae, C.

(eds) *Working with Young Children in Museums: Weaving Theory and Practice.* London: Routledge/Taylor & Francis Group, pp. 62–70.

Hodgins, B. D. (2019) 'Caring: Method as Affect, Obligation and Action', in Hodgins, B. D. (ed) *Feminist Research for 21st-Century Childhoods Common Worlds Methods*. London: Bloomsbury Academic.

Hohti, R. and Osgood, J. (2020) 'Pets That Have "Something Inside": The Material Politics of in/Animacy and Queer Kin within the Childhood Menagerie', *Genealogy*, 4(2), p. 38.

Holmes, R. and Jones, L. (2016) 'Flickering, Spilling and Diffusing Body/Knowledge in the Posthuman Early Years', in Taylor, C. .A. and Hughes, C. (eds) Posthuman Research Practices in Education. London: Palgrave Macmillan. Available at: <https://doi.org/10.1057/9781137453082_8>.

Honan, E. and Bright, D. (2016) 'Writing a Thesis Differently', *International Journal of Qualitative Studies in Education*, 29(5), pp. 731–743.

Huizinga, J. (1949) *Homo Ludens a Study of the Play-Element in Culture.* London: Routledge & Kegan Paul.

Hultman, K. and Lenz Taguchi, H. (2010) 'Challenging Anthropocentric Analysis of Visual Data: A Relational Materialist Methodological Approach to Educational Research', *International Journal of Qualitative Studies in Education*, 23(5), pp. 525–542.

Independent Editors. (2018) *Sexual Abuse, Exploitation of Women and Paedophile Priests: The Scandals That Have Rocked the Catholic Church, Independent Newsdesk.* Irish Independent Newspaper Newsdesk, 26 Aug 2018.

INFORM. (2020) 'Act Now: Re-imagining Early Childhood Education and Care in Ireland a Discussion Paper'. Submission for consideration for the Program for Government.

Ingold, T. (2010) 'The Textility of Making', *Cambridge Journal of Economics*, 34(1), pp. 91–102.

Ingold, T. (2013) *Making: Anthropology, Archaeology, Art and Architecture.* London: Routledge.

Ingold, T. (2016) 'On Human Correspondence', *Journal of the Royal Anthropological Institute (NS)*, (23), pp. 9–27.

Ingold, T. (2020) *Correspondences.* Cambridge, UK: Polity.

Irwin, R. L. (2013) 'Becoming a/r/tography', *Studies in Art Education*, 54(3), pp. 198–215.

Irwin, R. L. and De Cosson, A. (2004) *A/r/tography Rendering Self through Arts-based Living Inquiry.* Vancouver: Pacific Educational Press.

Jackson, A. Y. (2013) 'Posthumanist Data Analysis of Mangling Practices', *International Journal of Qualitative Studies in Education*, 26(6), pp. 741–748.

Jackson, A. Y. and Mazzei, L. A. (2012) *Thinking with Theory in Qualitative Research: Viewing Data Across Multiple Perspectives.* 1st edn. Abingdon, Oxon: Routledge.

Johansson, L. (2016) 'Post-qualitative Line of Flight and the Confabulative Conversation: A Methodological Ethnography', *International Journal of Qualitative Studies in Education*, 29(4), pp. 445–466. doi: 10.1080/09518398.2015.1053157.

John, P. (2011) 'Collage: Playing with Food?', *Teaching Young Children*, 4(3), p. 5.

Jones, L, MacLure, M, Holmes, R. and Macrae, C (2012) 'Children and Objects: Affection and Infection', *Early Years*, 32(1), pp. 49–60. doi: 10.1080/09575146.2011.593029.

Jordan, D. (2015) *The Dual Identity of the Artist-Teacher: What Does Teaching Do to the Artist-Teacher in a Contemporary Educational Context?* Dublin: Dublin City University.

Juffé, M. (2007) 'Lévinas as (Mis)reader of Spinoza', in Bugni, B. (trans) *Levinas Studies*, 2, pp. 153–173.

Kellog, R. (1959) *What Children Scribble and Why*. California: NP Publishers.

Kennedy, D. (2006) *Changing Conceptions of the Child from the Renaissance to Post-modernity: A Philosophy of Childhood*. Palo Alto: Edwin Mellen.

Kind, S. (2014) 'Material Encounters', *International Journal of Child, Youth and Family Studies*, 5(4.2), pp. 865–877.

Kind, S. (2018) 'Collective Improvisations: The Emergence of the Early Childhood Studio as an Event-Full Place', in Schulte, C. M. and Thompson Marme, C. M. (eds) *Communities of Practice: Art, Play, and Aesthetics in Early Childhood.* New York, NY: Springer.

Kind, S. and Argent, A. (2017) *Pedagogical Documentation in Early Years Practice: Seeing through Multiple Perspectives.* Edited by Fleet, A., Patterson, C. and Robertson, J., 1st edn. Thousand Oaks, CA: SAGE.

Kind, S. and Argent, A. (2019) 'Fabricating: Fabric Fluidities and Studio Encounters', in Hodgins, B. D. (ed.) *Feminist Research for 21st-Century Childhoods: Common Worlds Methods*. London: Bloomsbury Academic, pp. 35–42.

Kindler, A. M. (2010) 'Art and Art in Early Childhood: What Can Young Children Learn from Art Activities?', *International Art in Early Childhood Research Journal*, 2(1), pp. 1–14.

Knight, L. (2009) 'Desire and Rhizome: Affective Literacies in Early Childhood Education', in Masny, D. and Cole, D. R. (eds) *Multiple Literacies Theory: A Deleuzian Perspective*. Rotterdam: Sense Publishers, p. 12.

Knight, L. (2013) 'Not as It Seems: Using Deleuzian Concepts of the Imaginary to Rethink Children's Drawings', *Global Studies of Childhood*, 3(3), pp. 254–264.

Knight, L. (2018) 'Digital Aesthetics and Multidimensional Play in Early Childhood', in Schulte, C. M. and Thompson, C. M. (eds) *Communities of Practice: Art, Play, and Aesthetics in Early Childhood*. Cham: Springer International Publishing (Landscapes: The Arts, Aesthetics, and Education), pp. 133–151.

Knight, L. (2019) 'Playing: Inefficiently Mapping Human and Inhuman Play in Urban Commonplaces', in Hodgins, B. D. (ed.) *Feminist Research for 21st-Century Childhoods : Common Worlds Methods*. London: Bloomsbury Academic, pp. 139–145.

Knight, L. (2021) *Inefficient Mapping: A Protocol for Attuning to Phenomena*. Santa Barbara CA: Punctum Books.

Koro-Ljungberg, M., MacLure, M. and Ulmer, J. (2018) 'D… a…t…a…, Data++, Data, and Some Problematics', in Denzin, N. K. and Lincoln, Y. S. (eds) *The Sage Handbook of Qualitative Research*. 5th edn. Los Angeles: SAGE, p. 23.

Kuby, C. R. and Rowsell, J. (2017) 'Early Literacy and the Posthuman: Pedagogies and Methodologies', *Journal of Early Childhood Literacy*, 17(3), pp. 285–296.

Kucirkova, N. and Sakr, M. (2015) 'Child–Father Creative Text-Making at Home with Crayons, iPad Collage & PC', *Thinking Skills and Creativity*, 17, pp. 59–73.

Kummen, K. (2019) 'Crowing: Co-evolving Relationships', in Hodgins, B. D. (ed.) *Feminist Research for 21st-Century* Childhoods: Common Worlds Methods. London: Bloomsbury Academic.

Lather, P. (2013) 'Methodology-21: What Do We Do in the Afterward?', *International Journal of Qualitative Studies in Education*, 26(6), pp. 634–645.

Lather, P. and Pierre, E. A. S. (2013) 'Post-qualitative research', *International Journal of Qualitative Studies in Education*, 26(6), pp. 629–633.

Latour, B. (1996) 'On Actor-Network Theory: A Few Clarifications Plus More Than a Few Complications', *Soziale Welt*, 47, pp. 369–381.

Law, J. (2004) *After Method: Mess in Social Science Research*. London: Routledge.

Lebensztejn, J.-C. and Nagy, J. (2017) *Pissing Figures 1280–2014*. New York: David Zwirner Books (Ekphrasis).

Lenz Taguchi, H. (2010) *Going Beyond the Theory/Practice Divide in Early Childhood Education: Introducing an Intra-active Pedagogy*. London: Routledge (Contesting Early Childhood).

Lenz Taguchi, H. (2011) 'Investigating Learning, Participation and Becoming in Early Childhood Practices with a Relational Materialist Approach', *Global Studies of Childhood*, 1(1), pp. 36–50. doi: 10.2304/gsch.2011.1.1.36.

Lenz Taguchi, H. (2012) 'A Diffractive and Deleuzian Approach to Analysing Interview Data', *Feminist Theory*, 13(3), pp. 265–281.

Lenz Taguchi, H. (2016) 'Deleuzo-Guattarian Rhizomatics: Mapping the Desiring Forces and Connections between Educational Practices and the Neurosciences', in Taylor, C. A. and Hughes, C. (eds) *Posthuman Research Practices in Education*. Basingstoke: Palgrave Macmillan, pp. 37–57.

Lenz Taguchi, H. and St Pierre, E. A. (2017) 'Using Concept as Method in Educational and Social Science Inquiry', *Qualitative Inquiry*, 23(9), Sage Journals. doi: org/10.1177/1077800417732634.

Lester, S. (2020) *Everyday Playfulness: A New Approach to Children's Play and Adult Responses to It*. London: Jessica Kingsley Publishers.

Lewis, S. L. and Maslin, M. A. (2015) 'Defining the Anthropocene', *Nature*, 519(7542), pp. 171–180.

Lloyd, G. (1994) *Part of Nature: Self-knowledge in Spinoza's Ethics*. Ithaca: Cornell University Press.

Lodge, A. and Lynch, K. (2004) *Diversity at School*. Dublin: Institute of Public Administration for the Equality Authority.

Lord, B. (2010) *Spinoza's Ethics: An Edinburgh Philosophical Guide*. Edinburgh: Edinburgh University Press (Edinburgh Philosophical Guides Series).

Louv, R. (2005) *Last Child in the Woods: Saving Our Children from Nature-Deficit Disorder*. 1st edn. Chapel Hill, NC: Algonquin Books of Chapel Hill.

Loveless, N. (2019) *How to Make Art at the End of the World: A Manifesto for Research-Creation*. Durham: Duke University Press.

Lucero-Montaño, A. (2012) 'Spinoza's Ethics: Freedom and Determinism', *SSRN Electronic Journal* [Preprint]. doi: 10.2139/ssrn.2010571.

Lucretius and Ferguson-Smith. (1969) *Lucretius on the Nature of Things*. London: Sphere Books.

Lynch, K., Lyons, M. and Cantillon, S. (2007) 'Breaking Silence: Educating Citizens for Love, Care and Solidarity', *International Studies in Sociology of Education*, 17(1–2), pp. 1–19.

Lyons, A. (2020) *Oona*. Dublin: Liliput Press.

MacCormack, P. (2012) *Posthuman Ethics: Embodiment and Cultural Theory* (1st edn.). Routledge. Available at: <https://doi.org/10.4324/9781315601670>.

MacCormack, P. (2018) 'Posthuman Ethics', in Braidotti, R. and Hlavjova, M. (eds) *The Posthuman Glossery*. London: Bloomsbury Academic.

MacFarlane, R. (2016) 'Generation Anthropocene: How Humans Have Altered the Planet Forever', *The Guardian*, 1 April. Available at: <http://www.theguardian.com.books/2016/apr/01/generation-anthropocene-altered-planet-for-ever>.

MacLure, M., Holmes, R., Jones, L. and MacRae, C. (2010) 'Silence as Resistance to Analysis: Or, on Not Opening One's Mouth Properly', *Qualitative Inquiry*, 16(6), pp. 492–500.

MacLure, M. (2013a) 'Classification or Wonder: Coding as an Analytic Practice in Qualitative Research', in Coleman, R. and Ringrose, J. (eds) *Deleuze and Research Methodologies*. Edinburgh: Edinburgh University Press.

MacLure, M. (2013b) 'Researching without Representation? Language and Materiality in Post-qualitative Methodology', *International Journal of Qualitative Studies in Education*, 26(6), pp. 658–667.

MacRae, C. (2008) 'Representing Space: Katie's Horse and the Recalcitrant Object', *Contemporary Issues in Early Childhood*, 9(4), pp. 275–286.

MacRae, C. (2011) 'Making Payton's Rocket: Heterotopia and Lines of Flight', *International Journal of Art & Design Education*, 30(1), pp. 102–112.

MacRae, C. (2012) 'Encounters with a Life(Less) Baby Doll: Rethinking Relations of Agency through a Collectively Lived Moment', *Contemporary Issues in Early Childhood*, 13(2), pp. 120–131.

Macrae, C., Hackett, A., Holmes, R. and Jones, L. (2017) 'Vibrancy, Repetition and Movement: Posthuman Theories for Reconceptualising Young Children in Museums', *Children's Geographies*, 16(1), pp. 1–13.

MacRae, C. (2019) 'The Red Blanket: A Dance of Animacy', *Global Studies of Childhood*, 12(4), pp. 1–11.

MacRae, C. (2020) 'Tactful Hands and Vibrant Mattering in the Sand Tray', *Journal of Early Childhood Literacy*, 20(1), pp. 90–110.

MacRae, C. and Jones, L. (2020) 'A Philosophical Reflection on the "Leuven Scale" and Young Children's Expressions of Involvement', *International Journal of Qualitative Studies in Education*, 36(4), 36(4), pp. 1–13.

Malafouris, L. (2013) *How Things Shape the Mind: A Theory of Material Engagement*. Cambridge, MA: MIT Press.

Malaguzzi, L. (1993) 'For an Education Based on Relationships', *Young Children*, 49(1), pp. 9–12.

Malaguzzi, L. (1994) 'A Bill of Three Rights', *Innovations in Early Education: The International Reggio Exchange*, 2(1). Available at: <https://www.reggioalliance.org/downloads/v2.n1.rights.pdf>.

Malaguzzi, L. (1997) 'The Hundred Languages of Children', in Filippini, T. and Vecchi, V. (eds) *The Hundred Languages of Children*. Reggio Emilia, Italy: Reggio Children.

Malaguzzi, L. (2000) *The One Hundred Languages of Children*. Reggio Emilia: Reggio Children.

Malone, K. (2018) 'Reconfiguring the Child in the Anthropocene', in Malone, K. (ed.) *Children in the Anthropocene*. London: Palgrave Macmillan UK, pp. 247–257.

Malone, K. (2019) 'Re-turning Childhoodnature: A Diffractive Account of the Past Tracings of Childhoodnature as a Series of Theoretical Turns', in Latiner Raby,

R. and Valeau, E. J. (eds) *Handbook of Comparative Studies on Community Colleges and Global Counterparts*. Cham: Springer International Publishing (Springer International Handbooks of Education), pp. 1–30.

Manning, E. (2016) *The Minor Gesture*. Durham: Duke University Press (Thought in the Act).

Marmé Thompson, C. (2019) '"So You Will Remember Me as an Artist": Art-Making as a Way of Being in Early Childhood', in Sakr, M. and Osgood, J. (eds) *Postdevelopmental Approaches to Childhood Art*. London: Bloomsbury Academic, pp. 137–152.

Masny, D. (2018) 'Becoming Childhoodnature: Experimenting a Research Assemblage', in Cutter-Mackenzie, A., Malone, K. and Barratt Hacking, E. (eds) *Research Handbook on Childhoodnature*. Cham: Springer International Publishing, pp. 1–18.

Masny, D. and Cole, D. R. (2009) *Multiple Literacies Theory: A Deleuzian Perspective*. Rotterdam: Sense Publishers.

Massumi, B. (2002) *Parables for the Virtual: Movement, Affect, Sensation*. Durham, NC: Duke University Press.

Mayo, E. (1832) *Lessons on Objects as Given to Children between the Ages of Five and Eight, in a Pestalozzian School, at Cheam, Surrey*. 3rd edn. London: R. B. Seeley and W. Burnside.

Mazzei, L. A. (2013) 'A Voice without Organs: Interviewing in Posthumanist Research', *International Journal of Qualitative Studies in Education*, 26(6), pp. 732–740.

McCarthy, S. (2016) *Cost of Child Abuse Inquiry and Redress*. Comptroller and Auditor General, Special Report 96. Dublin, Ireland.

McClure, M. (2011) 'Child as Totem: Redressing the Myth of Inherent Creativity in Early Childhood', *Studies in Art Education*, 52(2), pp. 127–141.

McGinnity, F., Murray, A. and McNally, S. (2013) *Growing Up in Ireland – National Longitudinal Study of Children: Mothers' Return to Work and Childcare Choices for Infants in Ireland*. Dublin: Stationery Office.

McMillan, M. (1927) *The Life of Rachel McMillan*. London: J. M. Dent.

Mitchelmore, S. and Fleet, A. (2017) 'The Worlds of the Very Young: Seeing the Everyday in Small Pieces', in Fleet, A., Patterson, C. and Robertson, J. (eds) *Pedagogical Documentation in Early Years Practice: Seeing through Multiple Perspectives*. 1st edn. Thousand Oaks, CA: SAGE, pp. 61–72.

Montessori, M. (2014) *The Montessori Method*. New Brunswick, USA: Transaction Publishers.

Murris, K. (2016) *The Posthuman Child: Educational Transformation through Philosophy with Picture Books*. London: Routledge, Taylor & Francis Group.

Murris, K. (2017) 'Reading Two Rhizomatic Pedagogies Diffractively Through One Another: A Reggio Inspired Philosophy with Children for the Postdevelopmental Child', *Pedagogy, Culture & Society*, 25(4), pp. 1–20.

Myers, C. Y., Smith, K. A. and Tesar, M. (2017) 'Diffracting Mandates for Reflective Practices in Teacher Education and Development: Multiple Readings from Australia, New Zealand, and the United States', *Journal of Early Childhood Teacher Education*, 38(4), pp. 279–292.

Nancy, J.-L. (2015) *After Fukushima: The Equivalence of Catastrophes*. Translated by C. Mandell. New York: Fordham University Press.

NCCA. (2009) *Aistear: The Early Childhood Curriculum Framework*. Dublin: National Council for Curriculum and Assessment. Available at: <https://www.ncca.ie/media/1112/how-aistear-was-developed-research-papers.pdf>

NCCA. (2012) *Literacy in Early Childhood and Primary Education*. Dublin: National Council for Curriculum and Assessment.

NESF. (2005) *Early Childhood Care and Education*. Dublin: National Council for Curriculum and Assessment.

Nutbrown, C. (1994) *Threads of Thinking: Young Childrens Learning and the Role of Early Education*. London: Paul Chapman Pub.

Nutbrown, C., Clough, P. and Selbie, P. (2008) *Early Childhood Education: History, Philosophy, Experience*. Los Angeles: SAGE.

Odegard, N. (2012) 'When Matter Comes to Matter: Working Pedagogically with Junk Materials', *Education Inquiry*, 3(3), pp. 387–400.

Odegard, N. and Rossholt, N. (2016) 'In-betweens Spaces: Tales from a Remida', in Reinert, A. B. (ed.) *Becoming Earth: A Posthuman Turn in Educational Discourse Collapsing Nature/Culture Divides*. Rotterdam: Sense Publishers, pp. 53–63.

O'Donnell, A. (2015) 'Experimental Philosophy and Experimental Pedagogy: A Single Vision', in Hickey-Moody, A. and Page, Tara (eds) *Arts, Pedagogy and Cultural Resistance New Materialisms*. London: Littlefield International.

O'Donnell, A. (2016) 'For the Scavengers and the Gleaners: An Anti-heroic Vision of Education', *Other Education: The Journal of Educational Alternatives*, 5(2), pp. 17–38.

O'Donnell, A. (2017) 'Spinoza, Experimentation and Education: How Things Teach Us', *Educational Philosophy and Theory*, 50(9), pp. 819–829.

OECD. (2012) PISA 2009 Technical Report, PISA, OECD Publishing. Available at: <http://dx.doi.org/10.1787/9789264167872-en>.

Olsson, L. M. (2009) *Movement and Experimentation in Young Children's Learning: Deleuze and Guattari in Early Childhood Education*. London: Routledge (Contesting Early Childhood).

O'Neill-Butler, L. (2016) *Donna J. Haraway Speaks about Her Latest Book, Artforum.* Available at: <https://www.artforum.com/interviews/donna-j-haraway-speaks-about-her-latest-book-63147>

Osgood, J. (2019) '"You Can't Separate It from Anything": Glitter's Doings as Materialized Figurations of Childhood (and) Art', in *Postdevelopmental Approaches to Childhood Art*, . London: Bloomsbury Academic, pp. 111–136.

Osgood J. (2022) 'New Images of Thought in the Study of Childhood Drawing', in Trafí-Prats, L. and Schulte, C. M. (eds.) *Children: Global Posthumanist Perspectives and Materialist Theories*. Springer Nature: Switzerland, p. 176. Available at: <https://doi.org/10.1007/978-3-031-07143-0>.

O'Sullivan, S. (2006) *Art Encounters Deleuze and Guattari: Thoughts Beyond Representation.* London: Palgrave.

Pacini-Ketchabaw, V., Clark, V. and Hodgins, B. D. (2014) 'Thinking with Paint: Troubling Settler Colonialisms through Early Childhood Art Pedagogies', *International Journal of Child, Youth and Family Studies*, 5(4.2), pp. 751–781.

Pacini-Ketchabaw, V., Kind, S. and Kocher, L. L. M. (2017) *Encounters with Materials in Early Childhood Education*. New York: Routledge.

Pacini-Ketchabaw, V., Taylor, A. and Blaise, M. (2016) 'Decentring the Human in Multispecies Ethnographies', in C. A. and Hughes, C. (eds) *Posthuman Research Practices in Education*. Basingstoke: Palgrave Macmillan, pp. 149–167.

Page, J. (2018) 'Love, Care and Intimacy in Early Childhood Education and Care', *International Journal of Early Years Education*, 26(2), pp. 123–124. doi: 10.1080/09669760.2018.1459509.

Page, T. (2018) 'Teaching and Learning with Matter', *Arts*, 7(4), p. 82.

Papatheodorou, T. and Moyles, J. (2009) *Learning Together in the Early Years Exploring Relational Pedagogy*. London: Routledge.

Parikka, J. (2015) *The Anthrobscene*. Minneapolis: University of Minnesota Press.

Pedersen, H. and Pini, B. (2017) 'Educational Epistemologies and Methods in a More-than-Human World', *Educational Philosophy and Theory*, 49(11), pp. 1051–1054.

Pestalozzi, J. H. (2008) *How Gertrude Teaches Her Children*. Edited by Holland, L. E. and Turner, F. C. Translated by Cooke, E. Whitefish, MT: Kessinger Publishing.

Petersen, E. B. (2018) '"Data found us": A Critique of Some New Materialist Tropes in Educational Research', *Research in Education*, 101(1), pp. 5–16.

Piaget, J. (1952) *The Origin of Intelligence in Children*. Translated by Cook, M. New York: International Universities Press.

Piaget, J. (1971) *Mental Imagery in the Child: A Study of the Development of Imaginal Representation*. London: Routledge & Kegan Paul.

Pine, L. (1996) 'The Dissemination of Nazi Ideology and Family Values through School Textbooks', *History of Education*, 25(1), pp. 91–109.

Pope Edwards, C. (2002) 'Three Approaches from Europe: Waldorf, Montessori, and Reggio Emilia.', *Early Childhood Research & Practice*, 4(1), pp. 1–24.

Pope Edwards, C. (2003) '"Fine Designs" from Italy: Montessori Education and the Reggio Approach', *Faculty Publications, Department of Child, Youth, and Family Studies*, 20, p. 7.

Provenzo, E. F. (2009) 'Friedrich Froebel's Gifts: Connecting the Spiritual and Aesthetic to the Real World of Play and Learning', *American Journal of Play*, Summer, 2(1), pp. 85–99.

Puig de la Bellacasa, M. (2017) *Matters of Care: Speculative Ethics in More than Human Worlds*. Minneapolis: University of Minnesota Press (Posthumanities, 41).

Rautio, P. (2013) 'Children Who Carry Stones in Their Pockets: On Autotelic Material Practices in Everyday Life', *Children's Geographies*, 11(4), pp. 394–408.

Rautio, P. (2019) 'Theory That Cats Have about Swift Louseflies: A Diffractive Response', in Kuby, C. R., Spector, K. and Thiel, J. (eds) *Posthumanism and Literacy Education: Knowing/Becoming/Doing Literacies*. New York: Routledge (Expanding Literacies in Education Series).

Rautio, P. and Jokinen, P. (2015) 'Children's Relations to the More-Than-Human World Beyond Developmental Views', in Evans, B., Horton, J. and Skelton, T. (eds) *Play, Recreation, Health and Well Being*. Singapore: Springer, pp. 1–15.

Reggio Children. (2005) 'Remida: The Creative Recycling Centre in Reggio Emilia: An Interview with Elena Giacopini, Graziella Brighenti, Arturo Bertoldi and Alba Ferrari', *Innovations in Early Education*, 12(3), pp. 1–13.

Rekret, P. (2016) 'A Critique of New Materialism: Ethics and Ontology', *Subjectivity*, 9(3), pp. 225–245.

Richards, J. M. (1976) *Provision for the Arts: Report of an Enquiry Carried Out During 1974–75 Throughout the Twenty-Six Counties of the Republic of Ireland*. The Arts Council (Ireland) and the Gulbenkian Foundation.

de Rijke, V. (2019) '"It Might Get Messy, or Not Be Right": Scribble as Postdevelopmental Art', in Sakr, M. and Osgood, J. (eds) *Postdevelopmental Approaches to Childhood Art*. London: Bloomsbury.

Rinaldi, C. (2006) *In Dialogue with Reggio Emilia: Listening, Researching and Learning*. London: Routledge.

Rogers, Y., French, G. and EarlyArts UK (2013) *Early Childhood Arts Three Perspectives*. Dublin: The Arts Council.

Rooney, T. (2019) 'Sticking: Children and the Lively Matter of Sticks', in Hodgins, B. D. (ed.) *Feminist Research for 21st Century Childhoods*. London: Bloomsbury Academic, pp. 43–51.

Rotas, A. (2015) 'Ecologies of Practice: Teaching and Learning against the Obvious', in Snaza, N. and Weaver, J. A. (eds) *Posthumanism and Educational Research*. New York: Routledge, pp. 91–104.

Rotas, N. (2019) 'Mashing: A Practice that Makes Vision Felt', in Hodgins, B. D. (ed.) *Feminist Research for 21st-Century Childhoods: Common Worlds Methods*. London: Bloomsbury Academic.

Rousell, D. et al. (2018) 'Propositions for an Environmental Arts Pedagogy: A/r/tographic Experimentations with Movement and Materiality', in Latiner Raby, R. and Valeau, E. J. (eds) *Handbook of Comparative Studies on Community Colleges and Global Counterparts*. Cham: Springer International.

Sakr, M. and Osgood, J. (eds) (2019) *Postdevelopmental Approaches to Childhood Art*. London: Bloomsbury.

Sandvik, N. (2010) 'The Art of/in Educational Research: Assemblages at Work', *Reconceptualizing Educational Research Methodology*, 1(1), pp. 29–40.

Schulte, C. M. (2019) 'The Untimely Death of a Bird: A Posthuman Tale', in Kuby, C. R., Spector, K. and Thiel, J. (eds) *Posthumanism and Literacy Education*. New York: Routledge.

Schwarz, T. and Luckenbill, J. (2012) 'Let's Get Messy! Exploring Sensory and Art Activities with Infants and Toddlers', *Young Children*, 67(4), pp. 26–34.

Sehgal, M. (2014) 'Diffractive Propositions: Reading Alfred North Whitehead with Donna Haraway and Karen Barad', *Parallax*, 20(3), pp. 188–201.

Seigworth, G. J. and Gregg, M. (2010) 'An Inventory of Shimmers', in Gregg, M. & Seigworth, M. (eds) *The Affect Theory Reader*. Durham: Duke University Press.

Sellers, M. (2015) '… Working with (a) Rhizoanalysis … and Working (with) a Rhizoanalysis', *Complicity: An International Journal of Complexity and Education*, 12(1), pp. 6–31.

Sellers, M. A. (2009) 'Re(con)ceiving Children in Curriculum: Mapping (a) Milieu(s) of Becoming'. Available at: <https://espace.library.uq.edu.au/view/UQ:180106>

Sengupta, P. (2003) 'An Object Lesson in Colonial Pedagogy', *Comparative Studies in Society and History*, 45(1), pp. 96–121.

Sheldrake, M. (2020) *Entangled Life: How Fungi Make Our Worlds, Change Our Minds, and Shape Our Futures*. London: The Bodley Head.

Shields, R. (2015) 'Ludic Ontology Play's Relationship to Language, Cultural Forms, and Transformative Politics', *American Journal of Play*, 7(3), pp. 298–321.

Shotter, J. (2015) 'On Being Dialogical: An Ethics of "Attunement"', *Context*, 137, p. 4.

Sinner, A. (2017) 'Cultivating Researchful Dispositions: A Review of A/r/tographic Scholarship', *Journal of Visual Art Practice*, 16(1), pp. 1–22.

Snaza, N. (2013) 'Bewildering Education', *Journal of Curriculum and Pedagogy*, 10(1), pp. 38–54.

Snaza, N., Appelbaum, P., Bayne, S., Morris, M., Rotas, N., Sandlin, J., Wallin, J., Carlson, D. and Weaver, J. (2014) 'Toward a Posthumanist Education', *Journal of Curriculum Theorizing*, 30(2), p. 17.

Snaza, N. (2019) *Animate Literacies: Literature, Affect, and the Politics of Humanism*. Durham: Duke University Press (Thought in the Act).

Snaza, N. and Weaver, J. A. (2016) *Posthumanism and Educational Research*. New York: Routledge.

Soëtard, M. (1994) 'Johann Heinrich Pestalozzi', *Prospects: The Quarterly Review of Comparative Education (UNESCO: International Bureau of Education)*, XXIV(1/2), pp. 297–310.

Somerville, M. (2016) 'The Post-human I: Encountering "data" in New Materialism', *International Journal of Qualitative Studies in Education*, 29(9), pp. 1161–1172.

Somerville, M. (2018) 'Posthuman Theory and Practice in Early Years Learning', in Cutter-Mackenzie, A., Malone, K. and Barratt Hacking, E. (eds) *Research Handbook on Childhoodnature*. Cham: Springer (Springer International Handbooks of Education), pp. 1–25.

Sorensen, E. (2009) *The Materiality of Learning: Technology and Knowledge in Educational Practice*. New York: Cambridge University Press.

Spector, K., Thiel, J. and Kuby, C. R. (2019) 'Affect', in Kuby, C. R. (ed.) *Posthumanism and Literacy Education: Knowing/Becoming/Doing Literacies*. New York: Routledge (Expanding Literacies in Education Series).

Spinoza, B. (1996) *The Ethics*. Translated by Curley, E. London: Penguin.

Springay, S. (2016) '"Approximate-Rigorous Abstractions" Propositions of Activation for Posthumanist Research in Education', in Snaza, N. and Weaver, J. A. (eds) *Posthumanism and Educational Research*. New York, NY: Routledge, pp. 76–88.

Springay, S., Irwin, R. L. and Kind, S. (2008) 'A/R/Tographers and Living Inquiry', in Knowles, J. G. & Cole, A. L. (eds) *Handbook of the Arts in Qualitative Research: Perspectives, Methodologies, Examples, and Issues*. Los Angeles: SAGE, p. chap. 7.

St Pierre, E. A. (2011) 'Post Qualitative Research: The Critique and the Coming After', in Denzin, N. K. and Lincoln, Y. S. (eds) *Sage Handbook of Qualitative Inquiry*. 4th edn.Los Angeles: SAGE, pp. 611–635.

St Pierre, E. A. (2013) 'The Posts Continue: Becoming', *International Journal of Qualitative Studies in Education*, 26(6), pp. 646–657.

St Pierre, E. A. (2014) 'A Brief and Personal History of Post Qualitative Research: Toward "post inquiry"', *Journal of Curriculum Theorizing*, 30(2), pp. 2–19.

St Pierre, E. A. (2017) 'Deleuze and Guattari's Language for New Empirical Inquiry', *Educational Philosophy and Theory*, 49(11), pp. 1080–1089.

Steinhauer. (2017) *How Mierle Laderman Ukeles Turned Maintenance Work into Art, Hyperallergic*. Available at: <https://hyperallergic.com/355255/how-mierle-laderman-ukeles-turned-maintenance-work-into-art/>

Stengers, I. (2005) 'An Ecology of Practices', *Cultural Studies Review*, 11(1), pp. 183–196.

Stewart, K. (2007) *Ordinary Affects*. Durham, NC: Duke University Press.

Strom, K. J. (2020) *Learning from a 'Lost Year': An Autotheoretical Journey through Anxiety and Panic, Capacious Journal for Emerging Affect Inquiry*. Edited by Seigworth, G. J. and Arthur, M., 1(2), pp. 127–148. Available at: <http://capaciousjournal.com/article/learningfromalostyear/>

Sutton-Smith, B. (2006) 'Play and Ambiguity', in Salen, K. and Zimmerman, E. (eds) *The Game Design Reader a Rules of Play Anthology*. Cambridge, MA: The MIT Press, pp. 296–313.

Swim, T. J. and Freeman, R. (2004) 'Viewpoint. Time to Reflect: Using Food in Early Childhood Classrooms', *Young Children*, 59(6), pp. 18–22.

Syvitski, J. (2012) *Anthropocene: An Epoch of Our Making – IGBP, Global Change IGBP*. Available at: <http://www.igbp.net/news/features/features/anthropoceneanepochofourmaking.5.1081640c135c7c04eb480001082.html>.

Tarr, P. (1990) 'More than Movement: Scribbling Reassessed', *Visual Arts Research*, 16(1), pp. 83–89.

Tarr, P. (2001) 'Aesthetic Codes in Early Childhood Classrooms: What Art Educators Can Learn from Reggio Emilia', *Art Education*, 54(3), p. 33. doi: 10.2307/3193922.

Taylor, A. (2013) *Reconfiguring the Natures of Childhood*. London: Routledge.

Taylor, A. and Pacini-Ketchabaw, V. (2019) *The Common Worlds of Children and Animals: Relational Ethics for Entangled Lives*. London: Routledge.

Taylor, C. A. (2016) 'Edu-crafting a Cacophonous Ecology: Posthuman Research Practices for Education', in Taylor, C. A. and Hughes, C. (eds) *Posthuman Research Practices in Education*. Basingstoke: Palgrave Macmillan, pp. 5–24.

Taylor, C. A. (2018) 'What Can Bodies Do? En/gendering Body-Space Choreographies of Stillness, Movement and Flow in Post-16 Pedagogic Encounters', *International Journal of Educational Research*, 88, pp. 156–165.

Taylor, C. A. and Hughes, C. (eds) (2016) *Posthuman Research Practices in Education*. Basingstoke: Palgrave Macmillan.

Thiel, J., Kuby, C. R. and Spector, K. (2019) 'Agency', in Kuby, C. R., Spector, K. and Thiel, J. (eds) *Affect in Literacy Learning and Teaching*. New York: Routledge.

Thompson, C. M. (2015) 'Prosthetic Imaginings and Pedagogies of Early Childhood Art', *Qualitative Inquiry*, 21(6), pp. 554–561.

Thorp, A. (2020) 'On What Grounds', in Hackett, A., Holmes, R. and MacRae, C. (eds) *Working with Young Children in Museums: Weaving Theory and Practice*. London: Routledge/Taylor & Francis Group, pp. 187–194.

Todd, S. (2001) '"Bringing More Than I Contain": Ethics, Curriculum and the Pedagogical Demand for Altered Egos', *Journal of Curriculum Studies*, 33(4), pp. 431–450.

Todd, S. (2016) 'New Ethical Challenges within Environmental and Sustainability Education: A Response', *Environmental Education Research*, 22(6), pp. 842–844.

Todd, S., Jones, R. and O'Donnell, A. (2016) 'Shifting Education's Philosophical Imaginaries: Relations, Affects, Bodies, Materialities', *Gender and Education*, 28(2), pp. 187–194.

Todd, Z. (2015) 'Indigenizing the Anthropocene', in *Art in the Anthropocene: Encounters among Aesthetics, Politics, Environments and Epistemologies*. London: Open Humanities Press, pp. 241–255.

Tooze, A. (2020) 'We Are Living through the First Economic Crisis of the Anthropocene', *The Guardian*. Available at: <https://amp.theguardian.com/books/2020/may/07/we-are-living-through-the-first-economic-crisis-of-the-anthropocene>

Trafí-Prats, L. (2017) 'Learning with Children, Trees, and Art: For a Compositionist Visual Art-Based Research', *Studies in Art Education*, 58(4), pp. 325–334.

Trafí-Prats, L. (2019) 'Thinking Childhood Art with Care in an Ecology of Practices', in Osgood, J. and Sakr, M. (eds) *Postdevelopmental Approaches to Childhood Art*. London: Bloomsbury Academic, pp. 191–209.

Trafí-Prats, L. and Caton, L. (2020) 'Towards an Ethico-Aesthetic of Parenting: Sensing Ritornellos of Play with GoPro Data', *Genealogy*, 4(2), p. 34.

Tsing, A. L. (2015) *The Mushroom at the End of the World: On the Possibility of Life in Capitalist Ruins*. Princeton, NJ: Princeton University Press.

Ulmer, J. B. (2017) 'Posthumanism as Research Methodology: Inquiry in the Anthropocene', *International Journal of Qualitative Studies in Education*, 30(9), pp. 832–848.

UNESCO. (2006) *World Conference on Arts Education: Building Creative Capacities for the 21st Century*. Lisbon: Division of Arts and Cultural Enterprise of Sector for Culture.

Urban, M. (2016) 'At Sea: What Direction for Critical Early Childhood Research?', *Journal of Pedagogy*, 7(1), pp. 107–121.

Urban, M. (2018) '(D)Evaluation of Early Childhood Education and Care? A Critique of the OECD's International Early Learning Study', in Matthes, M. et al. (eds) *Improving the Quality of Childhood in Europe*. Brussels: Alliance for Children European Network Foundation, pp. 91–99.

Van Dooren, T. (2016) *Flight Ways: Life and Loss at the Edge of Extinction*. Paperback edition. New York: Columbia University Press (Critical Perspectives on Animals Theory, Culture, Science, and Law).

Vecchi, V., Chieli, F. and Reggio Children Srl (eds) (2008) *Children, Art, Artists: The Expressive Languages of Children, the Artistic Language of Alberto Burri ; [catalogue]*. 3. print. Reggio Emilia: Reggio Children.

Vehaba, A. (2018) 'Traversing Sites of Affective Immersion', *Capacious Journal for Emerging Affect Inquiry*. Edited by Seigworth, G. J. and Arthur, M., 1(2), pp. 127–148.

Vice News. (2017) *This Trans Artist Collected 200 Gallons of Urine to Protest Trump* [You tube short]. Vice News. Available at: <https://www.youtube.com/watch?v=p-TT7GoJ2iw&t=11s>.

Vygotsky, L. S. (1978a) 'Interaction between learning and development.' From: *Mind and Society* (pp. 79–91) Cambridge, M.A. Harvard University Press.

Vygotsky, L. S. (1978b) *Mind in Society: The Development of Higher Psychological Processes*. Translated by Cole, M. Cambridge: Harvard University Press.

Vygotsky, L. S. (1990) 'Imagination and Creativity in Childhood', *Journal of Russian and East European Psychology*, 28(1), pp. 84–96. doi: 10.2753/RPO1061-0405280184.

Weaver, J. A. and Snaza, N. (2017) 'Against Methodocentrism in Educational Research', *Educational Philosophy and Theory*, 49(11), pp. 1055–1065. doi: 10.1080/00131857.2016.1140015.

Wilson, B. (2003) 'Three Sites for Visual Cultural Pedagogy: Honoring Students Interests and Imagery', in *The International Journal of Arts Education*, 3(1).

Wilson, B. (2007) 'Art, Visual Culture, and Child/Adult Collaborative Images: Recognizing the Other-than.', *Visual Arts Research*, 33(65), pp. 6–20.

Wilson, F. R. (1999) *The Hand: How Its Use Shapes the Brain, Language, and Human Culture*. New York: Vintage Books.

Winner, E., Goldstein, T. R. and Vincent-Lancrin, S. (2013) *Art for Art's Sake?: The Impact of Arts Education*. Paris: OECD (Educational Research and Innovation).

Wolfe, C. (2010) *What Is Posthumanism?* London: University of Minnesota Press (Posthumanities; 8). Available at: <http://capitadiscovery.co.uk/ncad/items/63808>

Yunkaporta, T. (2019) *Sand Talk: How Indigenous Thinking Can Save the World*. Melbourne, VIC: Text Publishing.

Zuckerman, O. (2006) *Historical Overview and Classification of Traditional and Digital Learning Objects*. MIT Media Lab.

Index

New Disciplinary Perspectives on Education

Series Editors

Stephen Cowden is a Social Work educator and researcher with an interest in the relationships between pegagogy, social justice and equality.

Jones Irwin, Associate Professor in Philosophy and Education, Dublin City University

Published Volumes

Vol 1 Stephen Cowden and David Ridley (eds): The Practice of Equality: Jacques Rancière and Critical Pedagogy. 2019. 206 pages.
ISBN 978-1-78874-029-6

Vol 2 David Ridley: The Method of Democracy: John Dewey's Theory of Collective Intelligence. 2021. 230 pages.
ISBN 978-1-78997-337-2

Vol 3 Marina Avelar: Disrupting Education Policy: How New Philanthropy Works to Change Education. 2020. 222 pages.
ISBN 978-1-78707-688-4

Vol 4 Jones Irwin and Letterio Todaro (eds): Paulo Freire's Philosophy of Education in Contemporary Context: From Italy to the World. 2022. 216 pages.
ISBN 978-1-80079-689-8

Vol 5 Gordon Asher, Stephen Cowden, Shirin Housee and Alpesh Maisuria (eds): Critical Pedagogy and Emancipation: A Festschrift in Memory of Joyce Canaan. 2022. 400 pages.
ISBN 978-1-80079-692-8

Vol 6 Susan Bailey: The Levinasian Teacher. 2023. 188 pages.
ISBN 978-1-80079-781-9

Vol 7 Lucy Hill: Everyday Artfulness: A posthuman perspective on young children's play with materials. 2024. 262 pages.
ISBN 978-1-80079-784-0

www.ingramcontent.com/pod-product-compliance
Ingram Content Group UK Ltd.
Pitfield, Milton Keynes, MK11 3LW, UK
UKHW020657041026
15590UKWH00001B/327

9 781800 797840

Terotechnology

10th Conference on Terotechnology, 18-19 October 2017, Kielce, Poland

Agnieszka Szczotok[1], Jacek Pietraszek[2], Norbert Radek[3]

[1] Silesia University of Technology,
agnieszka.szczotok@polsl.pl, http://www.polsl.pl

[2] Cracow University of Technology, Faculty of Mechanical Engineering, Institute of Applied Informatics
jacek.pietraszek@mech.pk.edu.pl, http://www.mech.pk.edu.pl

[3] Kielce University of Technology, Centre for Laser Technologies of Metals
norrad@tu.kielce.pl, http://www.tu.kielce.pl

Peer review statement

All papers published in this volume of "Materials Research Proceedings" have been peer reviewed. The process of peer review was initiated and overseen by the above proceedings editors. All reviews were conducted by expert referees in accordance to Materials Research Forum LLC high standards.

Published under License by **Materials Research Forum LLC**
Millersville, PA 17551, USA

Published as part of the proceedings series
Materials Research Proceedings
Volume 5 (2018)

ISSN 2474-3941 (Print)
ISSN 2474-395X (Online)

ISBN 978-1-945291-80-7 (Print)
ISBN 978-1-945291-81-4 (eBook)

Distributed worldwide by

Materials Research Forum LLC
105 Springdale Lane
Millersville, PA 17551
USA
http://www.mrforum.com

Manufactured in the United State of America
10 9 8 7 6 5 4 3 2 1

Table of Contents

Preface

Terotechnology is the technology of installation, commissioning, maintenance, replacement and removal of plant machinery and equipment, of feedback on operation and design thereof, and on related subjects and practices.

It has been sixteen years since the first Conference on Terotechnology was held. The conference is still a venture of Kielce University of Technology, Polish Maintenance Society, Kielce Trade Fairs and Polish Society for Stereology. The first conference was held in 2001 and subsequent meetings were held in 2005-2009 in the annual cycle. In 2011, a two-year conference cycle was established.

The 10th International Conference on Terotechnology was held at the Kielce Trade Fair Centre, Kielce, POLAND on 18-19 October, 2017. Kielce is situated halfway between Kraków and Warsaw. The selected papers from the 10th Conference, having been reviewed by at least two experts, were prepared for this volume of *Materials Research Proceedings*. The papers were chosen on the basis of their quality and relevancy to the conference. The volume represents the recent advances in materials, technologies and methods.

Dozens of people meet at our conference: experienced researchers, young scientists and many practitioners from industry. A special feature of our Conference is a broad formula resulting from the extent of the concept of "terotechnology".

The aim of the conference is to present results and exchange experiences of research work in the area of broadly defined use of machinery. The conference is addressed to academic staff, research institutes, manufacturing and transportation, as well as the local government administration. The conference covers all areas of exploitation, namely: the theories of exploitation, industrial management, reliability, technical diagnostics, materials science, design of experiments, tribology and technical safety.

The Organizing Committee is grateful to deans and workers of the Centre for Laser Technologies of Metals of Kielce University of Technology, Faculty of Mechanical Engineering of Cracow University of Technology who helped with the organization of the 10th Conference and the edition of this volume. The Committee would also like to express their gratitude to all who have updated and reviewed the papers submitted to the conference, the scientific secretary for her editing work, and Materials Research Forum LLC for producing the volume.

Conference Organizers

Kielce University of Technology,
Polish Maintenance Society,
Kielce Trade Fairs,
Polish Society for Stereology

Organizing Committee

Bogdan Antoszewski – chairman
Mieczysław Scendo – co-chairman
Augustin Sládek – co-chairman
Ivo Hlavaty – co-chairman
Norbert Radek – scientific secretary
Dariusz Gontarski – organizing secretary
Jacek Pietraszek
Aneta Gądek-Moszczak
Agnieszka Szczotok
Piotr Sęk
Piotr Kurp
Izabela Pliszka
Szymon Tofil
Hubert Danielewski
Marta Kubi

Simulation of Cavitation Erosion in a Hydraulic Valve

DOMAGAŁA Mariusz[1, a *], MOMENI Hassan[2,b], DOMAGALA-FABIS Joanna[1,c]
FILO Grzegorz[1,d], KWIATKOWSKI Dominik[1,e]

[1]Cracow University of Technology, Institute of Applied Informatics, Al. Jana Pawla II 37
31-841 Krakow, Poland

[2]Western Norway University of Applied Sciences, Department of Mechanical and Marine Engineering, Postboks 7030, 5020 Bergen, Norway

[a]domagala@mech.pk.edu.pl, [b]Hassan.Momeni@hvl.no, [c]fabis@mech.pk.edu.pl,
[d]filo@mech.pk.edu.pl, [e]dkwiatkowski@pk.edu.pl

Keywords: Cavitation Erosion, CFD Simulation, Non-return Valve

Abstract. This paper presents a numerical simulation of cavitation of a non-return hydraulic valve. Cavitation is one of the most common reasons of hydraulic system failures or malfunctioning. Due to complex geometry and flow path which may rapidly change, hydraulic valves are exposed to a cavitation phenomenon. Additionally, hydraulic fluid always includes contaminants including dissolver air, which may accelerate nucleation formation. Despite the fact that cavitation is not a fully discovered phenomenon, an attempt of simulation of cavitation process was taken using computational fluid dynamics (CFD) tools.

Introduction

Cavitation as a phenomenon has been studied from the early beginning of XX century, started by Rayleigh. Despite extensive research and numerous investigations on cavitation, it has not been completely discovered yet. The most common explanation of this phenomenon is that cavitation arises during fluid flow when pressure falls below critical values, which is called saturation pressure. It might be observed during internal flow when fluids gather locally velocity causing local pressure drop where bubbles of fluid vapor and air are created. These bubbles collapse when fluid reaches areas with higher pressure causing bubbles implosions and high peak pressure around bubbles surroundings. Implosions and high pressure peaks lead to noise, mechanical vibrations and surface erosion. Attempts of mathematical description of cavitation and nucleation formation have been conducted for many years [1]. Research on cavitation in hydraulic components was conducted by simulations verified by experimental tests or in an experimental way [2], which is difficult due to problems with direct observation of the phenomenon. R. Amirante, et al. [3] and J. R. Valdés at al. [4] conducted numerical simulations and experimental tests on hydraulic valves.

Erosion caused by cavitation in a hydraulic valve is very well known. Long term exposure of flow to cavitation makes valve surfaces erode, which leads to improper operation of the valve and, in the consequence, malfunctioning of the hydraulic system. An example of cavitation erosion is presented in Fig. 1.

Simulation of cavitation in hydraulic liquids is not an easy task. Contrary to water, which is a single compound liquid, hydraulic fluids are a mixture of mineral oils, esters, corrosion inhibitors and other additives and therefore it is difficult to evaluate critical saturation pressure. This paper presents a simulation of cavitation in a no-return hydraulic ball type valve. To this purpose, a general purpose CFD code, Ansys CFX, was used with an implemented two phase flow with phase change.

Fig. 1. Eroded surface of a valve [5]

Cavitation model

As it was mentioned before, lots of studies have been conducted on cavitation and bubble formation. Generally, the tendency of fluid flow to cavitation may be expressed as [6]:

$$c_a = \frac{p-p_v}{0.5\rho U^2} \quad (1)$$

The Rayleigh-Plesset formula is one of the approaches which describe bubble dynamics:

$$R_b\frac{d^2R_b}{dt^2} + \frac{3}{2}\left(\frac{dR_b}{dt}\right)^2 + \frac{2\sigma}{\rho R_b} = \frac{p_v-p}{\rho} \quad (2)$$

After neglecting surface tension and term of second order, the above eq. has the following form:

$$\frac{dR_b}{dt} = \sqrt{\frac{2}{3}\left(\frac{p_v-p}{\rho}\right)} \quad (3)$$

The rate of changes of bubble volume is expressed by the following formula:

$$\frac{dV_b}{dt} = 4\pi {R_b}^2\sqrt{\frac{2}{3}\left(\frac{p_v-p}{\rho}\right)} \quad (4)$$

The rate of changes of bubble mass is expressed by:

$$\frac{dm_g}{dt} = 4\pi {R_b}^2\rho_g\sqrt{\frac{2}{3}\left(\frac{p_v-p}{\rho}\right)} \quad (5)$$

The number of bubbles *Nb* per unit volume rg is expressed by:

$$r_g = \frac{4}{3}\pi R_b^2 N_b \quad (6)$$

The total interphase mass transfer per unit volume is expressed by:

$$\dot{m}_{fg} = 3\frac{r_g\rho_g}{R_b}\sqrt{\frac{2}{3}\frac{p_v-p}{p_f}} \quad (7)$$

After including condensation to the above expression, it has the following form:

Materials Research Proceedings **5** (2018) 1-6 doi: http://dx.doi.org/10.21741/9781945291814-1

$$\dot{m}_{fg} = 3F\frac{r_g\rho_g}{R_b}\sqrt{\frac{2}{3}\frac{|p_v - p|}{p_f}}\,\mathrm{sgn}(p_v - p) \quad (8)$$

And finally, the vapor transport equation appears in the following form:

$$\frac{\partial}{\partial t}(\alpha\rho_v) + \nabla\cdot\left(\alpha\rho_v\mathbf{U}_f\right) = R_e - R_c \quad (9)$$

where:

R_b – bubble radius, p_v– vapor pressure, p – pressure of liquid surrounding the bubble, ρ–liquid density, ρ_g– vapor density, σ – surface tension, $\mathbf{U}_v$ – vapor phase velocity, α– vapor volume fraction, n– bubble number, R– phase change rate, f_v– vapor mass fraction, f_g– non-condensable gases, R_e, R_c- mass transfer source terms connected to the growth and collapse of the vapor bubbles, respectively.

Object of study

A non-return valve which is also called a check or a one way valve is the object of the study. The aim of such a valve is to provide flow only in one direction and secure against reverse flow.

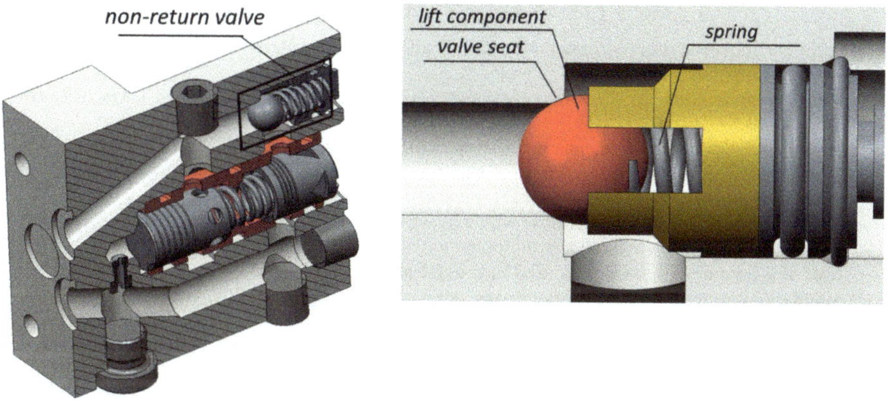

Fig. 2. Non-return valve assembled in a flow control valve

It consists a lift component, which has a spherical or conical shape and a spring which sets the opening pressure value. Such valves might be installed directly in-line or as a component of other valves.

CFD simulations

A flow simulation was conducted in the Ansys CFX general purpose CFD code, which uses the Finite Volume Method (FVM) for solving flow governing equations and applies a multiphase flow model with a phase change. For the numerical simulation, half of a valve model was used. Boundary conditions and a grid is presented in Fig. 3.

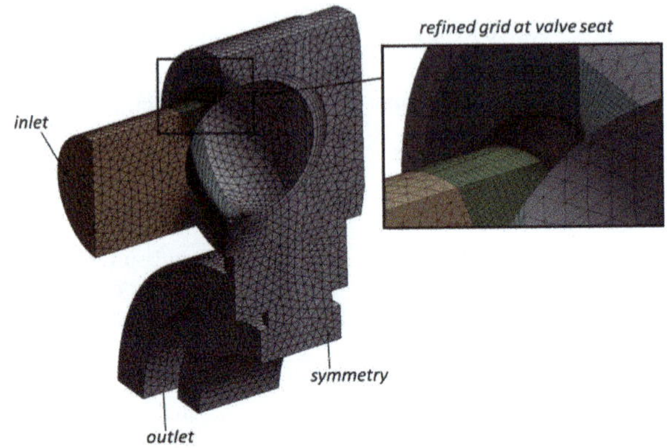

Fig. 3. CFD model: grid and boundary conditions

The CFD simulation was performed for a fixed ball position for steady state conditions and for the following assumptions:

- Fluid (hydraulic oil) has the following constant properties: density 880 [kg/m^3], viscosity υ=40 [mm^2/s]
- Flow is turbulent: k-ϖ turbulence model was used,
- Model is in thermodynamic equilibrium, heat transfer is not included.
- Saturation pressure is set to 0.2 [bar] [7],
- Simulation includes two phases flow with a phase change.
- Fluid is homogeneous without any dissolved gases.
- Half of a geometrical model was used in simulations.
- Simulations were performed for two cases, which differs the angle of the valve seat.

The CFD results, as a fluid velocity distribution for both cases at a symmetry plane are presented in Fig. 4.

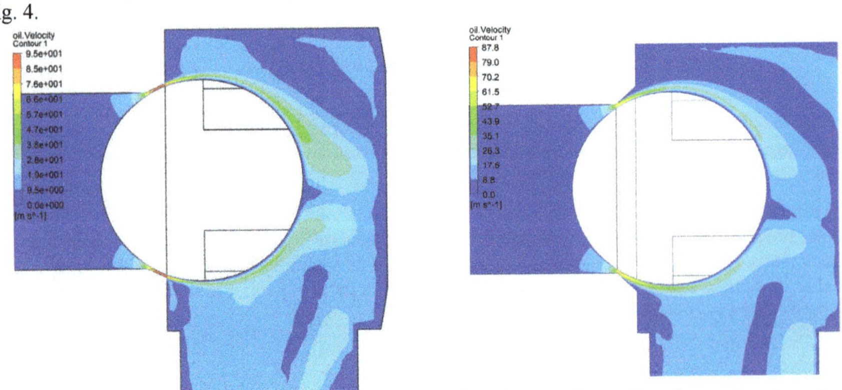

Fig. 4. Fluid velocity at a symmetry plane Velocity distribution shows that the angle of a valve seat influences the maximal velocity of the fluid.

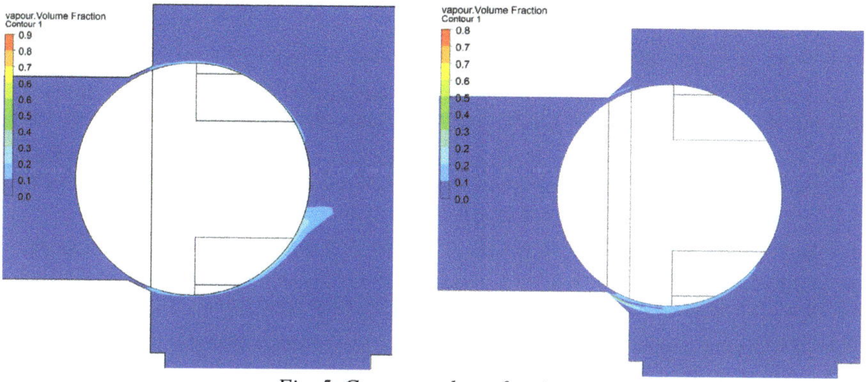

Fig. 5. Gaseous volume fraction

Figure 5 shows a gaseous volume fraction which allows the identification of areas where cavitation may occur. As it was expected, the area with a sudden velocity change is the most exposed to cavitation occurrence. However, the design parameters of the valve (angle of the seat) have influence on the place where cavitation may occur. Figure 6 shows path lines and areas where gaseous fraction may appear during fluid flow.

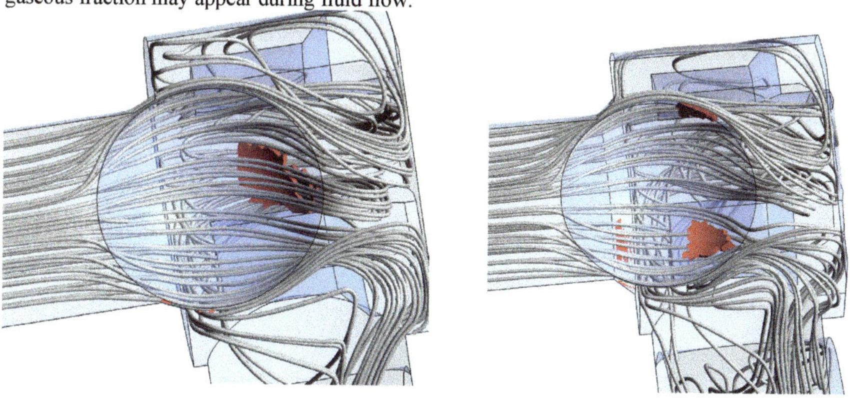

Fig. 6. Path lines and gaseous volume fraction

Summary

This paper presents a simulation of cavitation for one of the most commonly used hydraulic component, which is a ball type non-return valve. Simulations were performed for two cases, differing with valve seat geometry in the Ansys CFX code as a two phase flow with a phase change. The obtained results allowed us to indicate where cavitation may appear and what influence a valve design may have on cavitation phenomenon. The results presented in this paper showed that CFD codes may be recognized as tools for predicting cavitation erosion.

Materials Research Proceedings 5 (2018) 1-6 doi: http://dx.doi.org/10.21741/9781945291814-1

References

[1] Brennen C, E, Cavitation and buble dynamics. Oxford University Press (1995).

[2] Z. Kudzma, M. Stasiak, Studies of flow and cavitation of hydraulic lift valve, Archives of civil and mechanical engineering 15, 2015, 951-961. https://doi.org/10.1016/j.acme.2015.05.003

[3] R. Amirante, E. Distaso, P. Tamburrano, Experimental and numerical analysis of cavitation in hydraulic proportional directional valves, Energy Conversion and Management, 87 (2014) 208–219. https://doi.org/10.1016/j.enconman.2014.07.031

[4] J. R. Valdés, J. M. Rodríguez, R. Monge a, J. C. Peña, T. Pütz, Numerical simulation and experimental validation of the cavitating flow through a ball check valve, Energy Conversion and Management 78 (2014) 776–786. https://doi.org/10.1016/j.enconman.2013.11.038

[5] http://valvemagazine.com: Flashing and cavitation (information available on 15.02.2018)

[6] ANSYS CFX-Solver Theory Guide, Release 18.2, Ansys Inc, 2017

[7] M. Schümichen, F. Rüdiger, J. Fröhlich, Simulation of the cavitating flow in a model oil hydraulic spool valve using different model approaches 10th International Fluid Power Conference (10. IFK) March 8 - 10, 2016, Vol. 1, pp. 321-332

[8] J. Fabis-Domagala, Analysis of dfefects and failures of hydraulic gear pumps with the use of selected qualitative methods, Tribologia 2/2017, 33-38, ISSN 0208-7774

[9] M. Domagala, Simulation of cavitation in jet pumps, Technical Transactions, ISSN 1897-6328, 2013, 51-58

[10] E. Lisowski, G. Filo, CFD analysis of proportional flow control valve with an innovative opening shape, Energy Conversion and Management 123, 2016, 15-28. https://doi.org/10.1016/j.enconman.2016.06.025

Retention of a Diamond Particle in a Metallic Matrix

BOROWIECKA-JAMROZEK Joanna [1,a *], LACHOWSKI Jan [2,b]

[1]Kielce University of Technology, Faculty of Mechatronics and Mechanical Engineering, Al. 1000-lecia P.P. 7, 25-314 Kielce, Poland

[2]Kielce University of Technology, Faculty of Management and Computer Modelling, Al. 1000-lecia P.P. 7, 25-314 Kielce, Poland

[a]jamrozek@tu.kielce.pl, [b]jlach@tu.kielce.pl

Keywords: Diamond Tools, Sinters, Finite Element Method, Mathematical Modelling

Abstract. The paper deals with numerical and analytical modelling of retention of a diamond particle in a metallic matrix. The model of a diamond particle embedded in a metallic matrix was created using the Abaqus software. The analytical model of an elastic spherical particle in an elastic-plastic matrix was built. The aim of the paper was to determine the influence of the mechanical and thermal parameters of matrix materials on their retentive properties. The analysis has indicated mechanical parameters responsible for the retention of diamond particles in a metallic matrix. The pressure in the particle and the radius of plastic zone around particle were analyzed.

Introduction

Diamond segments (metal matrix composite containing diamond particles) of circular saws for cutting stone and ceramic materials are commonly made using the powder metallurgy process [1]. The process of the production of segments consists in mixing metallic matrix powder with diamond crystals and hot pressing them to nearly full density.

This paper discusses a mathematical model of a diamond particle in an elastic-plastic metallic matrix. The mechanical state of a particle embedded in a metal matrix results from the cooling of the sintered material required by the manufacturing process [2]. The coefficient of thermal expansion of the particle material is lower than that of the matrix and the particle is compressed by the contracting matrix. If the difference in the coefficients of thermal expansion is larger and there is a sufficient drop in temperature after hot pressing, a plastic zone around the particle is formed.

The term 'matrix retention' denotes the capacity of a metallic matrix material to retain diamond particles at the surface of a segment during cutting. This property is important in the case of matrices of diamond segments used for cutting materials.

The effect of mechanical properties of a matrix on its potential retentive capabilities is presented in this paper.

Numerical model of a diamond particle in a metallic matrix

Depending on the synthesis conditions, diamond crystallization leads to the formation of crystals with shapes ranging from a cube to an octahedron [3]. An intermediate shape of a diamond crystal is truncated octahedron. The 3D computer model of the truncated octahedron as a diamond crystal (Fig. 1a) was analyzed. The crystal size, i.e. the distance between the opposite square <100> facets (Fig. 1a), was assumed to be 350 μm. The calculations were performed for a deeply embedded particle as well as for the one protruding above the surface of the metal matrix (Fig. 1b).

The models of a diamond crystal embedded in a metallic matrix and protruding from its surface (Fig. 1b) were discussed in [3, 4]. Investigations were carried out by the finite elements method using the Abaqus software version 6.14 [5]. The values of the numerically calculated pressure inside a diamond particle was in agreement with those observed in the experiments [3, 6].

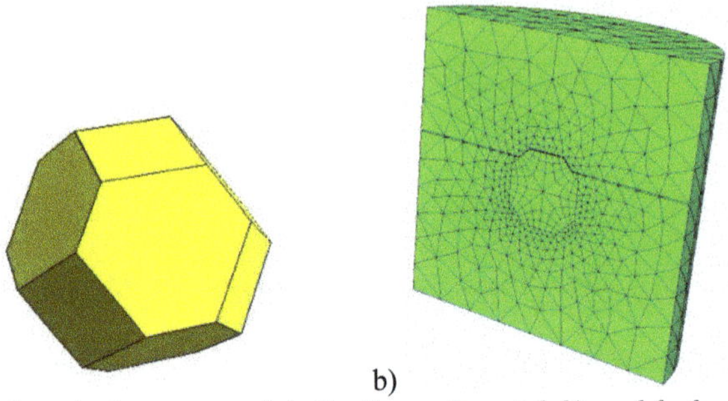

a) b)

Fig. 1. a) Truncated octahedron as a model of a diamond crystal, b) model of a diamond crystal inside the matrix and protruding above the matrix surface.

Mathematical model of a diamond particle in a metallic matrix

The truncated octahedron (Fig. 1a) is the most spherical object of all the semi-regular polyhedrons. Therefore, a spherical model of a diamond particle has been proposed. An analytical model of a spherical elastic particle in an elastic-ideally plastic matrix was presented in the paper [3]. The pressure inside the particle and the elastic energy of the particle are not essentially dependent on the particle shape, the particle is surrounded by the plastic zone with a relatively well defined radius (Fig. 2a) [3]. The mathematical model was verified by comparing the analytical results with the simulation results for the 3D numerical model of the spherical particle [3].

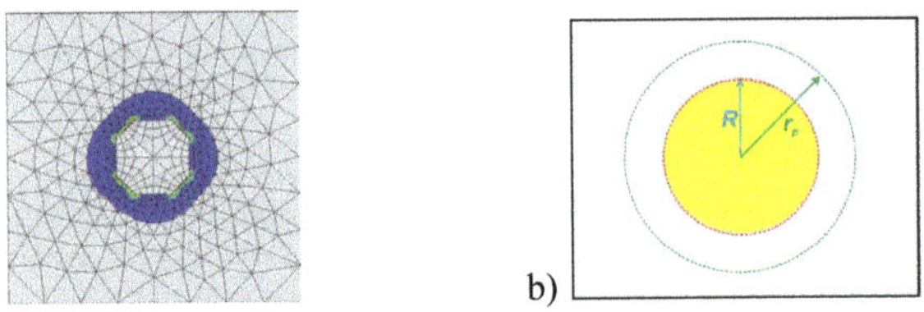

a) b)

Fig. 2. a) Plastic zone around the particle, b) model of the particle in the matrix,

Elastic stresses in the particle and in the matrix were determined using the Lame solution [7]; the plastic zone around the particle was determined from the theory of plasticity [8]. The radius of the plastic zone r_p can be calculated numerically using the following equation [3]:

$$(1-\nu_p)\frac{r_p^3}{R^3}-2\left((1-2\nu_m)-(1-2\nu_p)\frac{E_m}{E_p}\right)\left(\ln\left(\frac{r_p}{R}\right)+\frac{1}{3}\right)=\frac{E_m}{\sigma_0}(\alpha_m-\alpha_p)\Delta T \qquad (1)$$

where E is elastic modulus, ν is Poisson ratio, R is radius of the diamond particle, r_p is radius of the plastic zone around the particle (Fig. 2b), α_m and α_p are respectively the coefficients of thermal expansion of the matrix and the particle, ΔT is the decrease of temperature during cooling after hot pressing and σ_0 is the yield stress of the matrix. The p index denotes the particle parameters and the *m* index denotes the matrix parameters. The *p* pressure inside the particle is calculated from the following formula [3]:

$$p_p = 2\sigma_0 \ln\left(\frac{r_p}{R}\right) + \frac{2}{3}\sigma_0 \tag{2}$$

The stress components (the radial stress and the hoop stress) in the particle are equal to $-p_p$.

Retention of a diamond particle in a metallic matrix

A significant property of matrix material is diamond retention efficiency. Diamond particles are retained in the matrix as a result of mechanical and/or chemical bonding. Mechanical bonding is achieved through cooling that follows the process of hot pressing. Diamond particles are squeezed by the shrinking matrix. Mechanical bonding depends on the elastic and plastic properties of matrix material.

The mechanical properties of the matrices used in the proposed model are summarised in Table 1 [9]. It was assumed that materials were fully densified at hot pressing temperature and then cooled to the room temperature. Thus, the strain field generated in the diamond's surroundings was only an effect of thermal shrinkage of the metal matrix.

Table 1. Material data used for calculations [9]

	Diamond	Cobalt	Cobalt-Bronze
Composition		100%Co	54%Cu40%Co6%Sn
Hot pressing temperature [^{0}C]	-	850	750
Thermal linear expansion coefficient, [$m \cdot K^{-1}$] [10]	$2.9 \cdot 10^{-6}$	$13.36 \cdot 10^{-6}$	$15.5 \cdot 10^{-6}$
Young's modulus, [GPa]	1050	202	140
Poisson ratio	0.1	0.3	0.3
Yield strength, [MPa]	-	650	360
Tensile strength, [MPa]	-	900	520
Elongation, [%]	-	8.5	5.0

In order to verify the mathematical model, the analytical calculation results were compared with the simulation results in the 3D numerical model. The calculations were performed for two sinters: cobalt sinter and cobalt-bronze sinter (Table 1). The results of this analysis are shown in Table 2. The differences between numerical and analytical models are less than 10%.

The obtained values of pressure were correlated with the diamond retention index (DRI) of the matrices. The diamond retention index is the ratio of the total number of diamonds retained on the surface of segment to the total number of diamonds and pullout sites on the surface [11]. The DRI parameter was calculated on working surfaces of diamond segments after granite machining. The last row of Table 2 shows that the pressure in the particle can be taken as a matrix retention index.

Conclusions

Computer modeling enables the examination of the effect of different parameters on the potential retentive properties of the matrix in a very easy and cheap way. The presented results allow the following conclusions to be made:

1. potential retentive properties of the matrix strongly depend on the interactions taking place at the diamond-matrix interface,
2. the pressure inside the particle is not essentially dependent on the particle shape,
3. the particle is surrounded by the plastic zone with a relatively well defined radius.

The above parameters (2 and 3) could be used as the indexes diamond particle retention in the metal matrix.

Table 2. Results of the analysis

Matrix	DRI	Pressure inside particle, analytical model [MPa]	Pressure inside particle, numerical model [MPa]
Co	74	965	1045
CuSnCo	49	605	652
Ratio of Co/CuSnCo values	1.51	1.59	1.60

References

[1] J. Konstanty, Powder metallurgy diamond tools, Elsevier, Oxford, 2005.

[2] J. Borowiecka-Jamrozek, J. Konstanty, Microstructure and mechanical properties a new iron-base material used for the fabrication of sintered diamond tools, Advanced Materials Research 1052 (2014) 520–523. https://doi.org/10.4028/www.scientific.net/AMR.1052.520

[3] J. Lachowski, J. Borowiecka-Jamrozek, Modelling Thermomechanical Response of a Diamond Particle in a Metallic Matrix, Engineering Transactions, 65 (2017) 105–112.

[4] J. Borowiecka-Jamrozek, J. Lachowski, Modelling of retention of a diamond particle in matrices based on Fe and Cu, Procedia Engineering 177 (2017) 289 – 296. https://doi.org/10.1016/j.proeng.2017.02.227

[5] SIMULIA Dassault System, Abaqus Analysis User's Guide, ver. 6.14 (2014).

[6] D.A. Akyuz, Interface and microstructure in cobalt-based diamond tools containing chromium, PhD thesis, Ecole Polytechnique Federale de Lausanne, Lausanne, 1999, pp.105–108.

[7] L.D. Landau and E.M. Lifszic, Theory of Elasticity, 2nd ed. , Pergamon Press Ltd., Oxford, 1975.

[8] R. Hill, The Mathematical Theory of Plasticity, Clarendon Press, Oxford, (1950, reprinted 2009) pp. 97-101.

[9] A. Romański, J. Lachowski, H. Frydrych, Energy of plastic deformation as an estimator of the retentive properties of the metal matrix in diamond impregnated tools, Diamond Tooling Journal 1 (2009) 28-32.

[10] Spinger Handbook of Condensed Matter and Materials Data, eds. W. Martienssen & H. Warlimont, Springer Berlin Heidelberg 2005.

[11] A. Romański, Development of metal matrix for sintered diamond tools, AGH, Dissertations-Monographs, No 303, Kraków 2015 (in polish).

Materials Research Proceedings 5 (2018) 11-16 doi: http://dx.doi.org/10.21741/9781945291814-3

A Cause for Cracking of Head-Check Rails

MIKŁASZEWICZ Ireneusz[1,a*], CZARNECKI Marcin[1,b]

[1]Instytut Kolejnictwa, ul. Chłopickiego 50, 04-275 Warszawa, Polska

[a] imiklaszewicz@ikolej.pl

[b] marcinczarnecki@ikolej.pl

Keywords: Head Check of Rail, Rolling Load, Strain Hardening Surface, Chemical Composition of Steels

Abstract. The paper presents the reasons for the emergence of one of the main defects of railway rails, such as cracking, and factors affecting its increase. It was found that vehicle cooperation with the rail and the forming of the rolling surface caused by the load, as well as the resulting rail stresses caused by the dynamics of rolling stock are the cause of the head-check of the rails. The influence of impact resistance on rail cracks in the occurrence of these defects is also presented.

Introduction

Adapting to the increasing requirements of train safety and increasing speed, enforces modernization of the track superstructure and the use of good quality materials, including rails. The existing superstructure, especially on tracks with heavy traffic, requires the use of monitoring, which aims, among others, at assessing the condition of rails as one of the most important factors of safe transport.

One of the disadvantages of rails, which can appear during operation and have a negative impact on traffic safety, is the head check as shown in Fig. 1 and Fig. 2. It is a defect that arises on the running surface of the rail head and is characterized by the parallel arrangement of cracks at a specific angle to the direction of travel of the train. The defect arises mainly on the curves of the tracks near the edge of the rail head, then passing to the rolling surface. It was found that this defect also occurs on straight sections, however only where the rails are heavily loaded.

Carbon is an element that strongly affects the hardness of rails. The conducted tests of formation of head checks in rails made of steel with increased carbon content have shown that as the hardness of rails increases, the density of defects is increased, but the depth of their residualness decreases with a comparable transport load. Thus, the increase in hardness increases the resistance of the material to the rolling surface of the running surface of the rail head as well as the durability and wear of rails [1].

Therefore, as a result of the rail – wheel contact and crush of the running surface of the rail caused by a long exploitation in the track, the resulting crack becomes the beginning of the crack propagating inside the rail. It is then a fatigue type defect. It is visually detected. In contrast, fatigue defects of the rail with dislocation, vacancy and stress mechanisms, are created inside the head of the rail. They do not come to the surface and are detectable by ultrasonic testing [2]. At the same time, due to the difficulty in detecting this type of defect in the rails (Fig. 3), it can affect the safety of rail transport.

The rail manufacturing process used since the mid-eighties has caused a significant improvement in their quality. The introduced vacuum treatment significantly reduced the amount of gases in steel, i.e. oxygen, hydrogen and nitrogen, thereby reducing the number of non-metallic inclusions and hydrogen flakes as potential sources of microcracks in the rails. Thus, rails produced in the analyzed period had much fewer defects, especially those caused by non-metallic inclusions of the material. The problem was and is the defects appearing during operation.

From the analysis of publications concerning the formation of a defect in the operated rail, it appears that this is a process which is influenced by many factors. The basic ones, however, are rolling loads of the rail as well as residual stresses and stresses resulting from rolling stock dynamics.

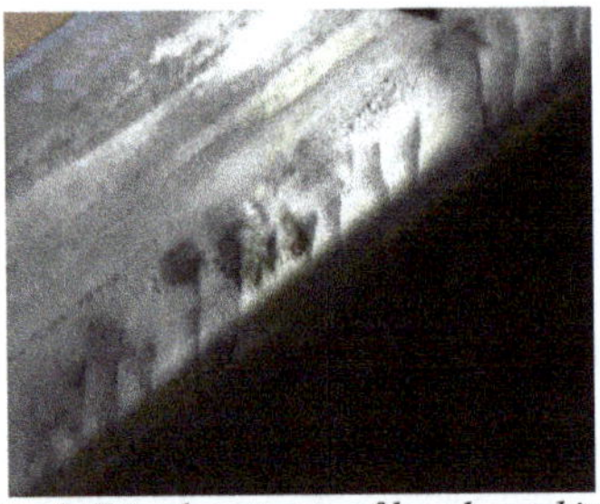

Fig. 1. Disadvantages of head cracking.

Fig. 2. A defect head check with cracking coming out.

Fig. 3. A defect head check was created inside the head of the rail.

Consequences of rolling loads

Rail steels in the R260 type have a high carbon content of 0.6 - 0.8%, and are classified as hard steels with low plasticity. Multiple cyclic pressures of the wagon wheels cause deformation in the surface part of the rail head, which results in crumbling and strengthening of the head surface of the rail. Measurements of HV5 microhardness of R260 rails exceeding even 1200 units, made on a strongly hardened surface of the rail head, indicated the local structure of the amorphous martensite texture - the so called white layer. [6]

Strengthening of the rail surface is progressive as the number of load cycles increases. Then strengthened zones are formed, passing from the surface deep into the material. The resulting structure is characterized by a structure similar to the layered system (Fig. 4, 5). The layered system is created by burnishing the material in the direction of the train. Grains in fortified layers are laid at a certain angle to the surface, depending on the material structure. As a result of increasing load cycles, the surface layer of the head first undergoes elastic deformation, then gradual wrinkling and strengthening. It is a phenomenon of plastic deformation that reaches the limit of material tensile strength. After exceeding this limit, the surface layer ruptures and microcracks arise [7, 8]. Cyclical structural stresses are an additional factor. This is the first stage of creating a defect in the rail head check. The further development of microcracks runs through the fatigue mechanism and is dependent on the number of cycles of rolling loads on the rail during track operation [3, 4].

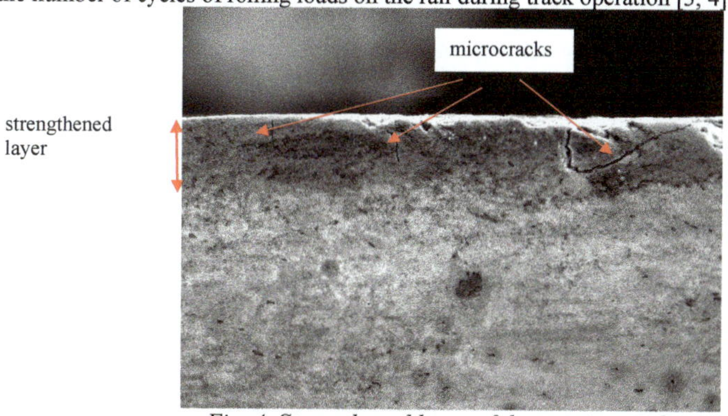

Fig. 4. Strengthened layer of the running surface on the cross-section of the rail head (SEM, 500x).

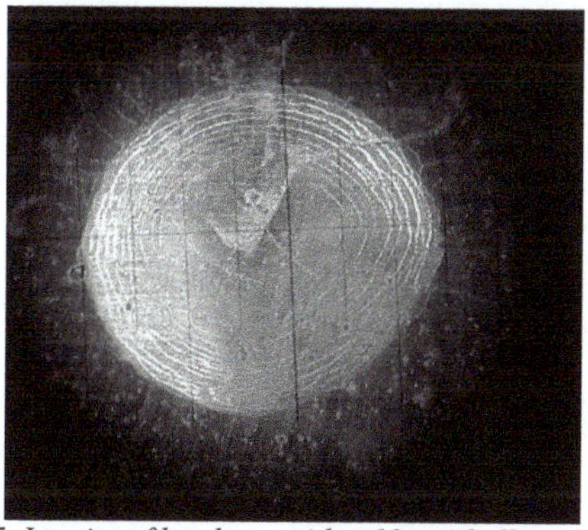

Fig. 5. Imprint of hardness with a 10 mm ball on the rail surface with a visible trace of crushed layers.

The above course of the defect formation occurs on straight sections, in the central part of the surface of the rail running head. The formation of this defect at the edge of the rail head located in the curve of the track causes additional lateral forces impacting the rail. Their impact depends on the size of the curve radius. The larger the radius, the smaller the number of head checks at the edges of the rail.

Parameter K_{1c} explicitly specifies the resistance requirement for R260 rails material for cracking after the crack is formed, according to EN 13674-1+A1: 2017, w (MPa $m^{1/2}$), and the speed of crack growth in the fatigue compartment, i.e. propagation determines the parameter da / dN expressed in m/Gc [5].

The material of the rails has a pearlitic structure with a trace of cementite content along the grain boundaries, as well as trace amounts of non-metallic inclusions in the form of oxides, silicates and sulphides produced in the smelting process. A large number of non-metallic inclusions contribute to the formation of microcracks under the influence of cyclic bus rails by weakening the cohesion of the material. In contrast, sulfide inclusions in the vicinity of other inclusions, arranged parallel to the plastic processing direction may also contribute to the formation of fatigue microcracks, but after a large number of load cycles.

One of the ways to eliminate surface imperfections of rails, including head checks, is the technological process of grinding the rail head. The most convenient period for removing cracking by grinding is the initial period of formation of this defect on the surface of the rail, when the head check has not sufficiently developed and is at a depth of about 0.5 - 1.0 mm. The use of rail grinding process, however, is associated with high economic costs, which makes it important to determine the right time to decide to grind the rails.

Impact resistance of rail steel

The rails in tracks exploited as a result of passing rolling stock are ~~subject~~ subjected to natural wear due to friction caused by rolling load and dynamic impact of the rail and wheel. Travels characterized by high dynamics, mainly caused by unevenness of the railway surface, are particularly dangerous. The parameter useful to determine the resistance of rail material to impact is impact strength.

Due to the increased number of rail cracks in winter and the need to replace them on the route, impact tests of rails were carried out at a temperature of -60°+ 40°C (Fig. 6), with a temperature increase of 20°C.

The research material was rails made and operated in the seventies marked - A1, in the nineties - A2 and made at the turn of the century - A3. The chemical compositions of the rails are given in Table 1. The impact strength was tested on KCU samples.

The presented chemical compositions of the rails show that the sample no. A1, contained the increased aluminum content binding in the previously produced rails, while samples no. A2 and A3 had a chemical composition in accordance with the standard PN-EN 13674-1+ A1: 2017 [5].

As a result of testing the impact strength of the rail material at subzero temperatures, no sudden reduction of impact strength or a clear brittleness threshold in the tested temperature range from +40 ° C to -60 ° C was observed.

Fig. 6 shows that, after 30 years of exploitation, the material of the A1 sample rail, has lower impact strength in the entire temperature range, proportionally lower in relation to the impact resistance obtained from the other rails of the A2 and A3 samples. The difference in impact resistance of rails at below zero temperatures is about 40% lower than at temperatures in summer. The size of the impact strength which determines the impact resistance of the material in the case of existing head checks, may indicate the possibility of rail cracks, especially at high fluctuations in the ambient temperature.

Materials Research Proceedings **5** (2018) 11-16
doi: http://dx.doi.org/10.21741/9781945291814-3

Table 1. The chemical composition of the tested rail steel in the R260 grade.

Sample number	Chemical composition, %											
	C	Mn	Si	P	S	Cr	Ni	Cu	Al	V	N	H(ppm)
A1.	0,75	1,17	0,27	0,033	0,009	0,01	0,02	0,01	0,019	0,010	0,010	-
A2	0,73	1,08	0,29	0,014	0,016	0,01	0,02	0,05	0,004	0,010	0,010	1,9
A3	0,73	0,90	0,39	0,020	0,014	0,05	0,02	0,08	0,004	ślad	ślad	1,6
R260 acc. EN-13674-1	0,60 0,82	0,65 1,25	0,13 0,60	max 0,030	max 0,030	≤0,15	max 0,10	max 0,15	max 0,004	max 0,030	max 0,01	max 2,5

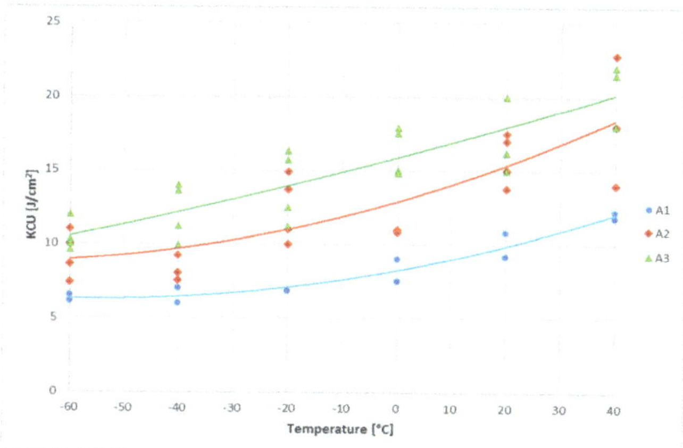

Fig. 6. The impact resistance on the R260 rails.

Summary

It was found that in order for the head check to be created in the rail, the condition must be fulfilled, i.e. the wheel-rail load action, resulting in a corresponding degree of surface strengthening of the rail head and the simultaneous action of stresses, mainly bending in the rail. These factors lead to the formation of micro-cracking of the rail head, and then as a result of repeated load-stress cycles caused by the dynamics of rolling stock, the development of a crack inside the rail head with fatigue mechanism. Non-metallic inclusions play a negative role in the formation of microcracks inside the material. However, currently due to the high technological level of rail production, the causes of defects will disappear, and in their place there will be defects caused by increased speed and greater load on the rails.

Analyzing the mechanism of formation crack rails, and especially the first stage of strengthening the surface layer in terms of plasticity with simultaneous stress, it should be stated that this is one of the most important stages of crack formation in the rail. It follows from the above statement that steel types should be used for the production of rails with a high elasticity and plasticity limit, while maintaining the current parameters of elongation and narrowing of steel. This will allow for a longer period of initial defect formation and will prolong the service life of the rails.

An even drop in impact strength of R260 rails was found with a drop in temperature over the entire temperature range. This has a negative effect on the strength of the material, in particular contributing to rails cracking in winter at below zero temperatures.

References

[1] G. Girsch, R. Heyder. Head – Hardened rail put to the test. Railway Gazette International, (2004).

[2] H. Bałuch. Trwałość i niezawodność eksploatacyjna nawierzchni kolejowej. Wydawnictwo Komunikacji i Łączności (1980) (in Polish).

[3] Roling Contact Fatigue in Rails: a Guide to Current Understanding and Practice, Railtrack PLC Guide-lines: RT/PWG/001, Issue 1, (13) (2001).

[4] Reaserch Programme Engineering. Management and Understanding of Rolling Contact Fatigue. WP1: Mechanisms of Crack Initiation. WP2: Crack Growth. Literature Survey. Rail Safety and Standards Board (2006).

[5] PN-EN 13674-1+A1:2017 Railway application – Track – Rail – Part 1: Vignole railway rails 46kg/m and above.

[6] Materiały Konferencji Naukowo – Technicznej Huty Katowice S.A., Rogoźnik (1999) 149-164 (in Polish).

[7] Z. Świderski, W. Gambin. Powstawanie przypowierzchniowych pęknięć kontaktowo zmęczeniowych w szynach kolejowych, IPPT Warszawa (1999) (in Polish).

[8] I. Mikłaszewicz, Odwęglenie a wady powierzchni główki szyny, Problemy Kolejnictwa, 165 (2014) 85-96 (in Polish).

Terotechnology 2017 Materials Research Forum LLC
Materials Research Proceedings **5** (2018) 17-24 doi: http://dx.doi.org/10.21741/9781945291814-4

Simulation of Particle Erosion in a Hydraulic Valve

DOMAGAŁA Mariusz[1, a *], MOMENI Hassan[2,b], DOMAGALA-FABIS Joanna[1,c], FILO Grzegorz[1,d], KRAWCZYK Mariusz[1,e], RAJDA Janusz[3,f]

[1]Cracow University of Technology, Institute of Applied Informatics, Al. Jana Pawla II 37, 31-841 Krakow, Poland

[2]Western Norway University of Applied Sciences, Department of Mechanical and Marine Engineering, Postboks 7030, 5020 Bergen, Norway

[3]PONAR Wadowice S.A., ul. W. Polskiego 29, 34-100 Wadowice, Poland

[a]domagala@mech.pk.edu.pl, [b]Hassan.Momeni@hvl.no, [c]fabis@mech.pk.edu.pl, [d]filo@mech.pk.edu.pl, [e]mariusz.krawczyk@mech.pk.edu.pl, [f]janusz.rajda@ponar-wadowice.pl

Keywords: Particle Erosion, CFD Simulation, Hydraulic Valve

Abstract. This paper presents an attempt of simulation particle erosion in a hydraulic flow control valve. A simulation of particle behavior during fluid flow was conducted in the Ansys CFX general purpose CFD code using two phases of flow: hydraulic fluid and solid particles. Simulations were performed for two types of particle material, steel and sand, and Finnie's model of erosion wear was applied. As a result of CFD simulations, particle trajectories were obtained as well as erosion rate on valve components for both particle materials. The obtained results made it possible to compare the influence of particle material on erosion wear in the flow control valve and components of valve which are exposed to particle erosion.

Introduction

Contamination of hydraulic fluid is one of the most common cause of hydraulic system failures. It is estimated that around 70% up to 90 % of total failures of hydraulic systems is caused by contamination of hydraulic fluid. Contaminants might be classified into two groups: particles (solid parts as a result of wear or dirt ingression) and chemical (water, air). Four sources of contamination might be distinguished:

1. Built in contamination: a result of assembly of hydraulic system.
2. Ingressed contamination, which can enter into fluid in a reservoir from air or from a hydraulic cylinder rod (sealing of a hydraulic cylinder cannot protect totally against particles) or during maintenance or repairing when a hydraulic line is disconnected.
3. Contamination of new hydraulic fluid, which might be caused by storage process or take place during manufacturing.
4. Contamination generated directly in a hydraulic system. Such contaminants may appear in the system as a result of abrasive wear, when loose particles are removed from metal and taken by the fluid. Another way in which contaminants can get into fluid is adhesive wear, when parts adhere and surfaces are damaged during relative motion due to the lack of oil film on moving parts. In parts under high structural loads, fatigue cracks may appear, which can be also a source of fluid contamination. Additionally, fluid which includes air and falls below saturation pressure during operation pressure may cause cavitation wear, which is also recognized as a source of internal contamination. Corrosive wear and erosion (particle in fluid erode surfaces of hydraulic components) are also internal contaminants.

Hydraulic fluids cannot be fully prevented against contaminations even by installing sophisticated filtration systems. However, hydraulic fluid may be classified for specific applications by controlling the size and amount of contaminants. Some applications of hydraulic systems do not require high cleanliness of working fluid like others. For example, hydraulic drive systems may include more particle contaminants than hydraulic control systems. Currently, the valid ISO 4406:1999 code defines cleanliness of hydraulic fluids as a number and size of particles in 1 [cm^3] of fluid. Figure 1 shows the difference between a selected class of cleanliness of hydraulic oil. The left picture shows fluid which contains 10e3-20e3 particles greater than 4 [μm] and 2.5e3-5e3 particles greater than 6 [μm] and 0.64e3-1.3e3 particles greater than 14 [μm]. The right picture shows hydraulic oil which includes 0.64e3-1.3e3 particles greater than 4 [μm], 80-160 particles greater than 6 [μm] and 10-20 particles greater than 14 [μm].

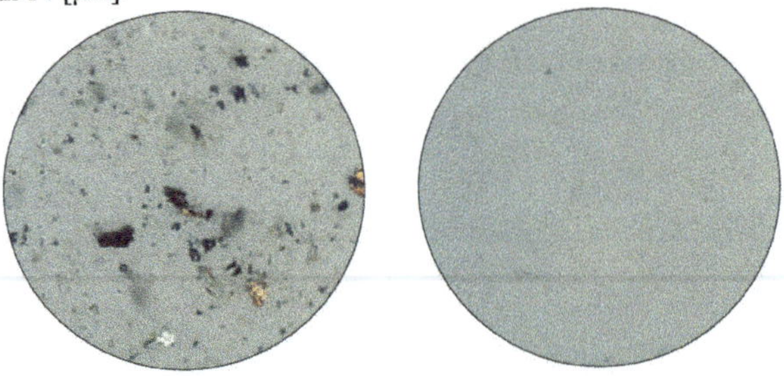

Fig. 1. Cleanliness of hydraulic oil according to the ISO 4406 code [1]
left: ISO 21/19/17, right: ISO16/14/11.

Solid particles in hydraulic oil during fluid flow gather momentum from fluid and may erode critical component surfaces which may lead to component failure or significant reduction of its lifetime. Therefore, prediction of wear caused by solid particles during flow is a more and more important feature. Erosion phenomenon is a complex task and has been investigated for decades, mostly by means of experimental tests [1], [2]. With the expansion of simulation tools and methods, including computational fluid dynamics (CFD), simulation of flow of fluid which includes solid contaminations (particles) is more and more often used. In case of hydraulic components, such attempts are also conducted. Y. Yaobaoa, Y. Jiayanga, G. Shengrongb [4] combined a CFD tool and MATLAB to resolve particle trajectory to predict erosion and worn profile of a hydraulic spool control valve.

This paper presents an attempt to predict erosion wear on a flow control valve using the Ansys CFX general purpose CFD code and an implemented erosion model.

Erosion model

The wear caused by the impact of solid particles on walls for almost all metals is a function of impact angle and particle velocity [5]:

$$E = kV_p^n f(\gamma) \tag{1}$$

where:

E - dimensionless mass (mass of eroded wall material divided by the mass of particle), V_p – particle velocity, n –exponent value which for metals is between 2.3 and 2.5, $f(\gamma)$ - function of impact angle.

The Finnie model is one of the models that describe the rate of wear [6]:

$$\mathrm{E} = \mathrm{k}\mathrm{V}_\mathrm{p}^2\mathrm{f}(\gamma) \tag{2}$$

where:

$$f(\gamma) = \frac{1}{3}cos^2(\gamma)\ if\ tan(\gamma) > \frac{1}{3} \tag{3}$$

$$f(\gamma) = sin(2\gamma) - 3sin^2(\gamma)\ if\ tan(\gamma) > \frac{1}{3} \tag{4}$$

Tabakoff and Grant is another model which describes the erosion rate [7]:

$$E = k_1 f(\gamma) V_p^2 cos^2(\gamma)\ [1 - R_T^2] + f(V_{PN}) \tag{5}$$

where:

$$f(\gamma) = \left[1 + k_2 k_{12} \sin\left(\gamma \frac{\frac{\pi}{2}}{\gamma_0}\right)\ \right]^2 \tag{6}$$

$$R_T = 1 - k_4 V_p \sin(\gamma) \tag{7}$$

$$f(V_{PN}) = k_3 \left(V_p \sin(\gamma)\right)^4 \tag{8}$$

$$k_2 = \begin{cases} 1.0\ if\ \gamma \leq 2\gamma_0 \\ 0\ if\ \gamma > 2\gamma_0 \end{cases} \tag{9}$$

k_1, k_{12}-constants

Simulation of particles in continuous fluid in the CFD is treated as two phases flow: discreet one (particles) and continuous one (fluid). Discreet particles are tracked into a fluid domain during fluid flow. The equation of a motion for a single particle according to Basset, Boussinesq and Oseen may be presented in the following way:

$$m_p \frac{dU_p}{dt} = F_D + F_B + F_R + F_{VM} + F_P + F_{BA} \tag{10}$$

where: F_D – drag force, F_B – buoyancy force, F_R – Coriolis force, F_{VM}- is inertia force of fluid occupied by particle (Virtual Mass), F_P – pressure force, F_{BA} – Basset force

The inertia force of fluid occupied by particle (Virtual Mass) is expressed by the following equation:

$$F_{VM} = \frac{C_{VM}}{2} m_F \left(\frac{dU_F}{dt} - \frac{dU_P}{dt}\right) \tag{10}$$

Transforming Eq. 10 we obtain:

$$\frac{dU_p}{dt} = \left(\frac{1}{m_p + \frac{C_{VM}}{2} m_F}\right)(F_D + F_B + F_R + F'_{VM} + F_P + F_{BA}) \tag{11}$$

where

Materials Research Forum LLC
doi: http://dx.doi.org/10.21741/9781945291814-4

$$F'_{VM} = \frac{C_{VM}}{2} m_F (U_F \nabla U_F) \tag{12}$$

Particle mass $m_P = \frac{\pi}{6} d_P^3 \rho_P$
Fluid mass $m_F = \frac{\pi}{6} d_P^3 \rho_F$
d_p is the particle diameter, ρ_P, ρ_F is density of particle and fluid respectively
Assuming that:

$$R_{VM} = \frac{m_P}{m_P + \frac{C_{VM}}{2} m_F} = \frac{\rho_P}{\rho_P + \frac{C_{VM}}{2} \rho_F} \tag{13}$$

Finally Eq. 10 has the following form:

$$\frac{dU_p}{dt} = \left(\frac{R_{VM}}{m_p}\right)(F_D + F_B + F_R + F'_{VM} + F_P + F_{BA}) \tag{14}$$

Object of study
The object of study is a flow control valve whose main function is to maintain constant flow rate independently on pressure on inlet and outlet port. The valve presented in Fig. 2 is manufactured by PONAR Wadowice as a proportional, solenoid control valve.

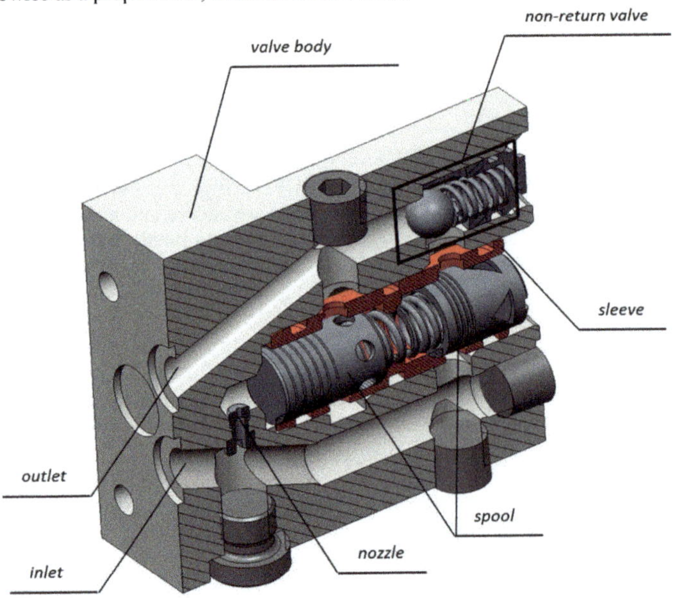

Fig. 2. Cross section of valve

Hydraulic oil flows at the inlet port to the valve, then through the holes in the sleeve inside and flows out to the outlet port by the holes at the spool. The position of the spool is determined by forces

Materials Research Proceedings **5** (2018) 17-24 doi: http://dx.doi.org/10.21741/9781945291814-4

generated on it during flow and pressure drop at the nozzle. For simplification, fixed positions of spool were assumed without taking into consideration dynamic behavior of the valve components.

CFD simulations

Flow simulation was conducted in the commercial Ansys CFX general purpose CFD code, which uses the Finite Volume Method (FVM) for solving flow governing equations. For the numerical simulation, half of a valve model was used. Boundary conditions and a grid is presented in Fig. 3.

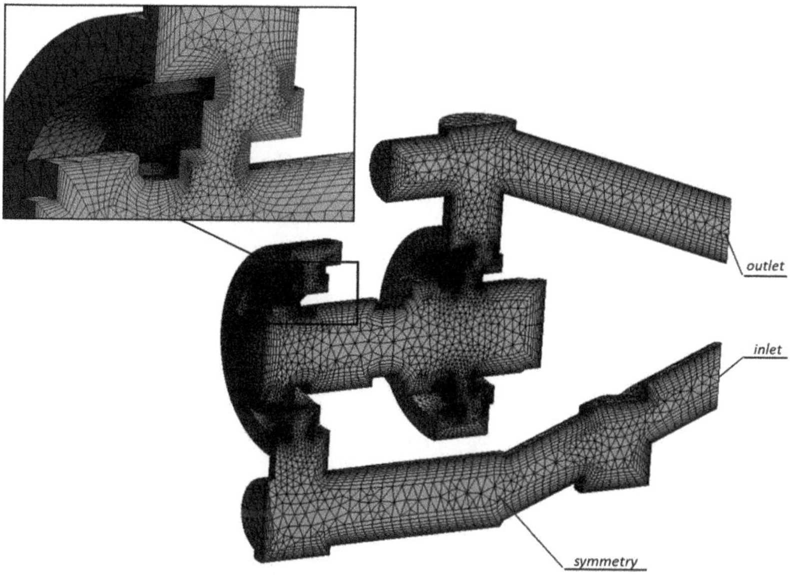

Fig. 3. CFD model: grid and boundary conditions

The CFD simulation was performed for a fixed component position for steady state conditions and for the following assumptions:

- Fluid (hydraulic oil) has the following constant properties: density 880 [kg/m^3], viscosity υ=40 [mm^2/s]
- Flow is turbulent: k-ϖ turbulence model was used,
- Model is in thermodynamic equilibrium, heat transfer is not included.
- Particles parameters have a uniform shape (spheres with a diameter of 20 [µm]) and constant properties, 20 particles in 1 [cm^3]
- Erosion model: Finnie.
- Interaction of particles and fluid is fully coupled.
- Simulations were performed for two particle materials: metal (steel) and sand.
- Particles are uniformly injected at the inlet to the valve.
- Half of a geometrical model was used in simulations.

The results of the CFD simulations are presented in the pictures below which show fluid velocity at a symmetry plane, trajectories of solid particles and erosion rate. Two types of solid particles were used in the CFD simulations to investigate erosion rate on the valve components for the same flow conditions. The results of steel particles are presented on the left side, sand particles on the right.

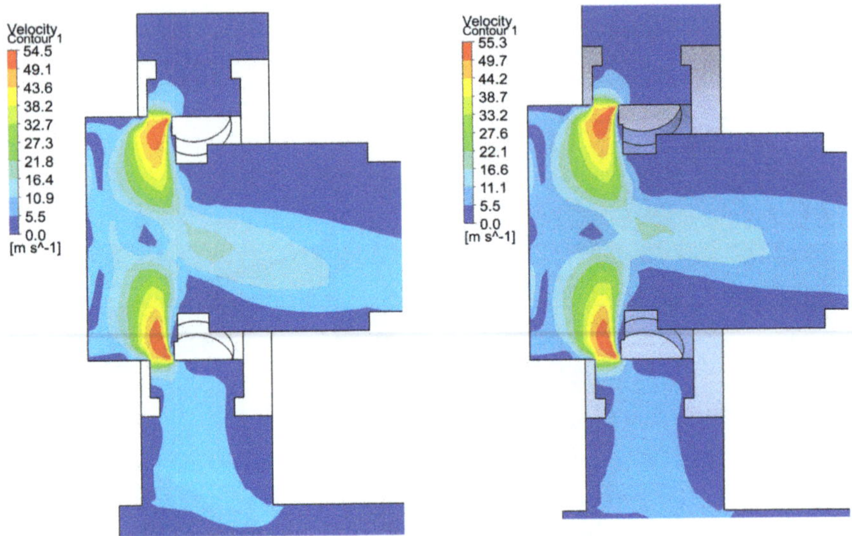

Fig. 4. Fluid velocity on a symmetry plane, right: steel particles, left: sand particles

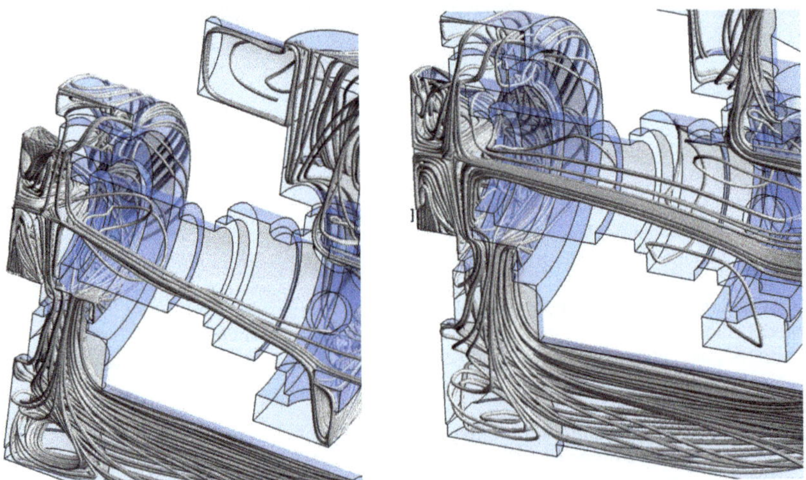

Fig. 5. Particle trajectory, right: steel particles, left: sand particles

The results of simulations show that, for given particle size and number, the flow of fluid is almost not disturbed. Fluid velocity distribution in both cases looks almost the same for both materials of particles (Fig. 4.). However, there is a difference with captured trajectories of particles between steel and sand (Fig. 5.). It was also found that surfaces which are exposed to erosion wear do not differ much in both cases, which is shown in Fig.6.

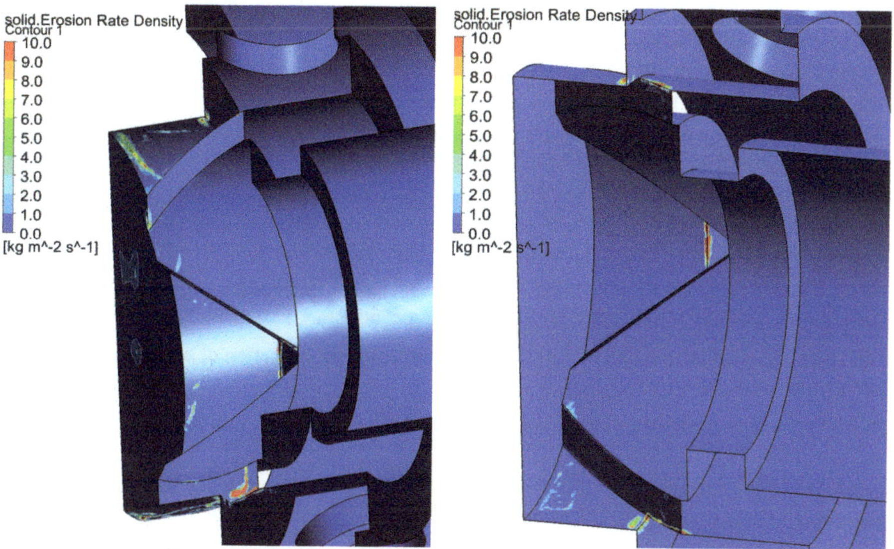

Fig. 6. Erosion rate, right: steel particles, left: sand particles

Summary

This paper is an attempt to predict particle erosion in a hydraulic flow control valve. A simulation of particle erosion is a complex task, however, simulation methods which have lately expanded their possibilities offer such capabilities. Among such methods, CFD tools seem to be very effective in a simulation of erosion wear. This work presents a simulation of flow of fluid containing solid particles in a flow control valve. Two materials of solid particles were investigated: metal (steel) and sand. The CFD simulations made it possible to obtain a trajectory of solid particles as well as erosion rate using Finnie's model. Simulations conducted for two materials of particles showed that, depending on particle material, various surfaces of valve components are exposed to erosion wear.

Materials Research Proceedings **5** (2018) 17-24 doi: http://dx.doi.org/10.21741/9781945291814-4

References

[1] Handbook of Hydraulic Filtration, Brochure Ref: FDHB289UK, Parker Hannifin Corporation 2006.

[2] A.T.Fry, P.Lovelock, N.Smith, M.Gee, A.J.Gant, An innovative high temperature solid particulate erosion testing system, Wear, Volumes 376–377, Part A, 15 April 2017, 458-467

[3] V.B.Nguyen C.Y.HLim Q.T.Trinh S.Sankaranarayanan Y.W.Zhang M.Gupta, Effect of impact angle and testing time on erosion of stainless steel at higher velocities, Wear Volume 321, 30 December 2014, 87-93

[4] Y. Yaobaoa, Y. Jiayanga, G. Shengrongb, Numerical study of solid particle erosion in hydraulic spool valves, Wear 392-393 (2017) 147-189. https://doi.org/10.1016/j.wear.2017.09.021

[5] Hutchings, I.M.,"Mechanical and metallurgical aspects of the erosion of metals",Proc. Conf. on Corrosion-Erosion of Coal Conversion System Materials, NACE (1979) 393.

[6] Dosanjh, S., and Humphrey, J.A.C., "The influence of turbulence on erosion by a particle laden fluid jet, Wear",V102, 1985, pp. 309-330. https://doi.org/10.1016/0043-1648(85)90175-9

[7] Dosa Tabakoff W., Grant G., Ball R., An experimental investigation of certain aerodynamics effect on erosion, AIAA Paper No. 74-639, 1974.

[8] J. Fabis-Domagala, Analysis of dfefects and failures of hydraulic gear pumps with the use of selected qualitative methods, Tribologia 2/2017, 33-38, ISSN 0208-7774

[9] H. Momeni, M. Domagala, CFD simulation of transport solid particles by jet pumps, Technical Transactions, ISSN 1897-6328, 2015, 185-191

[10] E. Lisowski, G. Filo, CFD analysis of proportional flow control valve with an innovative opening shape, Energy Conversion and Management 123, 2016, 15-28. https://doi.org/10.1016/j.enconman.2016.06.025

[11] E. Lisowski, G. Filo, J. Rajda, Analysis of Flow forces in the initial phase of throttle gap opening in a proportional control valve, Flow Measurement and Instrumentation, Volume 59, March 2018, 157-167. https://doi.org/10.1016/j.flowmeasinst.2017.12.011

Terotechnology 2017 Materials Research Forum LLC
Materials Research Proceedings **5** (2018) 25-30 doi: http://dx.doi.org/10.21741/9781945291814-5

Mechanically-Assisted Laser Forming of Flat Thin Beams made of Inconel 627 and Inconel 718 Alloys

KURP Piotr

Kielce University of Technology, Faculty of Mechatronics and Mechanical Engineering Department of Terotechnology and Industrial Laser Systems, Aleja Tysiąclecia Państwa Polskiego 7, PL- 25314 Kielce, Poland

pkurp@tu.kielce.pl

Keywords: Laser Forming, Mechanically-Assisted Laser Forming, Thin Beams

Abstract. The authors of this paper presented the results of research on mechanically-assisted laser forming of flat thin beams made of Inconel 627 and Inconel 718 alloys. The method of applying external forces to beams and the way of conducting this process will be presented. The deflections of beams depending on a given load will be presented. The research presented in this paper will be arranged to develop the technology of thin-walled spatial elements bending, e.g.: tubular elements, which are made of the tested materials.

Introduction

Laser forming is one of contactless forming methods of elements made from various materials. Due to the fact that there is no physical contact between the tool and the workpiece, it can also be carried out remotely. This technology was developed already in the 80s of the last century and is being developed all over the world, including at Center for Laser Technologies of Metals Kielce University of Technology and Polish Academy of Science (CLTM PŚk&PAN) [1], [2]. However, this is a time-consuming method. Therefore, the idea of hybrid forming, i.e. mechanically-assisted laser forming, appeared. The first tests of hybrid forming were carried out at CLTM PŚk&PAN in 2016 for thin-walled beams made of X5CrNi18-10 acid-proof steel [3].

The Inconel 625 and 718 nickel superalloys are widely used to manufacturing heat-resistant and high corrosion resistance elements [4]. Inter alia, they are used in the aviation industry as material for turboprop engine chamber components [5]. The research on hybrid mechanically-assisted laser forming will be presented below.

Research stage and methodology description

The research stage for mechanically-assisted laser forming of flat beams was made based on the TRUMPF TruFlow6000 CO_2 laser. Specimens were made of Inconel 625 and Inconel 718 alloys with dimensions of 200x20x1 mm (specimen laser scanning length L=130 mm - for Inconel 625 and L=150 mm - for Inconel 718). The specimens were covered with a technical absorber in the form of black enamel (to increase the absorption coefficient of laser radiation [6]). A laser segmented head generating rectangular laser beam with dimensions of 20x2 mm was used for machining. The beam scanned specimens surface in such a way that the length of the beam coincided with the width of the sample. The laser treatment parameters were selected in such a way that the surface temperature of the specimen was within the range of <800°C; 900°C>. The scanning parameters were as follows: laser power P=500W, beam speed v=200 mm/min. The F force was obtained by means of suitable weights mounted on the loose end of the specimen. The range of tested loads was the following: 110, 160, 210, 260, 310, 360, 410, 460, 510 and 560G. Fig.1 shows the diagram of loading and deflection measurement.

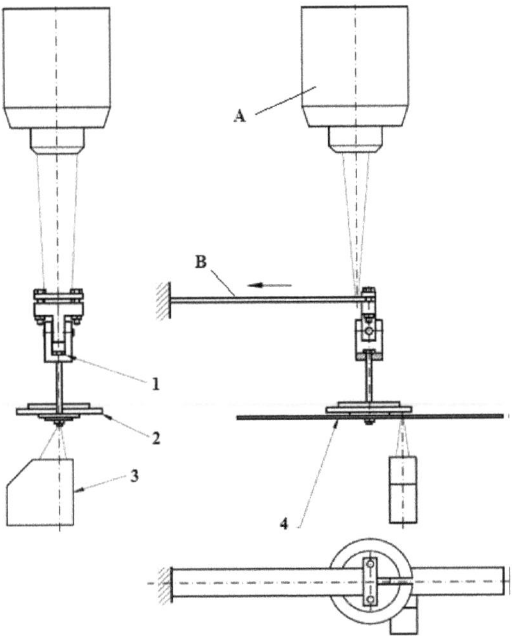

Fig. 1. Research stage diagram: A - laser segmented head, B - specimen tested, 1 - pivot suspension, 2 - weights (loads), 3 - deflection sensor, 4 - intermediate plate for deflection measurement.

Characteristics of the tested alloys and initial deflection values

The characteristics of the tested alloys are presented in Table 1 and Table 2 [7], [8].

Table 1. Mechanical and thermo-physical properties of Inconel 625 [7]

Constant	R_e [MPa]	R_m [MPa]	A [%]	E [GPa]	Thermal conductivity [W/m^{2}K]	Specific heat [J/kg°C]
Value	≥664	986 - 996	≥ 46,3	~200	9,8	410

Table 2. Mechanical and thermo-physical properties of Inconel 718 [8]

Constant	R_e [MPa]	R_m [MPa]	A [%]	E [GPa]	Thermal conductivity [W/m^{2}K]	Specific heat [J/kgK]
Value	830-1030*	1030-1275*	12-20*	~200	11,4	435

* - depending on the type of finishing treatment

Before the mechanically-assisted laser forming process was performed, the initial deflections of the sample for given loads were measured with an altimeter. The results of the measurements were compared with the theoretical values calculated from the dependencies below:

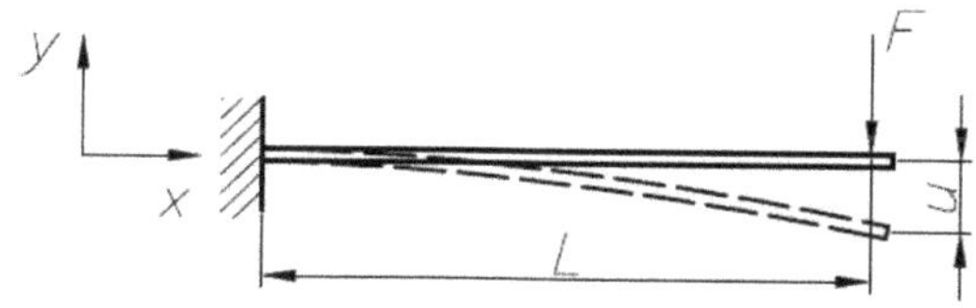

Fig. 2. Initial specimen deflection

The initial deflection of the *u* beam can be calculated from the following dependence:

$$u = \frac{F}{6 \cdot E \cdot I} \cdot x^3 - \frac{F \cdot L^2}{2 \cdot E \cdot I} \cdot x + \frac{F \cdot L^3}{3 \cdot E \cdot I} \tag{1}$$

The maximum deflection of the beam at the end of the interval is calculated for x = 0. Dependence (1) therefore takes the following form:

$$u = \frac{F \cdot L^3}{3 \cdot E \cdot I} \tag{2}$$

where:
E - modulus of longitudinal elasticity, I - inertia modulus of the beam cross-section:

$$I = \frac{g^3 \cdot s}{12} \tag{3}$$

Table 3 and Table 4 show the values for measured and calculated initial beam deflection.

Table 3. Initial deflection values for Inconel 625 alloy (L=130 mm)

load F, [G]	110	160	210	260	310	360	410	460	510	560
measured, [mm]	2,6	7,2	12,2	15,4	18,6	23,8	28,1	31,7	35,6	39,5
calculated, [mm]	3,6	5,2	6,8	8,4	10,0	11,6	13,2	14,9	16,5	18,1

Table 4. Initial deflection values for Inconel 718 alloy (L=150 mm)

load, [G]	110	160	210	260	310	360	410	460
measured, [mm]	5,8	8,5	10,9	13,9	17,8	19,1	22,8	25,4

Materials Research Forum LLC
doi: http://dx.doi.org/10.21741/9781945291814-5

calculated, [mm]	5,8	8,4	11,0	13,7	16,3	18,9	21,5	24,2

The experiment course and test results

The experiment course of mechanically-assisted laser forming for Inconel 625 and Inconel 718 alloys is shown in Fig. 3. The results of bending measurements during the process are shown in Fig. 4 and Fig. 5. The final bending values are tabulated in Table 5.

Fig. 3. Two extreme stages of mechanically-assisted laser thin beam forming for a total load of 360g (beginning and the finish of the process)

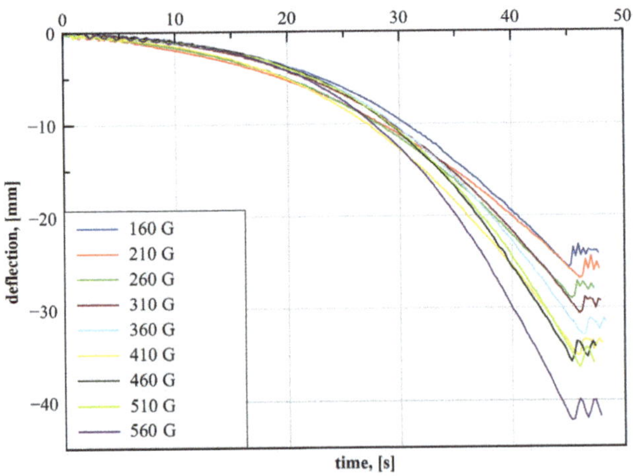

Fig. 4. Specimen deflection over time changes depending on the load. Values for Inconel 625 alloy.

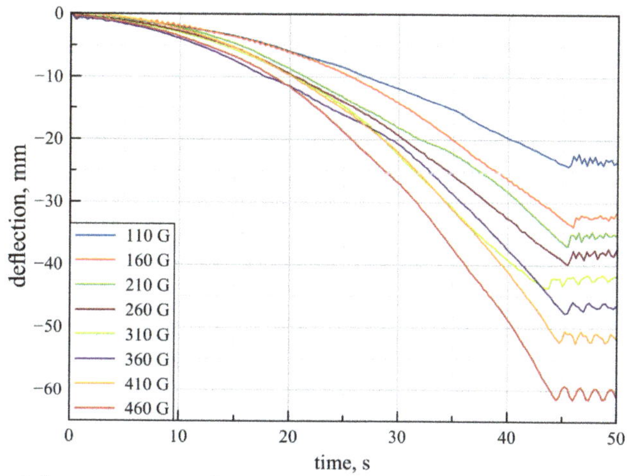

Fig. 5. Specimen deflection over time changes depending on the load. Values for Inconel 718 alloy

Table 5. Final deflection values of beam for given loads (tabular summary according to Fig. 4 and Fig. 5.)

load, [G]	110	160	210	260	310	360	410	460	510
(625) final deflection*, [mm]	-	23,05	25,01	27,02	28,91	32,21	34,01	35,11	36,15
(718) final deflection *, [mm]	23,5	32,5	35,0	37,5	42,5	47,0	51,5	61,0	-

* final deflection is plastic deflections after load removal

Summary

The test results indicate a good tendency for plastic deformation of the Inconel 625 and 718 nickel superalloy after reaching the appropriate temperature in element. The high content of alloying elements, in particular nickel, causes low thermal expansion of the material. However, by achieving the appropriate temperature of the beam, it changes the plastic properties of Inconel [9]. In addition, the bending process to the appropriate angle can be easily controlled by using additional, properly selected mechanical force. When supplying appropriate linear power density to the surface of formed specimens, which, however, must not liquefy the element, the force supporting the bending process may be small.

In the future research, structural studies will be carried out to determine structural changes in the material. The next stage of research will be mechanically-assisted laser forming of thin-walled elements in the form of pipes made of the above materials.

Acknowledgements

The research reported herein was supported by a grant from the National Centre for Research and Development (No. PBS3/A5/47/2015).

References

[1] D.L. Deacon, Material Degradation in Heavy Steel Plates Caused by Bending with a Laser. Master's thesis. Cambridge, USA: Massachusetts Institute of Technology, Dept. of Ocean Engineering, 1984.

[2] H. Frąckiewicz, Z. Mucha, W. Trąmpczyński, A. Baranowski, A. Cybulski, Sposób gięcia przedmiotów metalowych. Patent PL155358. Data zgłoszenia 26.11.1987.

[3] Z. Mucha, J. Widłaszewski, P. Kurp, K. Mulczyk, Mechanicallyassisted laser forming of thin beams. Proc. of SPIE 10159, Laser Technology 2016: Progress and Applications of Lasers, 2016, pp. 101590U-1–101590U-10.

[4] M. Scendo, N. Radek, J. Trela, Influence of laser treatment on the corrosive resistance of WC-Cu electrospark coatings, International Journal of Electrochemical Science, 8, 2013, pp. 9264-9277.

[5] J. Widłaszewski, M. Nowak, Z. Nowak, P. Kurp, Laser-assisted forming of thin-walled profiles. Metal Forming. XXVIII, 3, 2017, pp. 183–198.

[6] P. Kurp, Z. Mucha, K. Mulczyk, R. Gradoń, P. Trela, The influence of surface preparation on the absorption coefficient of laser radiation, Proc. of SPIE 10159, Laser Technology 2016: Progress and Applications of Lasers, 2016, pp. 101590M1-101590M8.

[7] Maher Limited, INCONEL® alloy 718 Data sheet, 2018 (http://www.maher.com/media/pdfs/718-datasheet.pdf)

[8] Special Metals Corporation, INCONEL® alloy 625 Data sheet, 2018 (http://www.specialmetals.com/assets/smc/documents/alloys/inconel/inconel-alloy-625.pdf)

[9] K.C. Miles, Recommended values of thermophysical properties for selected commercial alloys, Woodhead Publishing Ltd and ASM International, 2002. https://doi.org/10.1533/9781845690144

Fire Properties of Upholstery and Fire Resistance of the Complete Passenger Seat

RADZISZEWSKA-WOLIŃSKA Jolanta Maria [1, a *], KAŹMIERCZAK Adrian [2, b]

Instytut Kolejnictwa, 50, Chlopickiego Street, 04-275 Warsaw, Poland

[*a]jradziszewska-wolinska@ikolej.pl, [b]akazmierczak@ikolej.pl

Keywords: Furniture Calorimeter, Cone Calorimeter, Fire Properties, Smoke Density, EN 45545, Fire Safety in Rolling Stock, Fire Hazard, Fire Tests of Seats, Fire Test of Materials, Heat Release Rate

Abstract. The article discusses laboratory tests of fire properties of complete railway seats and upholstery systems used for their construction. The results obtained with the use of furniture and cone calorimeter were especially analyzed to determine the interrelations between them. It has been shown that the criteria for classification of upholstery systems and passenger seats need to be verified. High impact of proper selection of materials on the initiation and development of a possible fire was emphasized.

Introduction

Passenger seats in rolling stock constitute a significant part of the fire load density of vehicles. These elements are directly exposed to potential sources of ignition, such as acts of vandalism - arson, fire, heating system failures, as well as short circuits in electrical installations. Considering the history of fires in the rolling stock, which were the consequence of the ignition of the railway chair, it can be clearly stated that they pose a high risk to life and health of people. Humans are exposed to particular hazards in the direct area of an exposure to a thermal radiation and fumes containing toxic fire gases. The scale of danger increases many times if a fire occurs in the tunnel [1, 2]. That is why for many years now wagon seats and their component materials have been subjected to classification tests. Appropriate selection of rail seat components contributes significantly to the reduction of the fire risk level of the entire vehicle.

Requirements in the field of fire resistance

Selective selection of non-metal materials, as well as their location in the vehicle and production technology is a very important issue in designing rail vehicles. The above also applies to the component materials of the passenger seat.

The fire resistance characteristics of the tested materials under conditions simulating a flared fire make it possible to identify the hazards associated with them and take appropriate preventive measures. Investigations of flammable and smoke properties enable the elimination or modification already at the design stage of materials whose application may cause an unacceptable risk of events causing human and material losses. The research is aimed at obtaining materials that are difficult to ignite and limit the spread and development of fire. They do not eliminate the risk of danger. However, they allow controlling and minimizing them to a large extent. The requirements of the TSI (technical specifications for interoperability) [3, 4] and harmonized European standards are systematically verified.

The main threats resulting from the fire of a passenger seat include smoke density, gas toxicity and the amount of heat emission due to combustion. The listed hazards are included in the F requirements (furniture) (Table 2 according to EN 45545-2 + A1: 2015-12 [5]) for individual components and the complete seat. Table 1 presents a list of requirements applicable to elements

Materials Research Proceedings **5** (2018) 31-36 doi: http://dx.doi.org/10.21741/9781945291814-6

being the subject of this article, i.e. upholstery system (R21) and a complete passenger seat (R18). The parameter values vary depending on the hazard category (HL) for the rolling stock resulting from its operating category and the assigned railway infrastructure in which it moves.

Table 1. Statement of the R21 and R18 requirements according to EN 45545:2013+A1:2015 [5]

Requirements	Tests method	Parameter [Unit]	Requirement for:		
			HL1	HL2	HL3
R21 (upholstery on passenger seats and headboards)	ISO 5660-1: 25 [kWm^{-2}]	MARHE [kWm^{-2}]	≤ 75	≤ 50	≤ 50
	EN ISO 5659-2: 25 [kWm^{-2}]	D_s max. [dimensionless]	≤ 300	≤ 300	≤ 200
	EN ISO 5659-2: 25 [kWm^{-2}]	CIT_G [dimensionless]	≤ 1,2	≤ 0,9	≤ 0,75
R18 (complete passenger seat)	ISO/TR 9705-2	MARHE [kW]	≤ 75	≤ 50	≤ 20
	ISO/TR 9705-2	RHR Peak [kW]	≤ 350	≤ 350	≤ 350

Laboratory tests

Tests of complete passenger seats and the upholstery systems used for their construction were carried out on 6 representative samples designated successively as seat 1 to seat 6. The material composition of upholstery systems was determined analogously from 1 to 6 and shown below:

1. PUR foam, VIGOR upholstery,
2. NPD-100 foam, INNOVO upholstery fabric,
3. NPD-100 foam, fire kevlar para-aramid barriers, upholstery fabric,
4. NPD-100 foam, kevlar para-aramid fabric, INNOVO upholstery fabric,
5. NPD-100 foam, Kevlar para-aramid fireproof fabric, upholstery fabric,
6. PUR FRC foam, DARA ZWFR107 fireproof nonwoven, Kneitz upholstery fabric.

The scope of testing of upholstery systems included tests in accordance with Table 1. The following parameters were determined:

- Maximum Average Rate of Heat Emission (MARHE) according to ISO 5660-1:2015 [6] determined using a cone calorimeter at a radiation intensity of 25 kW/m^2,
- Maximum Specific Optical Density of smoke (Dsmax) according to EN ISO 5659-2:2013 [7], determined using a one-chamber test,
- Conventional Index of Toxicity (CIT_G) according to EN ISO 5659-2:2013 [7] and EN 45545-2: 2013 + A1:2015 [5] Annex C, determined using the FTIR analyzer connected to the smoke chamber.

A summary of the final results is included in Tab.2. The presented values represent the average of the tests for three samples/objects. In contrast, Fig. 1 shows the heat release rates (HRR) recorded during tests using a cone calorimeter. The graphs for the best (seat 6) and the worst upholstery system (seat 1) are shown. As can be seen, the upholstered systems with the fireproof layer applied have definitely achieved better values for most of the parameters determined. MARHE value in the range (44.2 to 48.27) kW/m^2, while for the systems without a fireblocker: (58.63 to 64.7) kW/m^2. Also, their later ignition and shorter burning time were clearly visible. It was also found that despite the different material composition of the studied systems, Ds max values are at a similar level within the limits of 218-242 and meet the requirements of HL1 and HL2. The toxic properties of combustion products are varied, however, all of them are below the acceptable value for the HL3 level, which is definitely lower than 0.75. Nevertheless, none of the upholstered systems tested met the fire requirements for the HL3 threat level.

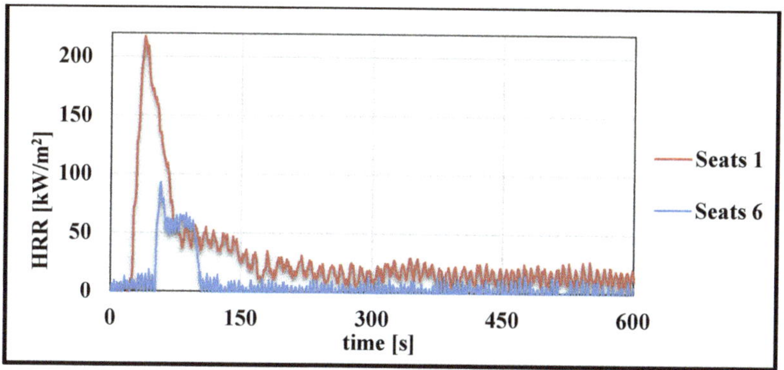

Fig.1. HRR chart during the test on a cone calorimeter

Fig. 2. Seat 1 on the test stand according to EN 45545-2:2013 + A1:2015 [5] and ISO/TR 9705-2:2001 [8] during the test.

Then, complete passenger seats were tested according to EN 45545-2:2013 + A1:2015 [5] Annex B, using a furniture calorimeter according to ISO/TR 9705-2:2001 [8]. This test stand makes it possible to determine parameters that are very important when assessing the level of fire safety. It is used in many research projects and certification tests [9 - 14]. MARHE (Maximum Average Rate of Heat Emission), RHR Peak (Rear Heat Release Peak), as well as TSR (Total Smoke Released) based on SPR (Smoke Production Rate) are the main parameters measured during the test..

During the test, a complete passenger seat placed under the hood with a forced air flow of 0.6m/s shall be treated for 180 seconds with a Belfagor propellant gas burner with a nominal power of 7 kW. Fig. 2 shows one of the tested seats on a furniture calorimeter.

Materials Research Forum LLC
doi: http://dx.doi.org/10.21741/9781945291814-6

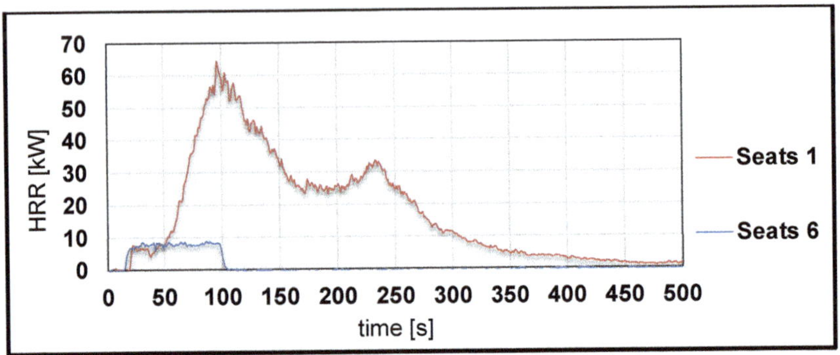

Fig.3. HRR chart during the study on a furniture calorimeter

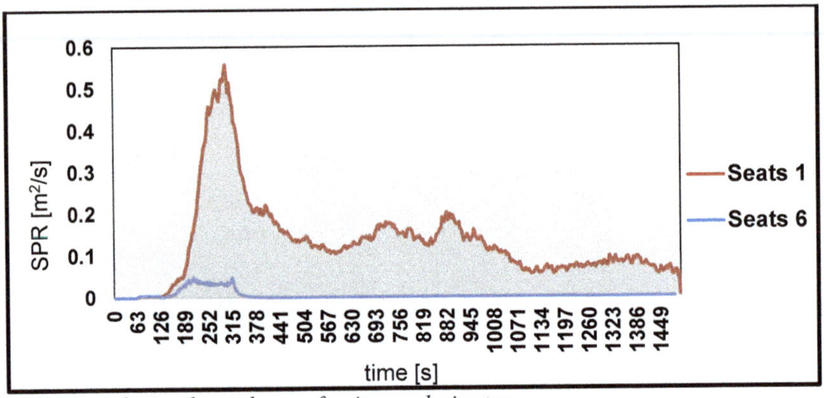

Fig.4. SPR chart during the study on a furniture calorimeter.

Table 2. List of research results obtained on furniture and cone calorimeter.

Parametr	Object					
	Seat 1	Seat 2	Seat 3	Seat 4	Seat 5	Seat 6
	EN 45545-2+A1:2015 Annex B (complete seat)					
RHR Peak [kW]	64.77	28.03	11.71	10.28	10.15	8.74
MARHE [kW]	39.0	14.62	8.83	7.91	8.27	7.67
Total Smoke Released [m^2]	196.0	24.3	8.4	7.9	7.5	6.2
	ISO 5660-1:25 kWm^{-2} (upholstery)					
MARHE [kWm^{-2}]	62.55	58.63	48.27	47.63	46.1	44.2
	EN ISO 5659-2:25 kWm^{-2} (upholstery)					
D_s max.	231.85	242.0	230.9	229.2	227.7	218.7
$CIT_{G(4)}$	0.284	0.09	0.12	0.11	0.08	0.08
$CIT_{G(8)}$	0.348	0.24	0.23	0.21	0.11	0.11

The obtained results, including the additionally determined TSR values, are also included in Tab.2. The heat HRR waveforms recorded during the tests are shown in Fig.3, and SPR waveforms in Fig.4.

The analysis of the results, also with the fire-retardant layer, shows that they achieved significantly better values of the parameters determined.

At the same time, all the complete seats with a fireproof layer and seat 2 without a fireblocker, met the HL3 requirements and reached the following values: MARHE: (7.67 - 14.62) kW and RHR Peak: (10.15 - 64.77) kW. On the other hand, seat 1, for which an intensive smoking process was observed with a flame of a maximum of 1.9 m in height and whose upholstery system was qualified only for HL1, met the requirements adopted for HL2.

As a part of the further analysis, the relationship between the MARHE values of upholstery systems and complete seats was determined. The graph in Figure 5 shows that this relationship is not linear and can be described by a 4th degree polynomial.

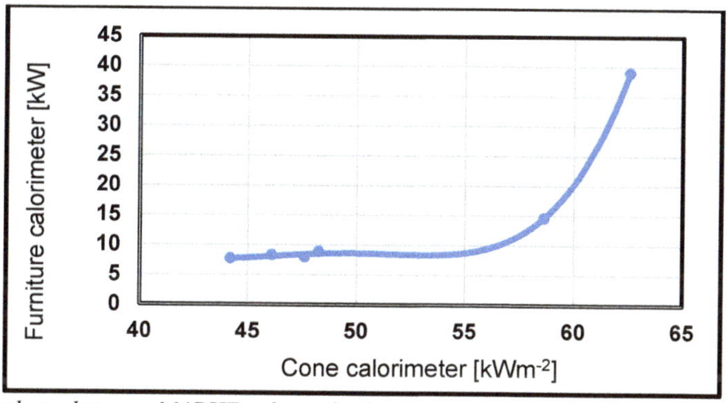

Fig. 5. Dependence between MARHE values of upholstery systems and MARHE of complete seats.

Summary

The tests showed that the seats, not protected by a fireproof fabric during combustion, are characterized by a much higher intensity of heat emission, as well as dense smoke, than the secured seats. The above phenomenon occurred also as part of tests on samples of the upholstery systems. The relationship between MARHE values of upholstery systems and MARHE of complete seats has the character of a 4- degree polynomial curve. In addition, there was no correlation in the classification of seats and their component materials. The above should be taken into account when verifying requirements as part of the ongoing revision process of EN 45545-2 + A1: 2015-12 [5], in which the Railway Research Institute participates. The proper selection of materials has a significant impact on the emergence and development of a possible fire, i.e. on the safety of travelers. Therefore, it is advisable to obtain compatibility of the criterial values of all parameters included in the classification of passenger seats.

References

[1] J. Radziszewska-Wolińska.- Passenger train fire in a tunnel, Tunnel Management International Ltd., Kempston, UK, vol. 5, No 4, 2002, ISBN 1 901808 20 3, 109-118

[2] J. Radziszewska-Wolińska, D. Milczarek Uniepalnienie materiałów niemetalowych a ich właściwości funkcjonalne, TTS 11-12 (2012), 56-59, (in Polish).

[3] COMMISSION REGULATION (EU) No 1303/2014 of 18 November 2014 concerning the technical specification for interoperability relating to 'safety in railway tunnels' of the rail system of the European Union.

[4] COMMISSION REGULATION (EU) No 1302/2014 of 18 November 2014 concerning a technical specification for interoperability relating to the 'rolling stock — locomotives and passenger rolling stock' subsystem of the rail system in the European Union.

[5] EN 45545-2+A1:2015-12 Fire protection of railway vehicles – Part 2: Requirement for fire behaviors of materials and components.

[6] ISO 5660-1:2015 Reaction-to-fire tests — Heat release, smoke production and mass loss rate — Part 1: Heat release rate (cone calorimeter method) and smoke production rate (dynamic measurement)

[7] EN ISO 5659-2:2013 Plastics- Smoke generation – Part 2: Determination of optical density by a single chamber test

[8] EN 45545-2:2013+A1:2015 Railway applications – Fire protection on railway vehicles-Part 2: Requirements for fire behaviour of materials and components

[9] ISO / TR 9705-2:2001 Reaction-to-fire tests -- Full-scale room tests for surface products -- Part 2: Technical background and guidance.

[10] EC, Firestarr Project, Final Report, European Commission (2001).

[11] M. Kumm, Carried Fire Load in Mass Transport Systems - A study of occurrence, allocation and fire behaviour of bags and luggage in metro and commuter trains in Stockholm, RESEARCH REPORT:4, Studies in Sustainable Technology, Mälardalen University, Västerås, Sweden (2010).

[12] Jie Zhu Xiao Ju Li Cheng Feng Mie - Combustion performance of flame-ignited high-speed train seats via full-scale tests, accepted manuscript in Case studies – Fire Safety, PII: S2214-398X(15)00007-2, DOI: http://dx.doi.org/doi:10.1016/j.csfs.2015.05.002, Reference: CSFS 17, Accepted date: 29-5 (2015).

[13] Hohenwarter D. - Experience Gained from Fire Tests According to EN 45545-2 and DIN 5510-2 for Testing of Seats, Railway Reports 60, 171 (2016) 27-38.

[14] Radziszewska-Wolińska J.M. – The Influence of the Ignition Source on Passenger Seat Burning, Railway Reports 60, 171 (2016) 67-78.

Terotechnology 2017 Materials Research Forum LLC
Materials Research Proceedings **5** (2018) 37-40 doi: http://dx.doi.org/10.21741/9781945291814-7

Impact of Laser Machining on the Structure and Properties of Tool Steels

ULEWICZ Robert[1, a*], SELEJDAK Jacek[2,b]

[1]Czestochowa University of Technology, Armii Krajowej 19B, Poland

[2]Czestochowa University of Technology, Armii Krajowej 19B, Poland

[a]*robert.ulewicz@wz.pcz.pl, [b] jselejdak@bud.pcz.pl

Keywords: Laser Heat Treatment, Tool Steel

Abstract. The paper presents the results of research of the impact of laser treatment on the properties of the surface layer of tool steel. Two grades of tool steel were qualified for the tests and they were subjected to a pulse laser and a continuous-mode laser treatment. In order to analyze the influence of laser beam impact on the properties of the surface layer of the studied material, metallographic investigations and microhardness profile tests were carried out in the surface layer.

Introduction

These laser machining technologies are used primarily in the machining of components from difficult-to-machine construction plastics, as well as in the manufacture of components with very complex shapes, whose production with the use of traditional method would be very labor-intensive and time-consuming. A characteristic feature of beam technologies is that materials are processed as a result of the influence of concentrated energy stream. By controlling the energy stream, we can process selected areas of a workpiece. The diameter of the spot of the influence of energy stream varies from a fraction of nanometers to several tens of centimeters [1, 2]. It depends on the basic physical phenomenon occurring in a given technology [3, 4]. Heating with a laser beam is used in the industry to modify the surface layer, e.g. hardening, tempering or modification of the microstructure [5, 6]. The remelting process also allows for changing the chemical composition of a surface layer, which is used in alloying cladding. Combined methods are also used, for example, in a thermo-chemical treatment in a fluidized bed and further laser treatment [7, 8]. Laser technologies associated with remelting require a higher density of radiation power (10^4-10^{10} W / cm^2) than the modification of a surface layer without remelting and the time of interaction of the beam with the material of 10^{-1}-10^{-9}s. High power lasers Nd: YAG and CO_2 with a continuous and pulse operation are mainly used [6]. The modification of a surface layer using laser technology is based on the process of rapid remelting and crystallization of a thin layer at the surface of the material [9, 10].

Research material

Two steel grades were used for the tests: tool steel for cold work 1.2379 and steel for hot work 1.2343. The chemical composition is shown in Table 1.

Table 1. The chemical composition of steel 1.2379 and 1.2343

steel 1.2379							
C	Mn	Si	P	S	Cr	Mo	V
1.77	0.29	0.16	0.02	0.02	11.84	0.85	0.75
steel 1.2343							
0.40	0.52	0.14	0.02	0.02	5.22	1.35	0.41

Materials Research Forum LLC
doi: http://dx.doi.org/10.21741/9781945291814-7

The microstructure of 1.2379 steel after softening annealing is shown in Figure 1. It consists of ledeburitic carbides, as well as of carbides of M_3C, MC and $M_{23}C_6$ type carbides in the ferrite matrix, whereas the steel 1.2343 microstructure is a ferrite alloy with separations of coagulated spheroidal alloy carbides. Visible banding is a result of the rolling process.

Machining parameters

Samples of steel 1.2379 in the softened state were given laser treatment with a pulsed laser with the following processing parameters: pulse energy 60 J, density energy 489.79 J/cm^2, frequency 48.5 GHz, power $2.91 \cdot 10^{12}$ W, power density $2.37 \cdot 10^{13}$ W/cm^2, exposure time $2.06 \cdot 10^{-11}$ s, area of impact of a single pulse 3.5 x 3.5 mm.

For the processing of steel 1.2343 in the softened state, a molecular laser with a wavelength of 10.6 μm and a power of 800W was used. Laser processing was performed by reducing the value of re-focus Δf and speed 12mm/s. The diameter of the beam equals 22.4 mm. The radiation was focused on a lens with a focal length of f = 96mm.

Research results and discussion

As a result of hardening with a pulsed laser of 1.2379 steel after softening annealing, a four-zone model of the surface layer structure was obtained.

In this model, we can distinguish three zones. The first zone (light-white) in which hard-to-digest hid acicular martenite was found and large amounts of residual austenite were found. The thickness of this zone depends on the heat dissipation rate from the surface layer and reaches 0.06 mm. Research confirms that this zone will be characterized by low hardness. When using this technology for machining tools, e.g. cutting knives machining will be necessary to remove this zone. The second zone contains martensite, primary carbides and residual austenite. The last third zone is a zone of heat influence, which was created as a result of the influence of heat coming from the laser pulse. Microscopic images of the zonal structure of the surface layer suggest that the energy distribution in the impulse is homogenous, which manifests itself with the same width of individual zones in a given sample. In the surface layer, we obtained a uniform martensite structure, which is related to the absence of the phenomenon of tempering of the material between successive areas of the laser pulse.

In the case of steel 1.2343, the top layer consists of three following spheres: a light white zone, created from the remelting material, a hardened warp zone and transition zone. All zones together have a total thickness of 162.5 μm, while the total width of the hardened zone at the surface is 1.15 mm. The light white zone has a fine-grained martensite structure that was created as a result of melting of the material. This zone reaches a depth of 37.5 μm. The second zone, which is a layer of a hardened matrix material, is 24.5 μm thick. Carbides, martensite and residual austenite are the structures of this zone. The transition zone between the hardened layer and the warp is about 100 μm.

In order to confirm the zone structure and determine the area of occurrence of individual zones, microhardness tests were carried out in the area of the surface layer. The microhardness tests were carried out in accordance with EN ISO 6507-3:2007 standards, hardness measurement with the Vickers method below HV 0.2. The load of 0.4903 N was used in the tests. At a distance of 0.5 mm from the edge of the sample in the middle of the pulse, a microhardness measurement was carried out for steel 1.2379 in a line perpendicular to the surface of the sample, passing through successive zones of heat influence. The profile of microhardness of the surface layer of steel 1.2379 is shown in Figure 1.

The microhardness tests confirmed the existence of a zoned structure of the surface layer of the material after the pulse laser processing of steel 1.2379. A slight increase in hardness to the level of approximately 550 HV005 is observed at the surface, in the next zone we observe an increase in hardness to the level of about 800 HV005, which confirms the occurrence of hid acicular martensite and residual austenite. The next zone is characterized by a decrease in hardness in relation to zone 2. The last zone is a zone of heat influence.

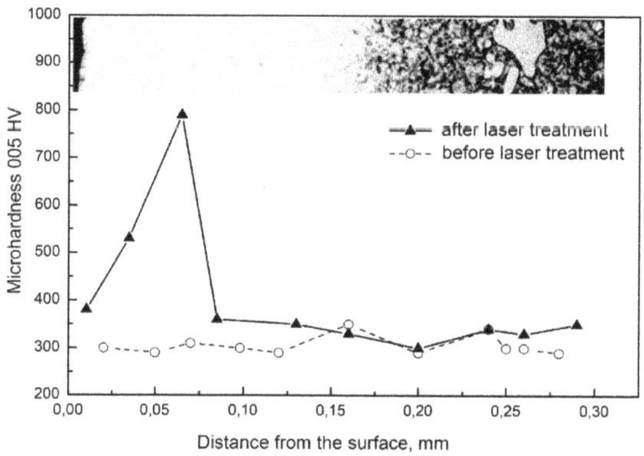

Fig.1. Profile of microhardness of steel 1.2379

In the case of steel 1.2343, a microhardness measurement was made perpendicular to the surface along the center line (laser beam impact paths). The conducted research confirmed the zonal structure, the first zone of fine-grain martensite has hardness of above 1000 HV005. In the second zone we observe a decrease in hardness. In this zone there are carbides, martensite and residual austenite. In the third zone, we still observe a decrease in hardness associated with a decrease in the share of martensite. The profile of microhardness of the surface layer of steel 1.2343 is shown in Figure 2.

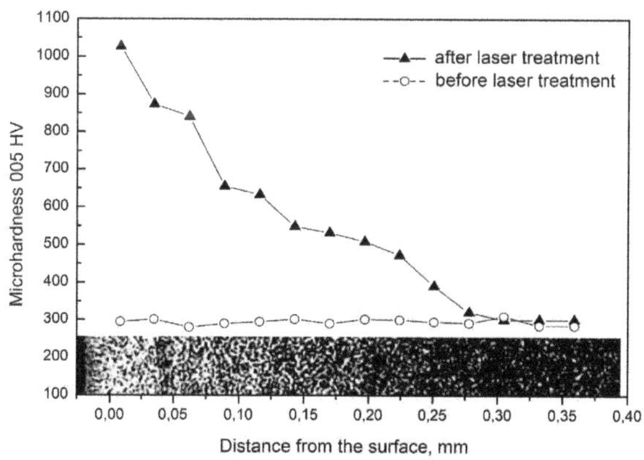

Fig. 2. Profile of microhardness of steel 1.2343

Summary

The amount of energy introduced into a material depends on the energy distribution in the beam, the time the beam interacts with the material and the absorption coefficient. The use of a pulsed laser enables obtaining an even distribution of energy in the beam, which was confirmed by metallographic and microhardness measurements. The interaction of the laser pulse ensures the formation of a homogeneous martensite structure in the top layer. As a consequence, we obtain a uniform state of stress, there is no phenomenon of material separation between the areas of impact of the impulse. Steel 1.2379 treated with pulsed laser has a zone structure described by a four-zone model. In case of steel 1.2343 and continuous laser treatment, there are thermal interactions with the stress fields. Also in this case there is a zone construction described by the three-zone model. The highest hardness was noted in the area of fine-grained martensite resulting from remelting. Between the paths, the phenomenon of tempering material was noted. To sum up, the properties of the surface layer of the material treated with a laser beam are characterized by a higher resistance to wear in relation to the starting material and in relation to conventional machining methods.

References

[1] P. Szataniak, R. Ulewicz, F. Novy, HSLA steels - comparison of cutting techniques, 23rd International Conference on Metallurgy and Materials, Metal (2014) 778-783.

[2] N. Radek, B. Antoszewski, I. Pliszka, J. Świderski, J. Pietraszek, Production of heterogeneous surfaces by ESD and LBM, Production Engineering Archives 13 (2016) 44-48. https://doi.org/10.30657/pea.2016.13.10

[3] M. Scendo, J. Trela, N. Radek, Influence of laser power on the corrosive resistance of WC-Cu coating, Surface & Coatings Technology 259 (2014) 401-407. https://doi.org/10.1016/j.surfcoat.2014.10.062

[4] M. Bramowicz, S. Kulesza, T. Lipiński, P. Szabracki, P. Piątkowski, Fractal analysis of AFM data characterizing strongly isotropic and anisotropic surface topography, Solid State Phenomena 203 (2013) 86-89. https://doi.org/10.4028/www.scientific.net/SSP.203-204.86

[5] A. Wrońska, A. Dudek, Characteristics of surface layer of sintered stainless steels after remelting using GTAW method, Archives of Civil and Mechanical Engineering, 14/3 (2014) 425-432. https://doi.org/10.1016/j.acme.2013.10.011

[6] YC. Zhang, L. Yang, TY. Chen, WH. Zhang, XW. Huang, J. Dai, Investigation on the optimized heat treatment procedure for laser fabricated IN718 alloy, Optics and Laser Technology 97 (2017) 172-179. https://doi.org/10.1016/j.optlastec.2017.06.027

[7] J. Lago, O. Bokuvka, F. Novy, The weld toe improvement of advanced HSLA steel by laser remelting, Materials Today-Proceedings, 3/4 (2016) 1037-1040. https://doi.org/10.1016/j.matpr.2016.03.043

[8] N. Radek, J. Pietraszek, B. Antoszewski, The Average Friction Coefficient of Laser Textured Surfaces of Silicon Carbide Identified by RSM Methodology, Adv. Mater. Res.-Switz. 874 (2014) 29-34.

[9] M. Sebek, L. Falat, F. Kovac, I. Petryshynets, P. Hornak, V. Girman, The Effects Of Laser Surface Hardening On Microstructural Characteristics And Wear Resistance Of Aisi H11 Hot Work Tool, Archives Of Metallurgy And Materials, 62/3 (2017) 1721-1726. https://doi.org/10.1515/amm-2017-0262

[10] E. Schmidova, P Savada, Behavior and Degradation Of Stabilized Corrosion Resistant Steel, 41. Mezinarodni Konference O Naterovych Hmotach, KNH (2010) 281-288.

Materials Research Proceedings 5 (2018) 41-46 doi: http://dx.doi.org/10.21741/9781945291814-8

On the Significance of Structure-Property Relationships

WOJNAR Leszek

Institute of Applied Informatics, Cracow University of Technology, al. Jana Pawła II 37, 31-864 Cracow, Poland

leszek.wojnar@mech.pk.edu.pl, leszek.wojnar@gmail.com

Keywords: Structure-Property Relationship, Microstructructure, Structure, Material Properties, Stereology

Abstract. The importance of deep knowledge of structural information in materials engineering is explained and discussed. Some basic relations, like Hall-Petch relation or linear dependency of hardness and phase composition are presented and discussed. Simultaneously, the difficulties in precise development of structure-property relationships are explained and selected examples of works devoted to structure-property relationships are discussed.

Introduction

If we analyze either the lectures or textbooks devoted to materials engineering, we always hear or read two terms: properties and structure. This is typical for old as well as for modern lectures and textbooks [1-3]. In commercial applications extremely important is also a price or production cost, but this item is not discussed in this paper. Properties describe and usually quantify the behavior of the materials under various loads and in various environments. So, properties constitute possibly the most important set of information for the potential user of the material. Usually, properties are discussed in conjunction with technological operations that allow for obtainment or modification of them.

The students are also forced to learn about the microstructure. Unfortunately, in most cases they have a lot of difficulty to answer a very simple question: Why microstructure is so important for the material? A typical answer is that microstructure decides what are the properties, but without any deeper analysis or understanding. Surprisingly, this is almost impossible to find clear and straightforward explanation of the importance of our knowledge about materials structure. Although this is not expressed in formal way, materials engineering is based on a fundamental theorem:

Two materials of identical (micro)structure have identical properties, irrespective of their origin and technological history.

The above presented theorem has no formal proof, but it is a quintessence of thousands years of experimental observations. A natural consequence of this theorem is looking for precise structure-property relationships. Having such relations coupled with the knowledge how technological processes modify (micro)structure, one can precisely choose technological operations leading to anticipated properties.

Selected basic structure-property relationships

A good example of structure-property relationship that has both theoretical background and satisfactory experimental verification is analysis of tensile strength of a dual-phase material (Fig. 1). We have two phases: phase A of high strength and phase B of low strength. The simplest case is a material with phases distributed in series. In this case the maximum possible tensile strength is equal to the strength of a weaker phase. It can be even lower if only the phase interface is very weak. Many layered materials used in civil engineering (for example, insulating foam covered by thin metallic or polymer sheets) can be good specimens. In these materials their strength in direction perpendicular to the external surface is very weak.

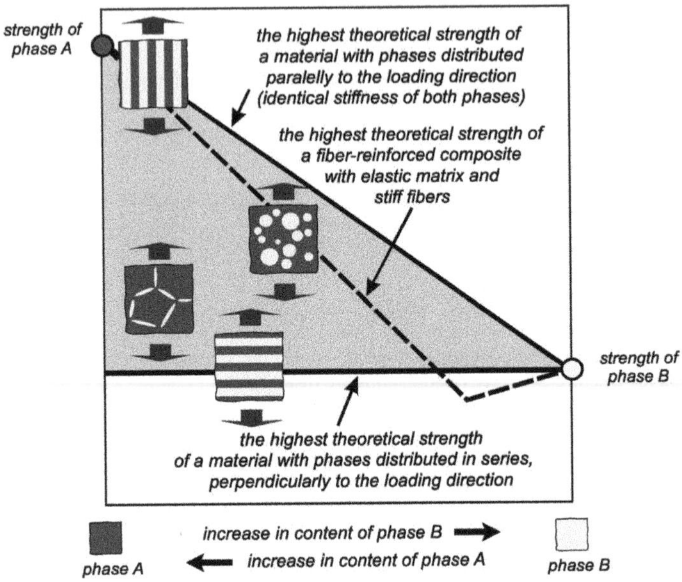

Fig. 1. Tensile strength of dual-phase materials. Details in the text.

Another extreme case is tensile strength of a material with phases distributed parallel to the loading direction (Fig. 1). If only the Young modulus is approximately equal in both phases, the final strength is equal to the weighted mean of values for both phases:

$$\sigma_T = V_V(A) * \sigma_T(A) + V_V(B) * \sigma_T(B) \tag{1}$$

where the symbols denote: tensile strength of the material, fraction of phase A, strength of phase A, fraction of phase B and strength of phase B, respectively.

The conclusion is that the material that is built of two phases has never strength higher than strength of these phases. However, this can be a good solution if one phase is resistant and very brittle and the second one has low strength but good plastic properties. Such a combination one can meet in pearlite in carbon steels (lamellar structure that consists of brittle cementite Fe_3C and ductile, soft ferrite) and many fiber-reinforced composites, for example glass fiber with polyester matrix. Unfortunately, in this case we have significant difference in the stiffness of both components (fiber are more stiff). In such a case practically all the loads are transmitted by fibers. On the other hand, in the case of very small amount of fibers they will be broken first and this leads to weakening the of matrix. In some cases a fiber-reinforced composite can be even less strong than the matrix (see dashed line in Fig. 1).

In the case of brittle fibers they have to be fully surrounded by a ductile matrix and therefore the optimum content of fibers is limited. In real composites the effect of fiber contents on the tensile strength is different than this one presented in simplified model (Fig.1). However, one can observe a local minimum in strength [3]. To conclude, even the simplified model presented above predicts the existence of optimum fiber content in fiber-reinforced composites.

One can meet also dual-phase structures with more complex arrangement of constituent phases. Considering one of them as a matrix, the second one can be distributed in a form of separated particles of different shape or in a form of more or less continuous network. The properties of such materials can be significantly different. A classic example is grey cast iron. Nodular graphite assures tensile strength comparable with steels, whereas flake graphite results in tensile strength 2-3 times lower.

Good agreement with the simple model presented in Fig. 1 demonstrates also carbon steel (part of the Fe-C equilibrium system with carbon contents up to approximately 1.2 %). In the case of ferrite-pearlite structure the tensile strength is proportional to the contents of pearlite, reaching maximum for pure pearlite (0.77 %C). Above this contents of carbon a network of brittle secondary cementite (Fe_3C) causes embrittlement of the whole material.

This is also noteworthy, that relation analogous to equation (1) describes well hardness of dual-phase materials. In the case of hardness, it can be also adapted for multiphase materials.

Another well documented relation bounds grain size and yield point. Hall [4] noticed that grain size establishes limit for slip space for dislocations responsible for plastic deformation of mild steels. Independently, Petch [5] obtained a similar relation, but his work concentrated more on brittle fracture. Their results are fully consistent and therefore this important relationship is named after both, Hall and Petch. The Hall-Petch relation illustrates experimental observation that yield point increases with decreasing grain size and can be expressed as:

$$\sigma = a + b * \sqrt{S_V} \tag{2}$$

where: σ - yield point of the material, a and b – constant values that depend on the material, S_V – specific surface of grain boundaries (ratio between surface area of grain boundaries and volume of the material).

The Hall-Petch relation also has limitations. In the case of extremely small grains (nan grains), smaller than approximately 1 micrometer, the yield point either remains constant or decreases with the decreasing grain size. So, this is opposite tendency if compared with equation (2) and therefore is called reverse Hall-Petch relation.

The above presented analysis demonstrates that even the simplest structure-property relations can highly non-linear and difficult to establish.

Stereology as a tool for microstructure quantification

In order to build analytical structure-property relations one need to quantify microstructure. In other words, we necessitate objective measures that allow to characterize microstructure by numbers. Appropriate solutions are given by stereology that defines both appropriate measures and tools for their evaluation. The principles of stereology are well described in the world famous monograph by Ervin E. Underwood [6]. In spite of big progress in measurement techniques only few procedures for microstructure quantification have been standardized. In metallography one can find ASTM standards for evaluation of grain size, non-metallic inclusions in steels or graphite in cast iron. The measurements are used mainly for quality control. Taking into account the theorem presented in *Introduction* to this paper this is postulated that materials showing similar results in quantitative microstructure control have also similar properties. This is noteworthy that from such quality control there is a long distance to development of structure-property relationship. Currently, quantification of microstructures is often automated thanks to image analysis equipment and software. In dedicated software appropriate procedures for typical structures and standardized methods are pre-defined and ready to use. However, in many cases one have to develop own procedures. Numerous solutions, described and explained step-by-step can be found in the monograph [7].

Development of structure-property relations on the basis of theoretical consideration is very difficult, as our analytical models are usually too simplified. Successful solutions are usually limited to a narrow group of materials, however differences between theoretical and real results can be surprisingly small. More common is the use of statistical analysis for establishing correlations between structural parameters and properties. One example can be evaluation of the effect of trabecular bone on compression strength of vertebral bodies in spine. On the basis of quantitative data from 3D micro tomography [8] the following relation can be evaluated:

$$R_T = 93.308\, V_V - 0.771\, N_V + 22.438\, degr - 2.77 \qquad (3)$$

where: R_T – compression strength [MPa], V_V – volume fraction of bone branches, N_V – relative number of junction points [1/mm^3] and *degr* – degradation coefficient; ratio between end points and junction points of bone branches.

The above presented equation gives very good correlation between the apparent and evaluated values with correlation coefficient r = 0.952 and mean relative error γ = 0.0952. Stereological parameters are used in this equation. However, it is difficult for physical interpretation, what is a weakness of statistical analysis. Nevertheless, closer analysis of this equation allows for postulating that the volume of bone branches is responsible for nearly 80 % of the bone strength.

Other examples

Shen and co-authors [9] demonstrated that the properties of the nanocomposites are determined by the intrinsic structure. The experimental results showed that the attapulgite silicates caused that the polymer network exhibits reinforcement in thermal and mechanical properties and slower relaxation in segments mobility compared with neat polyamides. The authors present many results of a wide spectrum of experiments. However, no clear structure-property relation has been established.

Ivanciuc and co-authors [10] presented a sophisticated method starting from molecular level analysis and obtained satisfactory approximation of numerous parameters, like, for example: boiling temperature, molar heat capacity, standard Gibbs energy, density and other quantities for alkanes. Their work was based on numerous computer programs used in chemistry.

Cheng and Ma [11] presented an extended work on properties of metallic glasses. They concluded that in these materials it is extremely difficult to investigate structure-property relationships as they have amorphous structure. Analysis of structure-property relationships is based on local non-homogeneity in distribution of single atoms and local structural organization. Some systematic relations were discovered, however they had rather qualitative character. This work demonstrated how both difficult and important analysis of the structure-property dependency is.

Ray and co-authors [12] studied rheological properties of organically modified layered silicate nanocomposites that are strongly influenced by their nanostructure and interfacial characteristics. The nanostructure was investigated by means of transmission electron microscopy and wide-angle X-ray diffraction. The images obtained using electron microscopy demonstrated how difficult such a structure is for any interpretation. Finally, after numerous experiments the authors demonstrated significant differences between two different biodegradable composites. This work also demonstrated that the internal structure of a material is decisive for its properties but precise quantitative characterisation of this effect is very difficult and requires extended further research. Deeper understanding of the structure-property relationship is in this case necessary for effective design of new nanocomposites or improvement in the already existing ones.

Gangopadhyay and co-authors [13] measured tribological properties of TiN-MoS_x composite coatings and various parameters related to the coating structure. Increase in MoS_x content was associated with decrease in grain size. From the structure-properties relationships viewpoint it was shown that the (Mo+S) content affects Vickers microhardness, critical load during scratch test and wear coefficient. However, each of these relations exhibited different character. Microhardness decreased with growing (Mo+S) content, critical load had a maximum corresponding to higher (Mo+S) values and wear coefficient had a minimum at lower (Mo+S) content. So, only one of these relations was monotonic.

This brief overview demonstrates that structure-property relationships are of interest for various groups of researchers, working on different materials and in different scale – from macro up to nanoscale. In all the cases deep understanding of the material structure is necessary and the structure-property relations has to be individually fitted to the discussed case.

Conclusions

Structure-property relationships are important for analysis of the properties and development of new or optimized materials of different character. A very complex nature of various processes and relations within materials causes that evaluation of analytical structure-property relations is very difficult. Usually these relations are nonlinear and often even discontinuous. Statistical analysis of the results can give correlation equations that can good results of prediction but simultaneously are almost useless for physical interpretation of the materials behavior. Numerous papers devoted to structure-property relations prove the importance of this problem. However, in most cases no clear analytical relation between structure and properties is possible to be evaluated and presented.

References

[1] L.A. Dobrzański, Podstawy nauki o materiałach I metaloznawstwao, Wydawnictwa Naukowo-Techniczne, Gliwice-Warszawa, 2002.

[2] A.G. Guy, Wprowadzenie do nauki o materiałach, Wydawnictwa Naukowo-Techniczne, Warszawa, 1977.

[3] O.H. Wyatt, D. Dew-Hughes, Wprowadzenie do inżynierii materiałowej. Metale, ceramika i tworzywa sztuczne, Wydawnictwa Naukowo-Techniczne, Warszawa, 1978.

[4] E.O. Hall, (1951), The Deformation and Ageing of Mild Steel: III Discussion of Results, Proc. Phys. Soc. London 64 (1951) 747–753. https://doi.org/10.1088/0370-1301/64/9/303

[5] N.J. Petch, N.J. (1953), The Cleavage Strength of Polycrystals, J. Iron Steel Inst. London 173 (1953) 25–28.

[6] E.E. Underwood, Quantitative stereology, Reading, MA, 1970.

[7] L. Wojnar, Image analysis. Applications in materials engineering, CRC Press LLC, Boca Raton, Fl, 1999.

[8] Opracowanie i optymalizacja metod diagnozowania ryzyka złamania w osteoporozie na podstawie analizy trójwymiarowych obrazów kości beleczkowej uzyskanych w warunkach in vivo, L. Wojnar, Ed., Politechnika Krakowska, Instytut Informatyki Stosowanej, Kraków 2012.

Materials Research Forum LLC
doi: http://dx.doi.org/10.21741/9781945291814-8

[9] L. Shen, Y. Lin, Q. Du, W. Zhong, Studies on structure–property relationship of polyamide-6/attapulgite nanocomposites, Composites Science and Technology, 66 (2006) 2242-2248. https://doi.org/10.1016/j.compscitech.2005.12.005

[10] O. Ivanciuc, T. Ivanciuc, D. Cabrol-Bass, A.T. Balaban, Evaluation of quantitative structure-property relationship models of structural descriptors derived from information-theory operators, J. Chem. Inf. Comput. Sci., 40 (2000) 631-643. https://doi.org/10.1021/ci9900884

[11] Y.Q. Cheng, E. Ma, Atomic level structure and structure-property relationship in metallic glasses, Progress in Materials Science, 56 (2011) 379-473. https://doi.org/10.1016/j.pmatsci.2010.12.002

[12] S.S. Ray, K. Okamoto, M. Okamoto, Structure-property relationship in biodegradable poly(butylene succinate)/layered silicate nanocomposites, Macromolecules 36 (203) 2355-2367.

[13] S. Gangopadhyay, R. Acharya, A.K. Chattopadhyay, S. Paul, Composition and structure-property relationship of low friction, wear resistant TiN-MoS_x composite coating deposited by pulsed closed-field unbalanced magnetron sputtering, Surface and Coatings Technology, 203 (209) 1565-1572. https://doi.org/10.1016/j.surfcoat.2008.11.004

The Tribological Properties of Titanium Carbonitride TiCN Coating Lubricated with Non-Toxic Cutting Fluid

MADEJ Monika[1, a], KOWALCZYK Joanna[1, b,*], OZIMINA Dariusz[1, c], MILEWSKI Krystian[2, d]

[1] Kielce University of Technology, Faculty of Mechatronics and Mechanical Engineering, Tysiąclecia PP 7, 25-314 Kielce

[2] Trzuskawica S.A., Sitkówka 24, 26-052 Nowiny

[a]mmadej@tu.kielce.pl, [b]jkowalczyk@tu.kielce.pl, [c]ozimina@tu.kielce.pl, [d]krystian.milewski@trzuskawica.pl

Keywords: Non-Toxic Cutting Fluid, Wear, Friction, TiCN Coating

Abstract: This paper presents the results of tribological studies of TiCN coatings formed on the HS6-5-2C steel by plasma-assisted chemical vapour deposition (PVD). The coating thickness was measured with the optical microscope. The surface texture analysis was performed before and after the tribological tests with a Talysurf CCI Lite optical profiler. The tribological properties were investigated using a T-01M tribotester for a ball-on-disc configuration in the sliding contact. The tests were carried out under lubricated friction conditions using lubricants. Cutting fluids and coatings have contributed to the reduction of resistance movement. The non-toxic cutting fluid has reached comparable coefficient of friction with a coolant containing mineral oil.

Introduction

Technological developments have created the need for coatings that meet demanding applications of advanced materials. Single layer and multilayer coatings, widely used especially on cutting tools [1-3], increase the quality of machined parts and enable high speed cutting [1]. Titanium carbonitride (TiCN) coating is characterized by high hardness and low friction, can be applied by physical deposition (PVD) at temperatures ranging from 300 to 450°C [3], exhibits high melting temperature [4], anti-wear behaviour, chemical stability and corrosion resistance [5, 6].

The effect of tools with coatings on the quality of surface treated and durability of the tool were examined [7, 8] to find that coated tools had lower wear than conventional high-speed steel tools. Two studies [9-11] describe experiments that involved machining with biodegradable, non-toxic coolants based on plant oils and oil-containing cutting fluids, in which the roughness of work materials and tool wear were compared. The results showed that the properties of plant oil coolants were comparable to or better than the properties of coolants containing mineral oil.

To reduce tool wear and improve the surface quality of a workpiece, cutting tool lubricants are used during machining for cooling, lubricating, transporting chips from the machining zone, reducing the coefficient of friction, and temporary protection against corrosion [12-14]. Lubricants are applied in over 80% of metalworking operations [15]. Widely used mineral oil-based coolants are usually toxic. Plant oils have become a renewable, hazard-free and easily biodegradable alternative, characterized by good lubricity, low volatility, higher fire resistance and good thermal stability. Currently, plant oil based metal working fluids are the area of extensive research [16-18].

Materials and methods

The tests were performed on a TiCN coating deposited on the HS6-5-2C high-speed tool steel by the PVD. The surface and cross-sections of the TiCN coating as well as its chemical analysis in the

Materials Research Forum LLC
doi: http://dx.doi.org/10.21741/9781945291814-9

micro-regions were carried out using a Jeol JSM 7100F field emission scanning electron microscope with an EDS analyser.
A Talysurf CCI Lite interferometer was used to analyse the geometric structure of the specimen and counter-specimen surfaces before and after the tribological tests [16].

The tribological tests were carried out on a T-01M ball (100Cr6 steel, Ø = 10 mm)-on-disk (HS6-5-2C steel without coating or with the TiCN coating, Ø = 38 mm) testing machine in accordance with the requirements of ASTM G 99. Friction tests were performed at a load of P = 50 N, a sliding speed of v = 0.1 m / s, along the friction path s = 1000 m, at the constant relative humidity of 40 ± 5% and ambient temperature T0 = 23 ± 1°C under technically dry friction, lubricated friction with a mineral oil-based coolant and with a non-toxic coolant, consisting of alkanolamine borate, a biodegradable polymer - zinc aspartate and DEMI demineralised water.

Test results
Figure 1a shows an image of the TiCN coating surface morphology. Figures 1b and 1c show the results of elemental analysis in the micro-regions of the coating.

Content, [% weight]	C	N	Ti	Fe	Total
	20,78	31,37	47,46	0,39	100,00

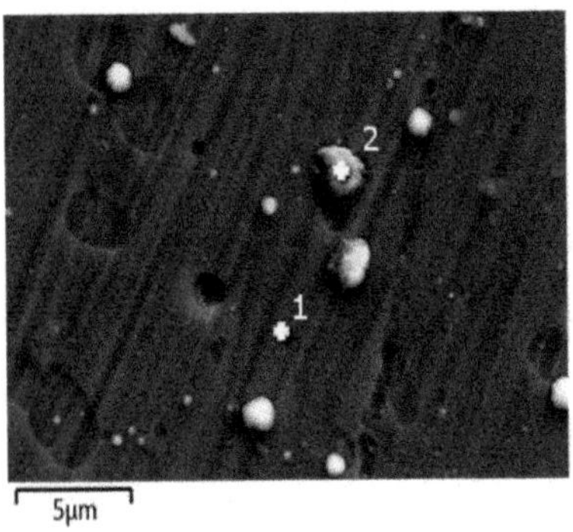

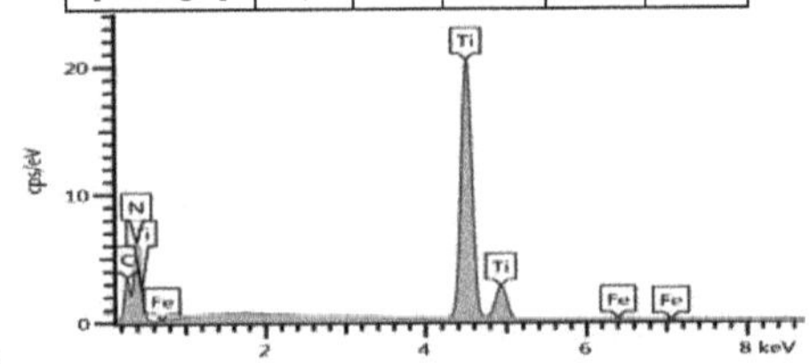

b)

Content, [% weight]	C	N	O	Si	Ti	Total
	12,95	38,13	3,40	0,12	45,90	100,00

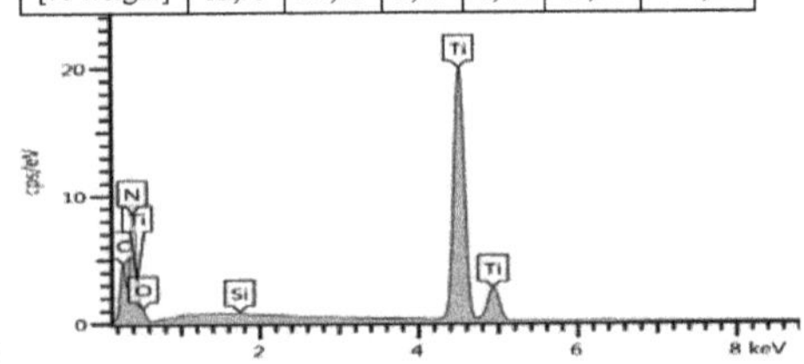

a) c)

Fig. 1. SEM: a) view of the disk area and the EDX radiation spectrum with the result of the chemical analysis for the TiCN coating: b) at point 1, c) at point 2.

The elemental analysis in the micro-region of the TiCN coating (Fig. 1b) showed the presence of titanium, nitrogen and carbon as components of the coating and iron from the substrate. In the image of the surface morphology (Fig. 1a), droplet shaped particles of various sizes were observed. They were most likely formed during the process of coating deposition as a result of titanium droplets splashing on the surface of the substrate. The analysis of the chemical composition showed that in addition to elements such as titanium, nitrogen and carbon, oxygen and silicon were also present (Fig. 1c). In the following testing stage, metallographic specimens were made, cross-sections of the coating were observed and its thickness of approx. 2 µm, was measured.

Materials Research Proceedings **5** (2018) 47-53 doi: http://dx.doi.org/10.21741/9781945291814-9

Figure 2 shows the force-penetration depth changes and the view of the impression in the TiCN coating on the HS6-5-2C steel after the nanohardness measurement. The nanohardness results are summarized in Table 1.

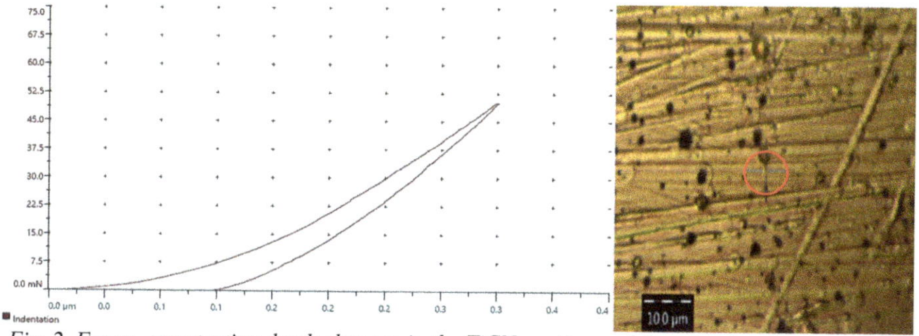

Fig. 2. Force - penetration depth changes in the TiCN coating and the microhardness impression.

Table 1. Results of nanohardness measurements for TiCN coating deposited on the HS6-5-2C steel.

Parameter	H_{IT}, [GPa]	E_{IT}, [GPa]	E*, [GPa]	H_{VIT}, [HV]	C_{IT}, [%]	F_{max}, [mN]	h_{max}, [mm]	S, [mN/mm]	A_p, [mm^2]	W_{elst}, [nJ]	W_{plast}, [nJ]	W_{total}, [nJ]
TiCN	38,18	311,57	342,38	3 535,70	0,36	49,99	0,32	352	1,31	4	1	5

The analysis of the loading-unloading curve indicates elastic properties of the TiCN coating, as can also be deduced from the values of the elastic modulus E_{IT} = 311.57 GPa and the reduced elastic modulus E* = 342.38 GPa. Hardness measured by instrumental indentation according to the Oliver and Pharr method equals 38.18 GPa and the Vickers hardness equals 3,535.70 HV. The TiCN coating is characterized by a small creep module C_{IT} - 0.36%, a small contact area A_p - 1.31 mm2, plasticity - 1 nJ and 4 times higher elasticity - 4 nJ.

Figure 3 shows the coefficients of friction recorded during technically dry friction and lubricated friction with a mineral oil-based coolant and coolants with zinc aspartate for the HS6-5-2C steel - 100Cr6 steel, and HS6-5-2C steel with the TiCN coating - 100Cr6 steel pairs.

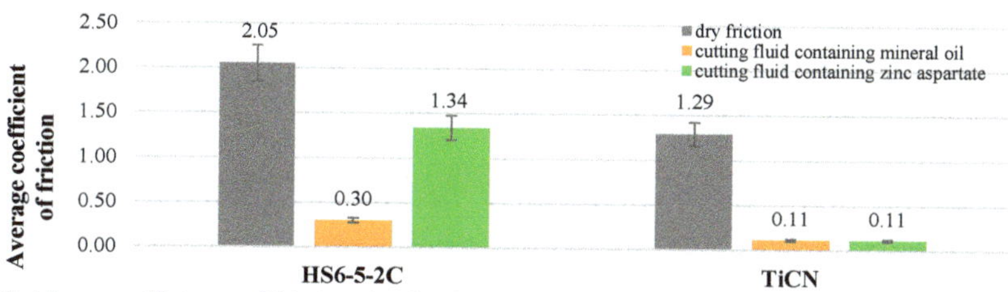

Fig. 3. Mean coefficients of friction for the friction pairs tested under technically dry friction and with the use of coolants.

The use of coolants reduced the values of mean coefficient of friction. The lowest mean friction coefficients of 0.11 were recorded for the TiCN coating under lubricating conditions both with mineral oil and zinc aspartate based working fluids. For the HS6-5-2C steel disk, the mineral oil-containing coolant turned out to be better. It can therefore be concluded that the use of zinc aspartate as a lubricant in combination with the TiCN coating does not reduce the coefficient of friction but reduces the effect of toxic agents contained in mineral oils.

Figures 4 ÷ 7 show the topography and primary profiles of disks after friction tests.

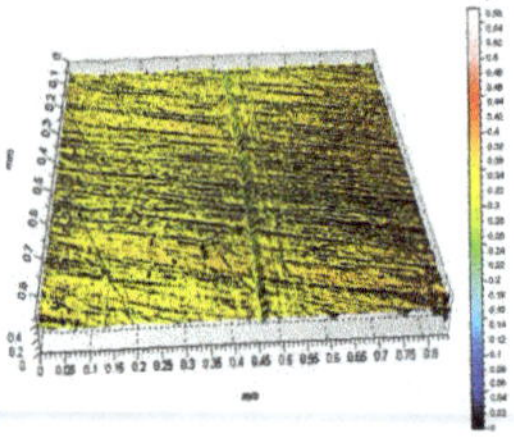
a)

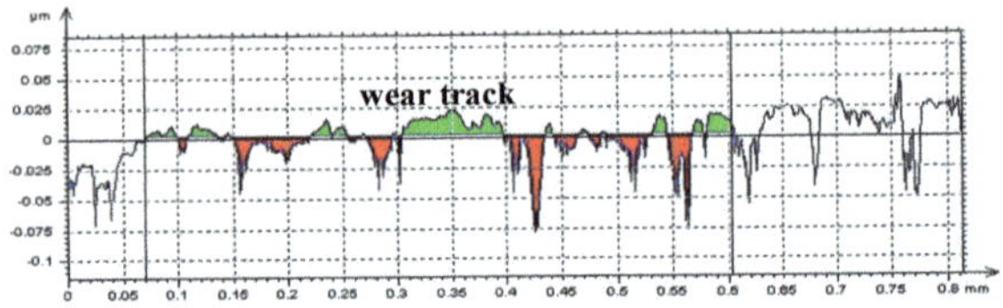

b)

Max. depth, [nm]	Max. height, [nm]	Hole field, [μm²]	Area outside, [μm²]
80,717	21,014	4,012	2,176

Fig. 4. Geometric structure of the HS6-5-2C steel disk: a) isometric image, b) primary profile after friction using a coolant containing mineral oil.

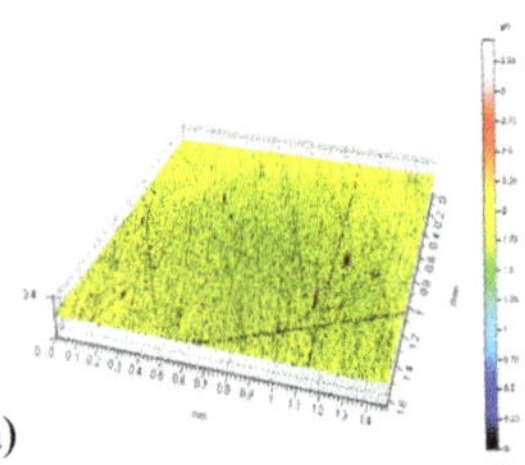
a)

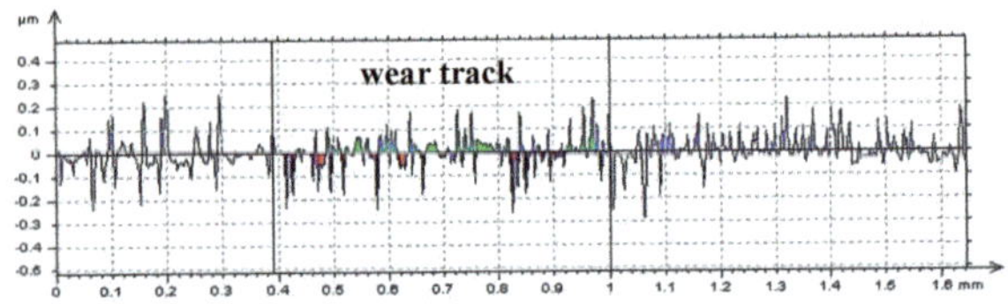

b)

Max. depth, [nm]	Max. height, [nm]	Hole field, [μm²]	Area outside, [μm²]
262,740 nm	226,250 nm	13,876 μm²	14,072 μm²

Fig. 5. Geometric structure of the HS6-5-2C steel disk with the TiCN coating: a) isometric image, b) primary profile after friction using a coolant containing mineral oil.

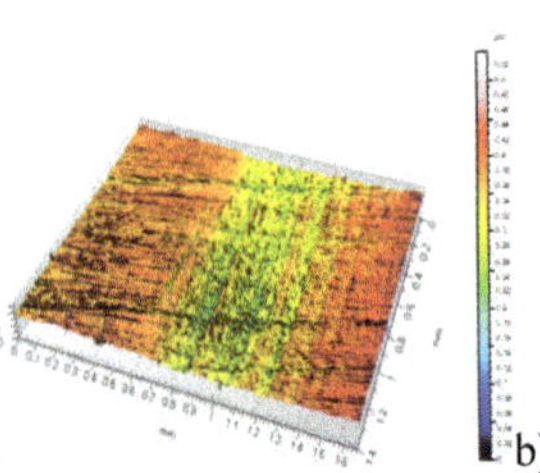
a)

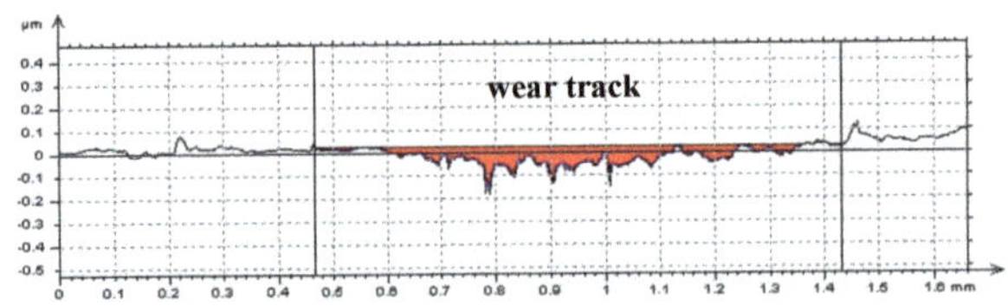

b)

Max. depth, [nm]	Max. height, [nm]	Hole field, [μm²]	Area outside, [μm²]
219,087 nm	10,152 nm	51,979 μm²	0,164 μm²

Fig. 6. Geometric structure of the HS6-5-2C steel disk: a) isometric image, b) primary profile after friction using a coolant containing zinc aspartate.

a) 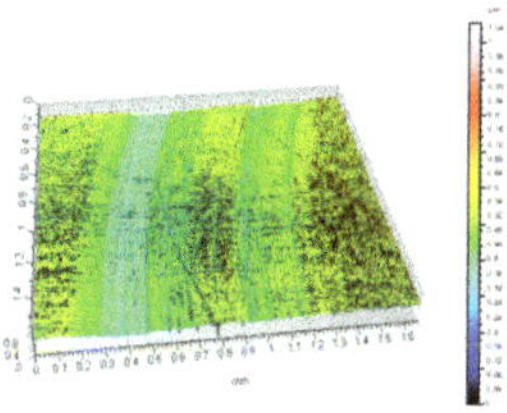

b) 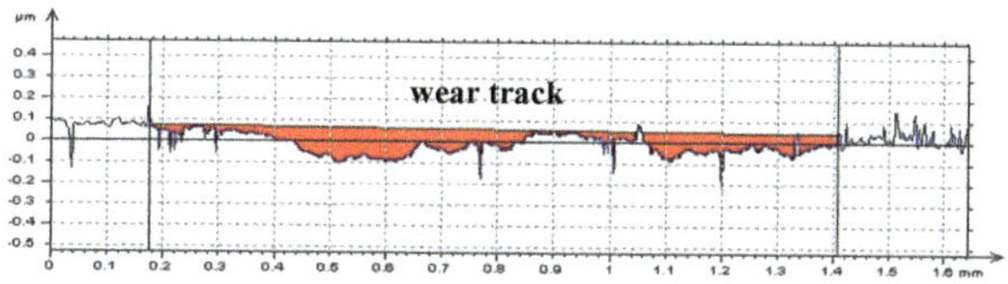

Max. depth, [nm]	Max. height, [nm]	Hole field, [μm^2]	Area outside, [μm^2]
268,490 nm	34,700 nm	99,094 μm^2	0,180 μm^2

Fig. 7. Geometric structure of the HS6-5-2C steel disk with the TiCN coating: a) isometric image, b) primary profile after friction using coolant containing zinc aspartate.

The analysis indicates that the wear of friction pair material was lower when the coolant containing mineral oil was used. After the tribological tests under lubricated conditions with the mineral oil coolant, build-ups and recesses formed on the disks, whereas after the tests with the zinc aspartate coolant only recesses were observed.

Table 2 shows the disk and ball surface geometric structure parameters before and after the tribological tests with the working fluids used.

Table 2. Parameters of the surface geometric structure

Test	DISC	Surface roughness parameters						
		Sa,[µm]	Sq,[µm]	Sp,[µm]	Sv,[µm]	Sz,[µm]	Ssk,[-]	Sku,[-]
BEFORE TESTS	HS6-5-2C	0,04	0,07	0,84	0,62	1,45	-1,03	8,94
	TiCN	0,23	0,29	0,67	1,11	1,78	-0,76	4,10
AFTER WET FRICTION WITH CUTTING FLUID CONTAINING MINERAL OIL	HS6-5-2C	0,02	0,02	0,14	0,27	0,40	-2,19	10,97
	TiCN	0,07	0,10	0,32	0,26	0,58	-0,41	4,06
AFTER WET FRICTION WITH CUTTING FLUID CONTAINING ZINC ASPARTATE	HS6-5-2C	0,03	0,03	0,23	0,31	0,54	-1,05	7,45
	TiCN	0,04	0,06	0,22	0,30	0,52	-1,33	8,00

The analysis of the results shows that the uncoated HS6-5-2C steel disk has the lowest values of amplitude Sa, Sq, Sp, Sv, Sz parameters after the tests carried out under lubrication with the mineral oil coolant. This indicates a smooth surface. The distribution of ordinates of this surface is characterized by high asymmetry and steepness. In turn, the lowest values of the amplitude parameters, i.e., a smoother surface of the disk with the TiCN coating, were observed after the tests using the zinc aspartate based coolant.

Summary

The thin, hard, elastic TiCN coating used in the study, obtained in the PVD process and working in tribosystems perfectly fulfills its function, i.e., it contributes to the low coefficient of friction, as was confirmed by the results of tribological tests, where the lowest values were obtained for the configuration with a TiCN coated disk after friction with lubrication using both mineral oil and zinc aspartate working fluids.

In the analysis of the obtained geometric structure of the disk surface after tests with a coolant containing mineral oil, a smoother surface was observed in the HS-5-2C steel disk, and after lubrication with a zinc aspartate coolant in the disc with the TiCN coating.

Considering the results obtained, the coolant containing zinc aspartate works better with the TiCN coating. This proves that the zinc aspartate based coolant is in some cases a better choice than the mineral oil based working fluid, as it can limit the use of toxic, commonly used cutting fluids.

References

[1] I. Peter, M. Rosso, Manufacturing, Composition, Properties and Application of Sintered Hard Metals, in: L. Dobrzański, Powder Metallurgy - Fundamentals and Case Studies, InTech, Croatia, 2017. https://doi.org/10.5772/66872

[2] J. Pietraszek, N. Radek, K. Bartkowiak, Advanced statistical refinement of surface layer's discretization in the case of electro-spark deposited carbide-ceramic coatings modified by a laser beam, Solid State Phenom. 197 (2013) 198-202. https://doi.org/10.4028/www.scientific.net/SSP.197.198

[3] Y. Iwai, Y. Nanjo, K. Okazaki, M. Tao, E. Sentoku, Application of Micro Slurry-Jet Erosion (MSE) for the Evaluation of Surface Properties of PVD TiN / TiCN Two-Layer Coatings, Tribology online. 12, 2 (2017) 49-57. https://doi.org/10.2474/trol.12.49

[4] Ch. P Rana, B. K Mishra, K. Abhishek, N. Patel, Comparing coating and uncoating on a cutting tools–a review, IJMTER. 4, 2 (2017) 85-91. https://doi.org/10.21884/IJMTER.2017.4057.KGR6P

[5] M. Madej, K. Marczewska-Boczkowska, D. Ozimina, Effect of tungsten on the durability of diamond-like carbon coatings in the chemical industry, Przemysł Chemiczny. 93, 4 (2014), 505-505.

[6] M. Madej, The effect of TiN and CrN interlayers on the tribological behavior of DLC coatings, Wear. 317, 1-2, (2014) 179-187.

[7] Jr J. de Paiva, R. Torres, F. Amorim, D. Covelli, M. Tauhiduzzaman, S. Veldhuis, G. Dosbaeva, G. Rabinovich, Frictional and wear performance of hard coatings during machining of superduplex stainless steel, The International Journal of Advanced Manufacturing Technology. 92, 1 – 4 (2017) 423 – 432.

[8] W. Tillmann, C. Schaak, D. Biermann, R. Aßmuth, S. Goeke, Robot based deposition of WC-Co HVOF coatings on HSS cutting tools as a substitution for solid cemented carbide cutting tools. IOP Conference Series: Materials Science and Engineering. 181 (2017) 1-9. https://doi.org/10.1088/1757-899X/181/1/012011

[9] R. Somashekaraiah, P. Suvin, D. Gnanadhas, S. Kailas, D. Chakravortty, Eco-Friendly, Non-Toxic Cutting Fluid for Sustainable Manufacturing and Machining Processes, Tribology Online. 11, 5 (2016) 556-567. https://doi.org/10.2474/trol.11.556

[10] J. Z. Zhang, P. Rao, M. Eckman, Experimental evaluation of a bio-based cutting fluid using multiple machining characteristics, IJME. 12, 2 (2012) 35-44.

[11] N. Ademoh, J. Didam, D. Garba, Investigation of Neem Seed Oil as an Altanative Metal Cutting Fluid, American Journal of Mechanical Engineering. 4, 5 (2016) 191-199.

[12] G. T Smith, Cutting Tool Technology.Industrial Handbook, Springer, London, 2008.

[13] M. Shah, V. Potdar, Study of the effect of Vegetable oil based cutting fluid on machining characteristics of AISI 316L Steel, Recent Trends in Mechanical Engineering, VVPIET. (2015) 1-6.

[14] S. Adekunle, A. Adebiyi, O. Durowoju, Performance evaluation of groundnut oil and melon oil as cutting fluids in machining operation acta technica corviniensis, Acta Technica Corviniensis – Bulletin of Engineering. 8 (2015) 97-100.

[15] R. D. Nageswara, R. R. Srikant, R. Ch. Srinivasa, Influence of Emulsifier Content on Properties and Durability of Cutting Fluids, Journal of the Brazilian Society of Mechanical Sciences and Engineering. XXIX, 4 (2007) 396-400.

[16] D. Ozimina, The exploitation of tribological systems. Tom I. The importance of scientific instrument in operation buildings, M48, Ed. TU, Kielce 2013.

[17] A. Petterson, High-performance base fluid for environmentally adapter lubricants, Tribology International. 40 (2007) 638-645. https://doi.org/10.1016/j.triboint.2005.11.016

[18] J. Manekotte, S. Kailas, Experimental investigation of coconut and palm oils as lubricants in four-stroke engine, Tribology Online. 6, 1 (2011) 76-82. https://doi.org/10.2474/trol.6.76

Materials Research Proceedings **5** (2018) 54-60 doi: http://dx.doi.org/10.21741/9781945291814-10

Analysis of the Causes of the Cracks in the Thermit Welds of the Tram Rails Type 60R2

KOWALCZYK Dariusz

Instytut Kolejnictwa Materials and Structure Laboratory, Chlopickiego Street 50, 04-275 Warsaw, Poland

dkowalczyk@ikolej.pl

Keywords: Rail Joints, 60 E1 Rail, 60R2 Rail, Rail Joint Cracks, Rail Joint Tests, FEM Analysis

Abstract. According to the PN EN 14578-1:2007 and PN EN 14578-2:2009 regulations and European standards, the requirements for the rail joints used in the rail roads point out that it is necessary to carry out research that confirms technology correctness and the quality of welded joints. The range of tests shall include the following: alignment, hardness, fatigue testing, macro and microstructure of the material, non-destructive magnetic particle examination (to identify cracks) and so on. As long as there are PN-EN 14811 + A1:2010 requirements for rails used in the tram infrastructure, no accurate regulations and requirements exist for tram rail joints and only the customer is able to specify the range of tests. In the light of fairly common problems associated with the rail and their joint cracks in the tramway tracks it seems to be reasonable to implement required research, in order to improve safety of the rail transport and reduce amounts of cracks occurred in the newly built and renovated tram roads, as well as minimize the costs of their maintenance during operation. In this article, examples of broken rail joints in the tram roads are shown and discussed and FEM simulation of loaded rails type 60E1 and 60R2 is analysed.

Introduction

One of the grooved rail's type 60R2 advantage is a possibility to build modern tram superstructures in common with road and, the so-called green track characterised by lower noise is becoming more and more popular. However, the obtained results of calculations indicate that, in the case of 60R2 rail, there are higher stresses in the rail foot area. FEM calculations and analysis of the cracked rail connections indicate that, due to a high concentration of stresses in the foot of 60R2 rail, rail road maintenance has to be of particularly high quality, and above all, perfect technologies of rails connection process. Rail connections should be made according to proven technology, requirements of the standards and relevant provisions of rules that demand keeping the registry of made welded connections. In this article there are presented difficulties that can occur in public transport (tram transport) as a result of rail connections damage or cracked rail.

In the case of road solutions used in the rail it is necessary to meet the following standards [1-3]: PN-EN 14587-1:2007 "Railway applications – Track – Flash butt welding of rails – Part 1: New R220, R260, R260Mn and R350HT" grade rails in a fixed plant and PN-EN 14587-2:2009 "Railway applications – Track – Flash butt welding of rails – Part 2: New R220, R260, R260Mn and R350HT grade rails by mobile welding machines at sites other than a fixed plant".

Rail connection techniques are designed to increase smooth running of trains (rail vehicle), and thus upgrade their speed (shortening the travel time) and increase the comfort of travelling [4, 5]. However, the construction of roads and railways of such characteristics requires connecting rails. Connections quality should guarantee safety and reliability during operation. They are obliged to carry out research on rail joints by accredited and independent research bodies which confirm the correctness of the execution of the work, technology and security solutions. In the case of trams and subways superstructure constructions such a range of research is not mandatory and only the customer himself defines the scope of requirements and expectations regarding studies, certificates, authorisations, etc. One of the most commonly used techniques for rails connecting in the tram superstructure is thermite welding. The basic technique is called SoWoS, i.e. welding without excess weld metal, with upper preheating of the rails ends. SoWoS technology consists of the following: preparation and setting contact rails for welding and seal forms, filing and setting the crucible, heating the rail ends, welding, cooling of connections, mould removing and post treatment [6-8].

Rails connections – joints and areas around them – are the most common places of the higher risk of cracks, defects, damage etc. This is due to thermal effects during the execution of the rails connections, in which emerges a heat affected zone that makes changes in the microstructure of the material. Connections quality closely depends on their technology. All important parameters of performed works should be written and recorded in the daily register of made welds [9, 10].

While analysing the documentation of the new tram tracks, lack of such records was stated and in many cases there were serious gaps that made it impossible to analyse and reconstruct process conditions and determine the causes of defective rail joints, and even the location of the performed works. For example, one of Daily Registers Made Welds did not contain such relevant information as ambient temperature, method and time of the rails ends preheating, pressure of the gases mixture, time of the molten metal pouring to the mould, temperature of the excess weld metal removing, method of the joint cooling, etc. Sometimes it happens that in the newly-built tracks there are many problems in the maintenance of track (even in less than 1 year) due to "cracked rails".

In the construction of new or upgraded tram tracks grooved rails are increasingly used, mainly the 60R2 type [11]. The advantage of this type of rails is the ability to make integrated road-rail track, green track with elastic sleepers fastening, green track on the concrete plate with elastic rails fastening system. In Fig. 1 there are samples of solutions using 60R2 rails.

Fig. 1 Integrated on-street tramways and green tramway with used rails 60R2 type.

Problems occurring during operation of the tramways

Problems occurring during operation related to the analysed cracks of the rails and their joints are connected with effects of many factors. The most important issue is the use of proper (proven in laboratory studies) technology of welding rails. In Fig. 2-4 there are examples of additional factors that may affect the durability of the rail connections resulting in a higher state of stresses in the rails, such as: irregular terrain, defects in rail joints, low quality of maintenance, improper construction of the track (sleepers, fastening systems), etc.

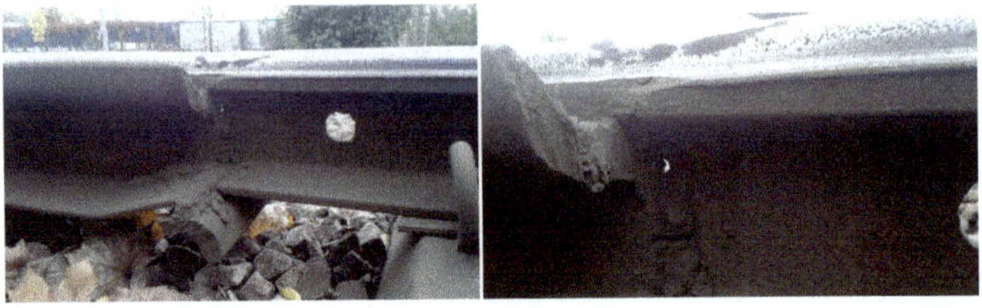

Fig. 2 Low quality of the rail joints – clearance, lack of material under rail head.

The following cases of defects, imperfections in the tram tracks come from different cities in Poland, and the dates come from the years 2016-2017 (Fig.2-4). The use of unproven techniques of rail joints and additional, described above, factors may cause cracks in the rail and their joints cracks. Examples of damage to the connections are shown in Fig. 4.

Fig. 3 Low quality of the tram tracks maintenance.

Fig. 4 Crack of the rail in the area of the rails joint.

Research problem

As shown above in the operation of the new tram tracks there are problems arising from cracks in the rails joints. The quality of rail joints depends on many factors: quality of the joints, track construction, residual stresses etc. One of the important issues to be taken into account is also the type of rail in the track, where rails joint, e. g. 60R2 are and 60E1. This issue has been analysed with FEM and its aim was to determine the distribution of stress fields, both rails in the same load condition (different rail profiles 60R2 and 60E1), as well as in the case of normal exploitation loads.

The results are given below. It should be noted that the calculation takes into account ideal conditions, without any imperfections associated with the construction of the track and residual stresses of the rail.

FEM calculation for the 60R2 and 60E1 rail types

The purpose of the FEM calculation, except for the issues described above, that concern rails welding and track constriction works, was to determine the distribution of the stress fields in the 60R2 and 60E1 rails and their joints.

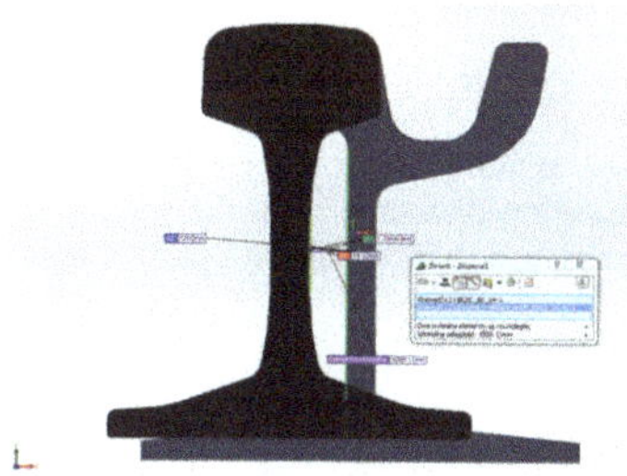

Fig. 5 Comparison of rail profiles 60R2 and 60E1 (profiles according to PN EN 13674-1:2011).

The presented (Fig. 5) layout of the rails profiles follows operating conditions, where appropriate geometry of the running surface of rails should be kept.

As can be seen from the scheme, distance between rail webs is about 15 cm, also rails 60R2 and 60E1 differ in the height and construction of the web and foot.

Models of the 60R2 and 60E1 rails are made in SolidWorks software. Boundary conditions are defined and the properties grid were built, as well as simulation and analysis of the results were done with software HyperWorks - version 14 of company ALTAIR and HyperView [12, 13] and with SolidWorks with Simulation pack.

Three simulations were performed for the 60R2 and 60E1 rails in the following load schemes: *Simulation 1*; fastened and supported rail has been deflected on the Y-axis at -1 mm; *Simulation 2*; fastened and supported rail has been deflected on the Y-axis at -1 mm and X-axis -1 mm; *Simulation 3*; fastened and supported rail has been loaded of 50 kN. Below are presented examples of stress fields distribution for selected loads, and summary results are presented in Table 1.

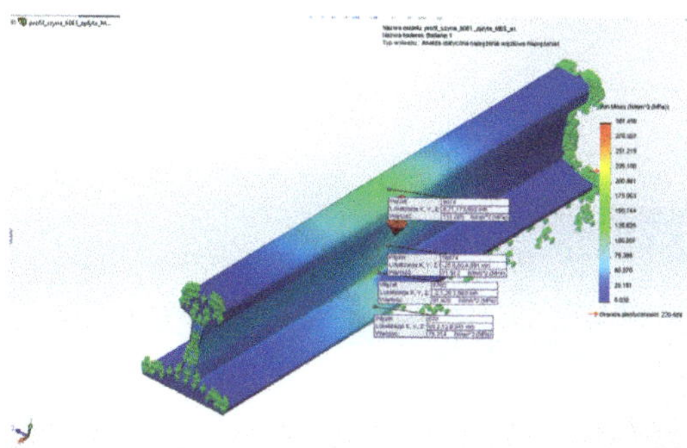

Fig. 6 Stress fields distribution according to von Misses in the rail 60E1 for simulation 1.

The results of FEM analysis no.1-3 are presented in Table 1.

Table 1. Maximum stress according to von Misses for each simulation.

FEM calculation option	Distortion direction	Stress [MPa]	
		Rail 60 E1 type	Rail 60 R2 type
Simulation 1	Axis Y -1 mm	91 (foot) 111 (head)	171 (foot) 104 (head)
Simulation 2	Axis Y -1 mm Axis X -1mm	211	305
Simulation 3	Axis Y – 50 kN	82	143

Simulation 4. Scheme of the symmetrical rails joint loading, joining with welds of 30 mm thickness. Symmetric way of fastening and support in the same distance of 250 mm from the joint. Applied material – steel R260 (rail) and PUR (under rail pad). Characteristics of materials are shown in Table 2.

Table 2. Material data used in the FEM calculation.

Material	Young's modulus [MPa]	Kirchhoff modulus [MPa]	Poisson's ratio	Density [g/cm^3]
Steel R260	210 000	80 769	0,3	7,85
PUR	250	100	0,42	0,95

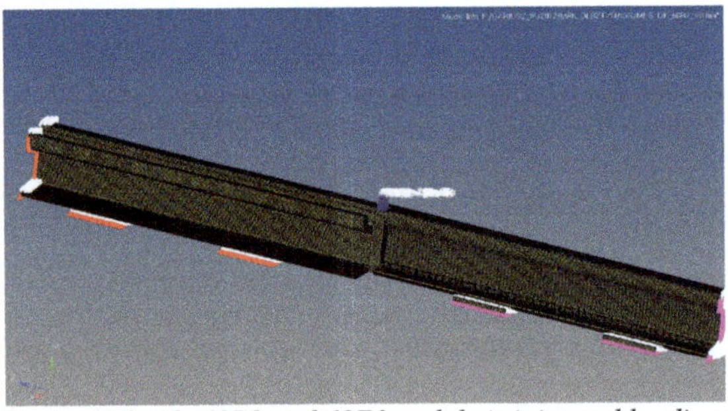

Fig. 7 Structure of rails 60R2 and 60E1 and their joint and loading scheme.

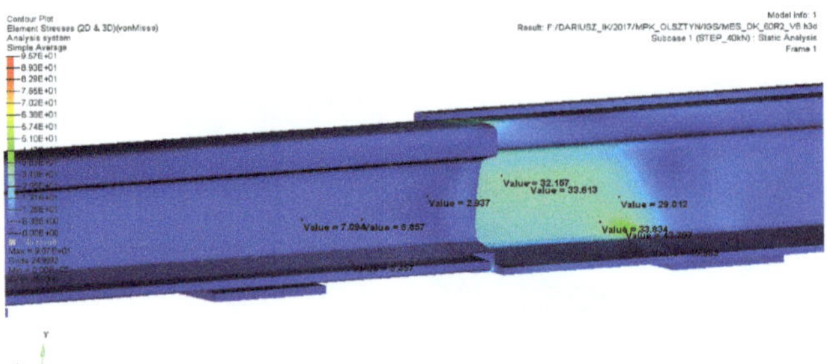

Fig. 8 Distribution of stress fields according to von Misses.

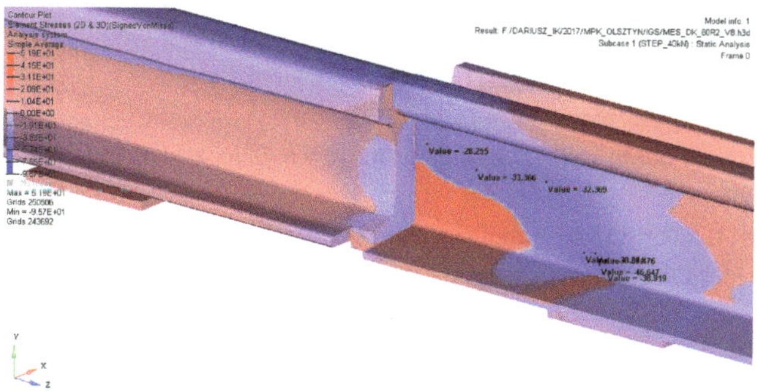

Fig. 9 Distribution of stress fields in the rails 60R2 and 60E1, and their joint.

Conclusions of the FEM analysis

For simulation: the 60E1 rails and 60R2 joint, which is 30 mm thick, while equiaxial and heads of rails layout is ensured, and with the use of symmetrical supports layout (defined the same edge condition) and the load of 40 kN placed at the joint (perpendicular to the rail head) it was found that for such a system in the area of-the 60R2 rail occurred higher stress than in the 60E1 rail for the same distance from the rail joint, and for the 60R2 rail occurred large differences in stress (+ 31 MPa, -50 MPa) in the close areas of the joint.

In the analysis of FEM residual stress has been ignored, which does not exist in reality.

Rail material after production process, transport and build into the track has its own residual stresses and in addition it is changed locally in the area rails joint made by the welding process (thermal impact, which also affects locally the structural changes of material).

The correct joint technology and other factors described in this article, as FEM calculation shows, are of crucial importance to the durability of the joint.

The FEM calculation showed that for symmetric load rail joint in the layout of the 60R2 and 60E1 rails, stress occurs in the 60R2 rail than 60E1. In addition, in the 60R2 rail web a big difference in the stress of compression and stretching occurs. Having regard to that fact and thermal stresses created during joint producing, it can be concluded that this area is particularly vulnerable to cracks. The load of 40 kN is not maximal but operational – at maximum tram axle loading it can be 55 kN. In addition, there should be taken into consideration track irregularities, its inclinations and curvatures, defects (lack of the ballast, absence of fastenings, etc.), and dynamic excess occurred during driving. In the adverse cases forces of about 70 kN can be present.

Summary

Problems with rail joints derive from the more difficult process of connecting the 60R2 rail type (profile less compact, more sensitive, higher stresses after thermite welding, in relation to the 60E1 rail profile). Due to the fact that there are no requirements for rail joints in the tram tracks, checking of them is provided during operation. Quite often there are joint cracks because the approved materials are not the only factor that guarantees high quality of the joints. Moreover, the technology is of great importance. The poor quality of the track works such as bad sleepers arrangement, bad construction of the track – lack of the ballast, improper operating fastening systems or their absence, etc. affect the increase of stress fields in the rail, and this can bring problems during operation of the track.

Materials Research Proceedings **5** (2018) 54-60 doi: http://dx.doi.org/10.21741/9781945291814-10

References

[1] PN-EN 14587-1:2007: Railway applications – Track – Flash butt welding of rails – Part 1: New R220, R260, R260Mn and R350HT grade rails in a fixed plant.

[2] PN-EN 14587-2:2009: Railway applications – Track – Flash butt welding of rails – Part 2: New R220, R260, R260Mn and R350HT grade rails by mobile welding machines at sites other than a fixed plant.

[3] PN-EN 14811:2006+A1:2010: Railway applications. Track. Special purpose rail. Grooved and associated construction.

[4] I. Miklaszewicz, Making and testing of the rail joints, Railway Research Institute, Warsaw, Railway Reports, 158, (2013) 35.

[5] B. Bogdaniuk, K. Towpik, Construction, modernisation and maintenance of the railway trucks, PKP, Warsaw, 2010.

[6] User ID-5 of the rails thermite welding, PKP, Warsaw, 2005.

[7] PN-EN 13674-1:2011: Railway applications. Track. Rail. Vignole railway rails 46 kg/m and above.

[8] PN-EN 14730-1+A1:2010: Railway applications. Track. Aluminothermic welding of rails. Approval of welding processes.

[9] PN-EN 14730-2:2006: Railway applications. Track. Aluminothermic welding of rails. Qualifications of aluminothermic welders, approval of contractors and acceptance of welds.

[10] Technical conditions WTWiO No ILK3d-518/1/08, joints making and acceptance of the new railway rails performed in the fixed plants. Requirements and tests, PKP, Warsaw, 2008.

[11] R. Wielgosz, Weld of railways rails, Technical Transactions, 106 (2-M) (2009), pp.7-19.

[12] Altair Engineering, Inc. Radioss and Optistruct Theory Manual 11.0 Version Jan, Large displacement finite Element Analysis (2011).

[13] Altair Engineering, Inc. OptiStruct 13.0 User's Guide.

Terotechnology 2017 Materials Research Forum LLC
Materials Research Proceedings **5** (2018) 61-66 doi: http://dx.doi.org/10.21741/9781945291814-11

Analysis of Possibilities of Using Polymeric Materials for Testing Prototypes of Harmonic Drive

PACANA Jacek[1, a *], PACANA Andrzej[1,b]

[1] Faculty of Mechanical Engineering and Aeronautics, Rzeszow University of Technology, Al. Powstancow Warszawy 12, 35-959 Rzeszow, Poland

[a] pacanaj@prz.edu.pl, [b] app@prz.edu.pl

Keywords: Harmonic Drive, Flexspline, Polymeric Materials, 3D Printing, FDM, MEM, Coordinate Measurements

Abstract. The main objective of the analysis was to assess the suitability of rapid prototyping methods to perform functional prototypes of a harmonic drive. Physical models of a gear wheel for the harmonic drive were made and geometric accuracy control was carried out therefor. Precise measurements of tooth profiles showed differences between models made with the use of 3D printing and corresponding theoretical outlines. The discrepancies were not decisive for the correctness of the work of the analyzed harmonic drive, since the key parameter affecting its kinematic accuracy is the scale's constancy.

Introduction

Harmonic drives are a special type of gear transmission, mainly due to an extremely complex way of load transfer. Their basic components are flexsplines, in the shape of a thin-walled sleeve with an external toothed ring, a circular spline and a wave generator. The basic diagram of such a gear is shown in Figure 1a.

The rotating generator deforms the flexspline, causing its moving in relation to the circular spline with internal teeth. Toothed rims of both wheels meet in the meshing areas, and pass each other outside of these areas [1, 2]. The load is transferred simultaneously by 20 ÷ 50% of the total number of teeth of the flexspline [3, 4]. These exceptional working conditions bring on the need for a very thorough stress analysis [5, 6] to improve the quality of this type of gears. Analytical calculations of this type of gear are not accurate, because the complexity of the problem forces the basing only on general assumptions. Numerical calculations carried out, for example, using the FEM finite element method are based on simplified models, so they also do not give precise results. Only bench tests allow for the exact determination of the strength and durability of a gear, as well as the assessment of the quality of its work or kinematic accuracy.

Experiments are carried out with the use of functional prototypes, often by rapid prototyping (RP) methods, from polymer materials. The use of the 3D printing method allows for fast and cheap production of all components of the gear. This is particularly important in the case of a flexspline of a harmonic drive, whose thin-walled construction causes a lot of technological problems during production with classical cutting methods. Therefore, the use of additive methods of rapid prototyping not only facilitates the creation of flexsplines, but opens up new opportunities to improve their design. The key issue, however, is the geometric accuracy of harmonic drives made by 3D printing.

Model making

The implementation of a functional prototype of a harmonic drive must be preceded by analytical calculations in order to determine its basic geometrical and strength parameters and to model its components in the selected CAD program. For the needs of gear production for harmonic drive, the MEM (Melting and Extrusion Modeling) method was chosen, i.e. FDM (Fused Deposition Modeling) with a single nozzle print head [7, 8]. It is an incremental shaping technology that allows for creating physical models with complex geometry, which is especially important in the case of a flexspline of a harmonic drive. The obtained surface quality is not comparable with other methods [9, 10], especially in mass production, but sufficient for prototyping. The ABS copolymer (acrylonitrile butadienestyrene) is one of the most often used materials in 3D printing, with strength and temperature parameters that allow it to be widely used both in industrial production and for prototyping [11, 12].

Measurements

A detailed analysis of geometric accuracy of models made with the use of 3D printing was performed using a Roland MDX-40A coordinate milling machine with a special ZSC-1 scanning head mounted (Fig.2b). The measurements were carried out using the contact method due to its high accuracy and the possibility of using models immediately after their manufacture.

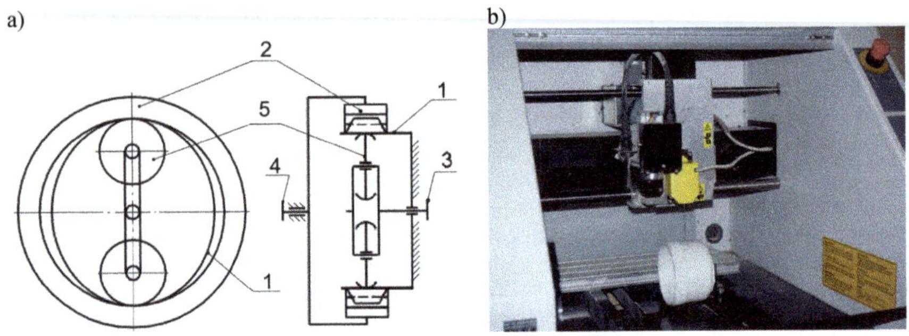

Fig. 1. Harmonic drive: a) scheme: 1-flexspline, 2-circular spline, 3-input shaft, 4-output shaft, 5-two-roll generator, b) measurement of the profile of the flexspline in a longitudinal section on the Roland MDX-40 cnc milling machine.

The gear wheels mounted in the tested harmonic drive have teeth which are about 1mm high. Therefore, a very small diameter of the scanning module measuring tip of 0.04mm is an additional factor determining the choice of this coordinate machine. The process of spatial models scanning consists in making measurements in the X, Y, Z coordinate system of the CNC machine when the head contacts the surface of the model. The scan result is obtained in the form of a points cloud, or after processing as a triangle grid. In this form it is saved for further processing and analysis.

Longitudinal section of the flexspline

Firstly, the outer surface of the flexspline was scanned in its longitudinal section passing through the middle of the inter-tooth groove. The view of the obtained profile along with the enlargement of characteristic areas is presented in Figure 3.

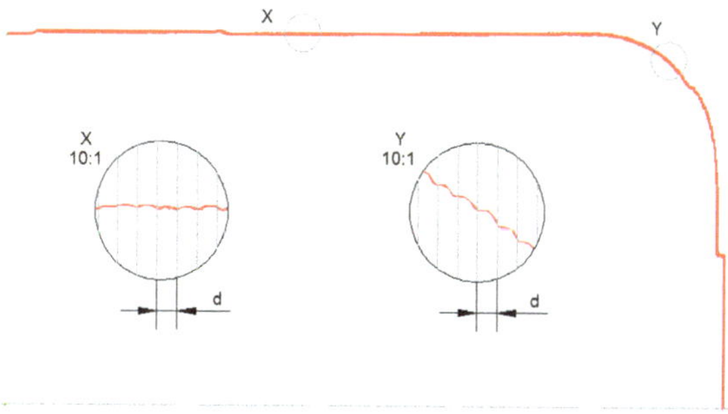

Fig. 3. The result of scanning of external profile of the flexspline.

The measured overall length of the flexspline equaled 69.87 mm and was slightly smaller than the nominal size of 70mm. However, it was about 0.48mm smaller from the value obtained during the initial measurements, made immediately after the model was printed. This reduction of overall dimensions is mainly a result of a processing shrinkage which in the case of ABS is up to 0.7%, i.e. for the analyzed model equals about 0.5 mm. Because the scanning of the gear wheel models was carried out long enough from their production, the reduction in length is a correct behavior resulting from the properties of the used material.

Figure 3 also shows the enlargement of the X and Y areas, related to the external outline of the flexspline of the harmonic drive obtained in the process of scanning. In the X area, the external surface of the model is cylindrical, and the outline obtained as a result of scanning is close to the linear one. Measured deviations from the average value do not exceed ± 0.25 mm, which should be considered as a very good result for the incremental method of model producing. However, when the external outline of the model ceases to be parallel to the direction of layer application, the quality of the external surface is getting worse. This is clearly visible on the Y magnification, where the profile of the flexspline body obtained in the measurements has a clearly stepped course. Individual layers can be recognized thereon and they are marked with lines to improve readability. The distances between successive lines equalled 0.1mm, which is exactly the same as the thickness of a single layer for 3D printing. In the case of the prototype of a flexspline, this has a negative effect only on the appearance of the external surface of a model, and does not affect its strength.

Teeth profiles in a cross-section

Checking the geometric accuracy of gear wheels should particularly concern areas that have a direct impact on their correct working. In the case of harmonic drives, it is very important to properly shape the small-module gear rim. When making 3D printing wheels with small teeth, the diameter of a nozzle that provides the molten material is a very important parameter. The minimum radius of rounding of all external corners on models depends thereon. In the case of gear wheels for harmonic drives, a problem may arise in the precise reproduction of a theoretical outline of teeth. Therefore, the models produced by the FDM / MEM incremental shaping method have been subjected to precise measurements also in this field. The shape of a tooth obtained on a physical model is significantly different from a theoretical one. To notice the differences, not only in relation to the correct outline, but even between successive teeth, it is enough to visually assess the profiles in a few-size magnification. The lack of sharp corners on models made with the FDM / MEM method is a result of its technical limitations. The machine used for 3D printing had a nozzle with a diameter of 0.4 mm.

Materials Research Proceedings **5** (2018) 61-66 doi: http://dx.doi.org/10.21741/9781945291814-11

With such limitations, it is not possible to obtain external corners with a radius smaller than the dimension of the hole in the nozzle supplying the material.

However, the assessment of the accuracy of gear wheels of harmonic drives is not based on the accuracy of the representation of the outline of individual teeth [13]. In these drives, wheel cooperation takes place due to a simultaneous meshing of even several pairs of teeth, which is why maintaining a constant scale is a more important criterion. It is also common to use outlines of teeth other than involute teeth in this type of gears, and their selection depends on production capacity [14]. There are also solutions with a rectilinear outline of the tooth's side, because at a small height and high correction factors used, the difference with regard to the used involute fragment is not significant. Experience suggests that flexspline defects occur most often in places where the thickness of a susceptible sleeve changes or in the grooves between teeth. Thus, the shape of a tooth does not have a significant impact on the strength of harmonic drives.

Averaged scan results for the grooves and tooth width as well as the scale of the teeth of both wheels are given in Table 1. They were compared with the reference values for the theoretical outline.

The results presented in Table 1 show a large convergence of measurement results with theoretical values. This is surprising, with respect to the previous preliminary assessment of the outlines of the teeth, whose profiles differed significantly.

Table 1. Selected tooth dimensions of the physical and theoretical model.

		Tooth's width [mm]	Groove's width [mm]	Scale [mm]
Susceptible wheel	Theoretical outline	1,552	1,055	2,607
	Scanning result	1,513	1,073	2,586
	Variation	***0,039***	***-0,018***	***0,021***
Rigid wheel	Theoretical outline	1,375	1,192	2,547
	Scanning result	1,369	1,214	2,583
	Variation	***0,006***	***-0,022***	***-0,016***

The discrepancies between the averaged tooth width values for the flexspline and circular spline are 0.039mm and 0.006mm, respectively. Also the widths of the grooves between the teeth, measured for both physical models, are very close to the theoretical values, and the differences are at the resolution level of the scanning method, so they can be ignored. Thus, the requirements regarding the most important geometrical parameters determining the kinematic accuracy of the harmonic drive have been met.

The analyzed harmonic drive is a special construction, where the teeth of the flexspline move along a complex trajectory on a shape which is influenced by simultaneous radial and peripheral displacements and the angle of twisting of the tooth axis. Therefore, when designing harmonic drive wheels, considerably greater interstices are expected in order to eliminate the phenomenon of interference. When the generator rotates by an angle of $0° \leq \varphi \leq 180°$, the tooth of the flexspline moves from one groove to the next. Figure 4 shows the subsequent positions of the flexspline's tooth which are a result of the generator rotation, for the theoretical outline.

Materials Research Proceedings **5** (2018) 61-66 doi: http://dx.doi.org/10.21741/9781945291814-11

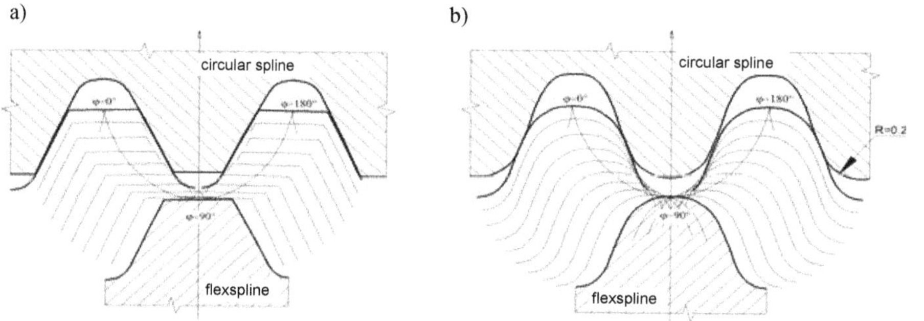

Fig. 4. Trajectory of the flexspline's tooth: a) for the theoretical outline, b) for the outline of the model obtained by 3D printing method.

While comparing the stages of tooth displacements of the theoretical and real outline, attention should be paid to a much greater space between the vertexes of the moving teeth in the second case. At the same time, for the angle of $\varphi=0^o$, in both cases there is a proper contact between the side surfaces of the teeth, because the measured differences in tooth width on the theoretical and physical models were not significant. Finally, it should be noted, keeping in mind the specific nature of harmonic drive working, that the deviations of the tooth outlines presented on the flexsplines do not adversely affect the quality of its work. Differences between the reference profile and the real one obtained using the FEM / MEM method concern mainly the rounding of tooth tips, which can be even a beneficial phenomenon. New technological possibilities, appearing along with methods of rapid prototyping, allow for the search of new tooth outlines that can positively influence the improvement of work quality and load capacity of harmonic drives.

Summary

The 3D printing method allows for a quick and easy preparation of components of a functional prototype of a harmonic drive. The additive building process of a model is particularly advantageous in the case of a flexspline with an extremely complex shape. The conducted contact measurements, using a coordinate machine, enabled the analysis of the geometric accuracy of gear wheels manufactured using the FDM/MEM method. A slight discrepancy between the physical models and their theoretical counterparts is visible. The problems concerned the mapping of the outer corners of the prototypes made with the 3D printing method. However, in the case of the analyzed harmonic drive, where even several tooth pairs stay in the meshing, the accuracy of a single tooth profile is not a key issue. Constancy of the scale or width of subsequent grooves and teeth are much more important parameters. These parameters were also checked by measuring the rim gear in the cross-section. The obtained results indicate high accuracy of gear wheel models in this range. The determined average values of the width of subsequent teeth and scale differ by no more than 2% from the theoretical values. This allows for a very good evaluation of the accuracy of physical models of gears made using 3D printing. Positive assessment in the area of key geometric parameters allows for the use of a prototype of a harmonic drive in bench tests.

References

[1] X. Chen, Y. Liu, J. Xing, S. Lin, W. Xu., The parametric design of double-circular-arc tooth profile and its influence on the functional backlash of harmonic driver, Mechanism and Machine Theory, no. 73 (2014) 1-24.

[2] O. Kayabaşi, F. Erzincanli, Shape optimization of tooth profile of a flexspline for a harmonic drive by finite element modeling, Materials and Design, no. 28 (2007) 441-447. https://doi.org/10.1016/j.matdes.2005.09.009

[3] Z. Xinjie, Y. Changxiang: Application of precision harmonic gear drive in focusing mechanism of space camera, Proceedings of the SPIE, vol. 10 (2010).

[4] H. S. Jeon, S. H. Oh, A study on stress and vibration analysis of a steel and hybrid flexspline for harmonic drive, Composite Structures, no. 47 (1999).

[5] J. Pacana, W. Witkowski, J. Mucha, FEM analysis of stress distribution in the hermetic harmonic drive flexspline, Strength of Materials, vol. 49, no. 3 (2017) 388-398. https://doi.org/10.1007/s11223-017-9879-z

[6] Z. Ren, J. Flasker, Numerical stress analysis of harmonic gear drive flexspline, Special Gears – modeling, product development and prospect of application. Warszawa (1996) 133-146.

[7] P. Dudek, FDM 3D printing technology in manufacturing composite elements, Archives of Metallurgy and Materials, vol. 58, issue 4 (2013) 1415-1418. https://doi.org/10.2478/amm-2013-0186

[8] G. Budzik, Possibilities of utilizing 3DP technology for foundry mould making, Archives of Foundry Engineering, Vol. 7, Issue 2 (2007) 65-68.

[9] R. Ulewicz, Practical aplication of Quality tools in the cast iron foundry, Manufacturing Technology, vol. 14, Issue 1 (2014) 104-111.

[10] L. Parv, Cost of product functions using analysis of value, Production engineering Archives, 12/3 (2016) 14-18.

[11] A. Gadek-Moszczak, N. Radek, S. Wronsk, J. Tarasiuk, Application the 3D Image Analysis Techniques for Assessment the Quality of Material Surface Layer Before and After Laser Treatment, TEROTECHNOLOGY, Advanced Materials Research, 874 (2014) 133-138.

[12] A. Pacana, L. Bednárová, J. Pacana, I. Liberko, A. Woźny, D. Malindžak, Effect of selected factors of the production proces of stretch film for its resistance to puncture. Przemysł Chemiczny. 12 (2014) 2263- 2265.

[13] T. Dziubek, M. Oleksy, Zastosowanie systemu optycznego ATOS II w technikach szybkiego prototypowania modeli kół zębatych otrzymywanych na bazie żywicy epoksydowej, Polimery nr 1 (2017) 44-51. https://doi.org/10.14314/polimery.2017.044

[14] M. Cader, R. Oliwa, O. Markowska, G. Budzik, Otrzymywanie prototypów części chwytaka robota mobilnego z materiałów polimerowych z wykorzystaniem technologii wytwarzania przyrostowego. Cz. I. Właściwości mechaniczne i stałe materiałowe próbek z kopolimeru akrylonitryl-butadien-styren, Polimery, T. 62, nr 1 (2017) 27-35.

Terotechnology 2017 Materials Research Forum LLC
Materials Research Proceedings **5** (2018) 67-72 doi: http://dx.doi.org/10.21741/9781945291814-12

The Influence of Friction Coefficient on Forward Slip in Experimental Research on Cold Longitudinal Flat Rolling

MIŁEK Tomasz

Department of Metal Science and Materials Technologies, Faculty of Mechatronics and Mechanical Engineering, Kielce University of Technology, Al. Tysiąclecia Państwa Polskiego 7, 25-314 Kielce, Poland

tmatm@tu.kielce.pl

Keywords: Longitudinal Rolling, Friction Factor, Forward Slip, EN AW-6063 Aluminium

Abstract. The paper presents experimental results that concern cold longitudinal rolling of flat bars made from EN AW-6063 aluminium for different conditions. The investigations aimed at determining the impact of friction factors on the forward slip. The friction factors were determined by the method of the roll bite angle. The forward slips were calculated from Fink's, Drezden's and Vinogradov's formulae. On the basis of investigations of cold longitudinal rolling of flat bars made from EN AW-6063 aluminium, it was found that the forward slip increases with an increase in friction factor.

Introduction

Rolling is the most extensively used metal forming process and it accounts for almost 90% of metal produced by forming. In this process, the material to be rolled is drawn by means of friction into the gap between two revolving rolls. Compressive forces applied by the rolls reduce the thickness of the material or change its cross-sectional area. The rolling process belongs to the compressive deformation processes and has been classified on the basis of kinematics, tool and workpiece geometry. Based on kinematics, the rolling process can be classified as longitudinal, cross and skewed [1-2].

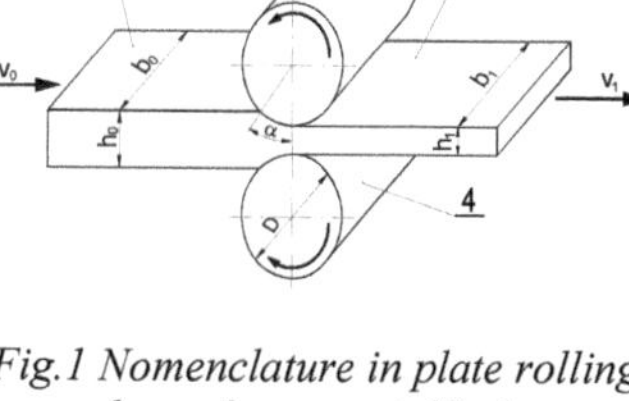

Fig.1 Nomenclature in plate rolling, where: 1 – material before deformation, 2 – material after deformation, 3,4 – mill rolls

In addition to conventional methods of rolling, there are also special methods, as the WPM method, for example. The most essential features of this method, which distinguish it from other known methods, involve the use of two circular segments as the tools and the adoption of kinematics that allows one-directional rolling with a symmetrical system of forces relative to the axis of the rolled material. The WPM method of cold rolling allowed cylindrical gears, involute splines and other circular profiles to be formed [3].

As can be seen from Fig. 1, in longitudinal rolling, the workpiece moves through the rolling gap perpendicular to the axis of the rolls, without rotation about the workpiece axis [1-2]. In the exit zone of deformation, the horizontal component of the roll circumferential velocity is less than the workpiece velocity. This phenomenon is called a forward slip [1]. The values of a forward slip were calculated from Fink's, Drezden's and Vinogradov's formulae [1, 4]. The forward slip is given by the equation of Fink below (1) [1, 4]:

$$S_w = \frac{[h_1 + D(1-\cos\gamma)]\cos\gamma}{h_1} - 1 \qquad (1)$$

where: h_1 – height oft he material after deformation (mm), D – the diameter of a mill roll (mm), γ - parting plane angle is given by the Ekelund formula (2) [4]:

$$\gamma = \sqrt{\frac{\Delta h}{2D}} - \frac{\Delta h}{2D\rho} \qquad (2)$$

where: ρ – the friction angle can be determined by formula (3) [4]:

$$\cos\rho = \frac{1}{\sqrt{1+\mu^2}} \qquad (3)$$

where: μ – the friction coefficient is (4) [4]:

$$\mu = \sqrt{\frac{1}{\left(1-\frac{\Delta h_{max}}{D}\right)} - 1} \qquad (4)$$

where: Δh_{max} - the high reduction for maximum deformation.
The following formula can be used to obtain the forward slip of the Drezden theory (5) [1,4] :

$$S_w = \frac{R}{h_1}\gamma^2 \qquad (5)$$

The forward slip may be estimated from Vinogradov's formula given below (6)[1,4]:

$$S_w = \frac{\left[b_0 + \Delta b\left(1-\frac{\gamma}{\alpha_{ch}}\right)\right][h_1 + D(1-\cos\gamma)]\cos\gamma}{(b_0 + \Delta b)h_1} \qquad (6)$$

where: b_0 – width oft he material before deformation (mm), Δb – spreading of the material (mm), α_{ch} – roll bit angle can be determined by (7) [4]:

$$\alpha_{ch} = \arccos\left(1-\frac{\Delta h}{D}\right) \qquad (7)$$

Cold longitudinal rolling process of flat bars and the design of the tooling have been reported in some studies [5-11]. Those covered both experimental and computer modeling investigations.

The paper presents experimental results that concern cold longitudinal rolling of flat bars made from EN AW-6063 aluminium for different conditions. The investigations aimed at determining the impact of friction factors on the forward slip. The friction factor was determined by formula (4) and forward slips were calculated by means of Fink's, Drezden's and Vinogradov's formulae (1), (5), (6).

Materials Research Proceedings **5** (2018) 67-72 doi: http://dx.doi.org/10.21741/9781945291814-12

EN AW-6063 aluminium was selected as the testing material in these investigations due to its good formability and wide applications in industry [12-13].

Materials

Samples made from EN AW-6063 aluminium constituted the material for experimental investigations into cold longitudinal rolling process. Chemical composition of material [12] is shown in Table 1. The mechanical properties of the material, determined by static tensile testing, are presented in Table 2.

Table 1 Chemical composition of EN AW-6063 aluminium [%] [12]

Mg	Si	Fe	Ti	Zn	Cr	Mn	Cu	Unspecified other elements		Al. minimum
								Each	Total	
0.45 - 0.9	0.20 - 0.6	max. 0.35	max. 0.10	max. 0.10	max. 0.10	max. 0.10	max. 0.10	max. 0.05	max. 0.15	rem

Table 2 Mechanical properties of specimens made from EN AW-6063 aluminium

R_m, [MPa]	$A_{11.3}$, [%]	Z, [%]
260	13.8	11.2

The tensile test was conducted on a LabTest5.20SP1 testing machine (LABORTECH firm) with the force of 20 kN. The machine was calibrated by PN-EN ISO 7500-1:2005 and meets metrological requirements for class 0.5. The mechanical properties indicate a strain hardening of the material after cold work. Flow curve of the sample made from from EN AW-6063 aluminium is presented in Fig. 2.

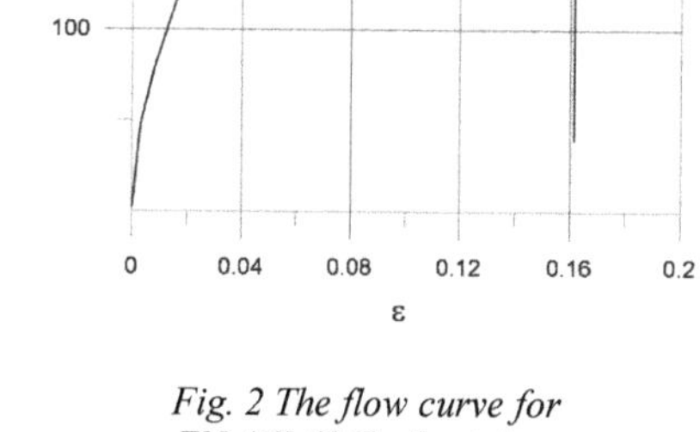

Fig. 2 The flow curve for EN AW-6063 aluminium

The samples used in experimental investigations into cold longitudinal rolling were made of segments of flat bars with the high of h_0 = 5 mm, width of b_0 = 20 mm and length of l_0 = 40 mm. The longitudinal rolling process is conducted at a special stand – a DUO-100 rolling mill. The rolling mill has two mill rolls at diameter D equals 102 mm. The samples were rolled in different conditions. Commonly used lubricants, such as extraction naphtha, kerosene and mineral oil were chosen for the experiment [14]. In addition, motor and rapeseed oils were used.

Extraction naphtha is a complex mixture of hydrocarbons obtained by the treatment of petroleum fraction with hydrogen in the presence of a catalyst. It consists of of hydrocarbons with number of carbon atoms in the range from C_4 to C_{11} whose boiling point ranges from from -20°C to 190°C. It is used as a solvent in the paint and varnish industry as well as in metal and rubber industry. Extraction naphtha is applied to production of adhesives. It is used for cleaning and degreasing in dry cleaners and tanneries as well as in service workshop. Extraction naphtha is a colourless liquid with a flash point below 0 °C. Ist density at 15 ^{0}C ranges from from 0,62 to 0,88 g/cm^3. It can be dissolved in most organic solvents, hydrocarbons, alcohols, ethers, carbon disulphide, carbon tetrachloride and chloroform [15].

Kerosene is a complex mixture of hydrocarbons obtained during the process of removing acidic substances. It consists of hydrocarbons with a number of carbon atoms in the range from C_9 to C_{16}. The boiling point of kerosene is from approx. 150 °C to 290 °C. Kerosene is a cleaner used to clean metal machine parts, especially rolling bearings (to remove lubricants). It has anti-corrosive properties. Its flash point is above 61 ^{0}C and density at 15 ^{0}C ranges from from 0,75 to 0,86 g/cm^3 [15].

The LHL-46 mineral oil (Hydrol LHL-46) is a mixture of basic mineral oils and refining additives. It is applied in low and medium-loaded systems of force transmission as well as hydraulic drive and control. Its relative density is approx. 0.85÷0.90 g/cm^3. The minimum of flash point equals 190 ^{0}C and maximum flow temperature equals -12 ^{0}C. The LHL-46 oil dissolves in hydrocarbons but is not soluble in water [14,16].

The evolution 700STI 10W40 motor oil (brand of the ELF) is a semi-synthetic oil which is a combination of base synthetic and mineral oils as well as anti-wear additives. It is recomended for all gasoline and diesel engines in cars. Its density at 15^{0}C equals 863 kg/m^3 and flash point equals 232 ^{0}C. The oil is insoluble in water and ist viscosity at 40 ^{0}C equals 96.81 mm^2/s [17].

The rapeseed oil is a deeply refined oil with high oxidative stability. It is applied in the food industry or as a raw material for carrying out chemical changes. The rapeseed oil is used as diesel fuel, either as biodiesel, straight in heated fuel systems, or blended with petroleum distillates for power motor vehicles. Its density at 20^{0}C equals 0.92 g/cm^3 and viscosity at 40 ^{0}C equals 35.40 mm^2/s [18].

Experimental results

The friction factors were determined by the method of the roll bite angle. This method involves rolling samples at the maximum degree of deformation. On the basis of geometric parameters of samples after deformation measurements, the values of roll bite angles, friction factors and forward slips were calculated from formulae (1)÷(7).

The deformation ratios of material in paper were defined as relative strain ε: on the high ε_{wh}, on the width ε_{wb}, the length ε_{wl}, respectively. The maximum values of deformation for different conditions of cold longitudial rolling are presented in Table 3. Nomenclature in table formulae is shown in Fig. 1.

Table 3. The maximum values of relative strain for samples made from EN AW-6063 aluminium for different conditions of cold longitudinal rolling

Relative strain	Lubrication:				
	extraction naphtha	kerosene	mineral oil	motor oil	rape oil
$\varepsilon_{wh} = \frac{\Delta h}{h_0}$	0.75	0.81	0.45	0.51	0.62
$\varepsilon_{wb} = \frac{\Delta b}{b_0}$	0.16	0.33	0.06	0.09	0.15
$\varepsilon_{wl} = \frac{\Delta l}{l_0}$	2.54	3.60	0.71	0.86	1.39

For comparison, values of roll bite angles, friction factors and forward slips are shown in Table 4. For calculation of the forward slip formulae, the following have been considered: Fink, Drezden and Vinogradov. Of these formulae, those of Fink and Drezden do not take into account spreading of Δb material and its influence on the forward slip. However, in all these formulae the γ parting plane angle is calculated neglecting the influence of spreading.

The influence of friction coefficient on the forward slip in cold longitudinal flat rolling for specimens made from EN AW-6063 aluminium is shown in the diagram (Fig. 3). The calculated forward slip increases with an increase in μ friction factor. The maximum value of the forward slip was obtained for friction factor $\mu = 0.29$ (kerosene as a lubricant of rolls) and minimum for $\mu = 0.22$ (mineral oil as a lubricant).

Table 4. Values of roll bite angles, friction factor and forward slips calculated in different conditions of rolling

Lubrication:	α_{ch} [^{0}]	μ	S_w from Fink's formula (1), [%]	S_w from Drezden's formula (5), [%]	S_w from Vinogradov's formula (6), [%]
extraction naphtha	15,58	0.29	27.30	27.60	19.30
kerosene	16,30	0.28	23.50	18.40	19.20
rapeseed oil	14,21	0.25	8.86	10.49	6.78
motor oil	12,81	0.23	7.20	7.42	4.83
mineral oil	12,16	0.22	5.13	5.39	5.71

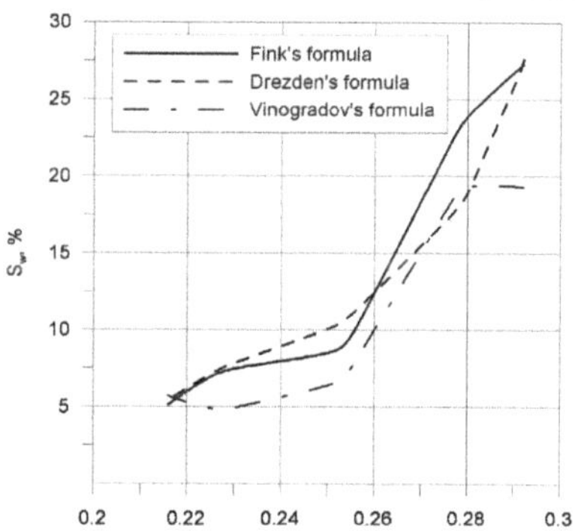

Fig. 3 Influence of friction coefficient on the forward slip in cold longitudinal flat rolling for specimens made from EN AW-6063 aluminium.

Summary

On the basis of investigations carried out into cold longitudinal rolling of flat bars made from EN AW-6063 aluminium, the following can be stated:

1. Values of forward slips calculated by means of Fink's, Drezden's and Vinogradov's formulae (1), (5), (6) did not differ much at the same μ friction factor, and in all the considered cases they were the greatest for Drezden's formula (5), whereas the lowest values were obtained for Vinogradov's formula (6).

2. Forward slip increases with an increase in μ friction factor. The maximum value of the forward slip was obtained for friction factor μ = 0.29 (kerosene as a lubricant of rolls) and minimum for μ = 0.22 (mineral oil).
3. Fink's (1), Drezden's (5) and Vinogradov's (6) formulae, used for calculating forward slip in longitudinal rolling can be applied in engineering practice. Their knowledge is necessary to determine the workpiece velocity in the exit zone of deformation.

References

[1] Wusatowski Z.: Fundamentals of rolling, Śląsk, Katowice, 1969.

[2] Lange K.: Handbook of metal forming, McGraw-Hill Book Company, 1985.

[3] Miłek T., Kopacz Z.: Evaluation of the possibility of diameter reduction in Mt 1020 steel tubes by rolling on the WPM-120 cold profile eccentric rolling machine, 24th Int. Conf. on Met. and Mat., Brno (2015) 377-382.

[4] Sińczak J. et al.: Metal forming processes, AKAPIT, Kraków 2001 (in Polish).

[5] Kowalik M., Trzepieciński T.: Experimental and numerical study of longitudinal cold rolling of the shafts, Steel Res. Int. se 14th Int. conf. Metal Forming (2012) 63-66.

[6] Engler O., Schäfer C., et al.: Flexible rolling of aluminium alloy sheet—Process optimization and control of materials properties, J. Mater. Process. Technol. 229 (2016) 139-148. https://doi.org/10.1016/j.jmatprotec.2015.09.010

[7] Tabary P.T., Sutcliffe M.P.F., Porral F., et al.: Measurements of friction in cold metal rolling, ASME J. Tribology 118 (1996) 629-636. https://doi.org/10.1115/1.2831584

[8] Gjønnes, L., Andersson B.: Mechanisms of surface deformation during cold rolling of aluminium, J. Mater. Sci. 33 (1998) 2469-2476. https://doi.org/10.1023/A:1004328513471

[9] Gerasimov D., Biba N. Stebunov S., Kadach M.: Implementation of a dual mesh method for longitudinal rolling in QForm V8, Mater. Sci. Forum 854 (2016) 158-162. https://doi.org/10.4028/www.scientific.net/MSF.854.158

[10] Ściężor W., Mamala A., Kwaśniewski P.: Analysis of properties of selected aluminium alloys, obtained by twin roll casting method and subjected to cold rolling process, Key Eng. Mat. 641 (2015) 202-209. https://doi.org/10.4028/www.scientific.net/KEM.641.202

[11] Grydin O., Bondarenko S., Stolbchenko M., et al.: Rolling of flat Aluminum strips with tailored mechanical properties, Mater. Sci. Forum 854 (2016) 87-92. https://doi.org/10.4028/www.scientific.net/MSF.854.87

[12] J.R. Davis, Aluminum and Aluminum Alloys, ASM Specialty Handbook, ASM International, Ohio, 1993.

[13] Hirsch J.: Aluminium in innovative light-weight car design, Mater. Trans. 52 (2011) 818-824. https://doi.org/10.2320/matertrans.L-MZ201132

[14] Czarny R.: Plastic lubricants, WNT. Warszawa, 2004 (in Polish).

[15] Information on http://www.chem-rozlew.com/

[16] Information on http://www.orlenoil.pl/

[17] Information on http://www.total.com.pl/

[18] Information on http://www.basiccomponents.com.pl/

Materials Research Proceedings 5 (2018) 73-77 doi: http://dx.doi.org/10.21741/9781945291814-13

Detergents for Washing of Rolling Stock

KOWALIK Pawel [1, a*], MILCZAREK Danuta [1, b]

[1] Instytut Kolejnictwa (Railway Research Institute), Materials & Structure Laboratory, Chlopickiego 50 Street, 04-275 Warsaw, Poland

[*a] kowalik@ikolej.pl, [b] dmilczarek@ikolej.pl

Keywords: Rolling Stock, Detergents, Clearing, Corrosivity, Acidic and Alkaline Agents

Abstract The article discusses acidic and alkaline cleaning agents and types of impurities removed. Their influence on paint coatings and construction elements of rolling stock is also described. It shows what contaminants occur during the operation of rolling stock exploitation and materials of carriage elements to be washed. The scope of the test and requirements for cleaning preparations for washing of rolling stock are also presented. Then, test results of the examined preparations are discussed along with the interpretation and conclusions.

Introduction

During rolling stock washing, its structural elements are exposed to aggressive substances found in the cleaning agents. Wagon boxes and chassis components are exposed to these substances. The type of preparation used depends on the impurities. The aim is to ensure that the composition of washing agents is so that their impact on the construction elements of rolling stock is as small as possible and preparation effectively removes impurities.

Rolling stock contaminants

Rolling stock contaminants can be divided into two groups:

- originating from dust deposited on the surface of rolling stock – these are mainly calcium, potassium and silicon compounds;
- arising depending on the type of traction used: iron and copper oxides, soot, oil and grease particles. Iron oxides (73%) and oils and lubricants (17%) predominate for diesel traction and iron oxides (83-90%) and mineral powders (10-16%) in the case of electric traction.

Materials of carriage elements exposed to action of cleaning agents

The external part of wagons contains elements made from steel (shell) covered with the paint coating, glass (panes), rubber (gaskets and rollers in gangways), plastics, cast iron and non-ferrous metals, mainly aluminum alloys. About 85% of the external surface of the wagons are paint coatings and about 12% are panes [1]. Paint materials are the most susceptible to dirt, among them phthalic coatings, which are relatively soft and not very resistant to acids, bases, oxidizing agents and organic solvents. The dirt is the most difficult to be removed. The most favorable in this respect are polyurethane and epoxy-ester coatings, the latter being between phthalic and polyurethane. In case of other materials, some problems are caused by some window frames made of an aluminum alloys, on which impurities accumulate easily and are difficult to remove.

Types of cleaning agents

Currently two types of cleaning agents are used: acidic (pH < 7) and alkaline (pH > 7) which remove various impurities. The use of inert (pH = 7) or slightly alkaline (pH = 7-8) agents does not give positive results. Washing preparations primarily consist of the following:

- acids or bases;
- surfactants;
- active fillers (e.g. corrosion inhibitors).

Materials Research Proceedings 5 (2018) 73-77 doi: http://dx.doi.org/10.21741/9781945291814-13

Acids and bases dissolve metal oxides and greases which make contaminations. Surfactants as surface active substances lessen water surface tension and stabilize double layer on the interface [2]. They contain two parts: hydrophilic and hydrophobic. The hydrophilic part (polar) is soluble in water and polar solvents. The hydrophobic part most often creates acidic or basic residue which dissolves action in water and polar solvents cause the formation of salt. Double action of surface active substances makes liquids containing surfactants exhibit foaming properties. Organic compounds with the polar structure of a molecule are the most often used surfactants:

- ionic – creating anions and cations as a result of dissociation in water, e. g. salts of higher carboxylic and sulphonic acids, higher amines and amine salts;
- nonionic compounds – which have in their structure polar groups that are unable to dissociate in water, e. g. higher polyhydric alcohols [3].

The selection of surfactants has a significant impact on the effectiveness of cleaning, because the lowering of water surface tension favors washing away of dirt and potential-forming compounds (acids and bases) dissolve it.

Active fillers support the action of surfactants. Their main functions are as follows: softening water (they remove calcium and magnesium ions), solution buffering, anti-corrosive action, binding of heavy metals and dispersing (breaking down into smaller particles) dirt. Most often they are chromates and organic substances absorbing on the oxide. Water cleaners have good effectiveness, are easy to use, non-flammable and relatively cheap.

Requirements for cleaning agents intended for rolling stock

The cleaning preparation should be effective and should:

- have an appropriate chemical composition (at most two acidic or basic substances with the concentration of 10-25%);
- be resistant to freezing;
- be stable after dilution;
- do not cause significant corrosion of steel and aluminum;
- do not exhibit aggressive action on paint coatings and rubber components of rail vehicle equipment.

Detailed requirements are specified in the Normative Document ND 001/07 "Cleaning agents for external and internal cleaning of rolling stock" [4].

Examination of cleaning agents for washing of rolling stock

The results of tests carried out on acidic and alkaline cleaning agents in the Railway Research Institute within the last 5 years, are presented below. The hydrochloric, sulfuric and oxalic acid were the main components of acidic cleaning preparations. Sodium hydroxide and potassium hydroxide were the predominant components in case of alkaline agents. The scope of research included among others the following tests:

- determination of pH value according to ISO 1262:2004 [5];
- determination of corrosive action on steel and aluminum according to ND 001/07 [4];
- determination of the impact on hardness change of material intended for doors and windows gaskets (rubber) according to PS-C-04238:1980 [6];
- determination of tensile strength of materials intended for doors and windows gaskets (rubber) according to ISO 37:2007 [7];
- determination of aggressive action on paint coatings according to ISO 2812-3:2012 [8].

Tests results

The results of test conducted on acidic and alkaline cleaning agents are depicted below accordingly in Table 1 and Table 2.

Materials Research Proceedings **5** (2018) 73-77 doi: http://dx.doi.org/10.21741/9781945291814-13

Table 1. Results of the tests of acidic cleaning agents (detailed results are in the possession of Materials & Structure Laboratory – Railway Research Institute)

Type of examination	Agent 1*	Agent 2*	Agent 3*	Agent 4**	Agent 5*	Agent 6***	Value required by ND 001/07
pH value [pH]	2.6	1.0	1.6	1.4	1.0	1.0	≥ 1
aggressive action on steel [g/m^2]	35.4	29.9	14.4	40.1	14 7	14.1	≤ 120
aggressive action on aluminum [g/m^2]	19.7	19.8	1.8	15.5	5.7	6.9	≤ 20
aggressive action on rubber elements – impact on change of rubber hardness [%]	1.6	0.0	0.0	1.6	4.5	4,5	≤ 15%
aggressive action on rubber elements – impact on change of tensile strength of rubber [%]	0.0	4.,1	1.7	1.1	8.0	8.3	≤ 15%
aggressive action on paint coatings I–IV	I: no change II: no change III: light dulling IV: no change	I: no change II: no change III: light dulling IV: no change	I: no change II: no change III: no change IV: no change	I: no change II: no change III: light dulling IV: no change	I: no change II: no change III: light dulling IV: no change	I: no change II: no change III: light dulling IV: no change	permissible light dulling or slight discoloration

*agents on a basis of hydrochloric acid, **agent on a basis of sulfuric acid, ***agent on a basis of oxalic acid
I – polyurethane coating (green), II – epoxy-acrylic coating (blue), III – phthalate coating (blue), IV – epoxy-polyurethane coating (blue).

Discussing the results and conclusions

The examined preparations have acidic and alkaline pH. Knowledge of pH value of cleaning preparations is important when predicting processes such as metals and paint coatings corrosion.

Aggressive (corrosive) effect on steel and aluminum

Iron and aluminum alloys are the most important construction materials of a rail vehicle exposed to corrosive action of washing agents. Corrosion cause transformation of metals and their alloys in nonmetallic compounds (oxides and hydroxides) and if not inhibited corrosion may lead to weakening of the quality of carriage boxes construction.

Corrosive processes of steel depend on the concentration of hydrogen ions [H+] in the electrolyte acting on it. The course of this phenomenon is presented by the equations of reactions [9]:

$Fe \rightarrow Fe^{2+} + 2e^-$ (iron ionization under water impact) (1)

$2e^- + 2H^+ \rightarrow 2H$ (hydrogen ions capture electrons) (2)

$2H + 1/2\ O_2 \rightarrow H_2O$ (reaction of hydrogen atoms with oxygen from the air) (3)

$Fe + H_2O + 1/2O_2 \rightarrow Fe^{2+} + 2OH^-$ (summary course of the reaction) (4)

In an alkaline environment the above reaction proceeds to a minimal extent, because in accordance with Le Chatelier rule OH^- ions shift the equilibrium state of the reaction (4) to the left. That is why in the presence of a base the corrosion process is ceased.

Table 2. Results of the tests of new alkaline cleaning agents (detailed results are in the possession of Materials & Structure Laboratory – Railway Research Institute)

Type of examination	Agent 7**	Agent 8*	Agent 9*	Agent 10**	Value required by ND 001/07
pH value [pH]	13.0	13.0	11.0	12.3	≤ 13
aggressive action on steel [g/m^2]	1.2	0.9	5.1	0.9	≤ 120
aggressive action on aluminum [g/m^2]	10.8	12.1	6.4	12.5	≤ 20
aggressive action on rubber elements – impact on change of rubber hardness [%]	2.9	1.6	1.6	3.1	≤ 15%
aggressive action on rubber elements – impact on change of tensile strength of rubber [%]	18.3	10.8	13.2	2.9	≤ 15%
aggressive action on paint coatings I–IV	I: no change II: light dulling III: dulling IV: no change	I: no change II: no change III: dulling IV: no change	I: no change II: no change III: dulling IV: no change	I: no change II: light dulling III: dulling IV: no change	permissible light dulling or slight discoloration

*agents on a basis of sodium hydroxide, **agents on a basis of potassium hydroxide
I – polyurethane coating (green), II – epoxy-acrylic coating (blue), III – phthalate coating (blue), IV – epoxy-polyurethane coating (blue).

In an acidic environment the state of equilibrium of the reaction (1) shifts to the right towards the formation of Fe^{2+} ions as a result of electrons consumption in reduction of hydrogen in acidic electrolyte. Favoring the formation of Fe^{2+} ions develops the corrosion process. Therefore, acidic cleaning agents are more aggressive to steel than alkaline preparations.

The above analysis is in agreement with the obtained test results. The examined alkaline preparations showed significantly lower aggressiveness to steel than acidic preparations.

Aluminum, on the other hand, is an amphoteric element that reacts with both acids and bases. In a natural environment it spontaneously covers the surface with an oxide layer that protects against corrosion:

$$2Al + O_2 \rightarrow Al_2O_3 \qquad (5)$$

Only the damage of the resulting protective layer, under the influence of some acids and strong bases solutions, may induce further course of corrosion process. This is the reason why it is difficult to predict in which environment the corrosion process will be more intense.

This was confirmed in the test results. Similar values of mass losses of aluminum samples were obtained in case of acidic preparations (agents 1, 2 and 4) as well as for alkaline preparations (agents 7, 8 and 10).

Aggressive action on paint coatings. The tested agents, both acidic and alkaline, showed no aggressive action on the surface of the tested coating samples. No changes were observed after a 24 h action of cleaning products. The exception was phthalic coating (blue), where there was a light dulling on the surface of the tested coating under the influence of acidic preparations and dulling under the influence of alkaline preparations. The study showed that phthalic coatings are indeed less resistant to action of strong acids and strong bases solutions than the other paint coatings.

Aggressive action on rubber elements. The obtained results of the experiments show that both alkaline and acidic preparations do not significantly change the hardness of rubber, constituting structural elements of carriages.

It is different in when it comes to changes of the tensile strength of rubber. Acidic agents showed little effect here, within the limit permitted by ND 001/07 [4]. Alkaline agents showed various impact (preparations 7 and 10). The main component of both preparations was potassium hydroxide of similar concentrations (2-5%). Despite this fact, agent 7 changed the tensile strength of rubber by 18,3% and agent 10 only by 2,9%. Agent 7 does not meet the requirements of ND 001/07 [4]. Such a big difference is probably caused by the influence of other ingredients in both preparations.

Summary

The examination of currently used and new acidic and alkaline cleaning preparations for rolling stock showed slight aggressive action (apart from agent 7) for structural elements of rolling stock and meet the requirements of ND 001/07 [4]. It is assessed that introduction for use of preparations systematically modified by manufacturers influenced the following: lengthening the life of used railway vehicles, reducing pressure on environment and reducing the exploitation costs.

References

[1] M. Moczarski Washing of passengers carriages in a depot and the environmental hazard, Railway Reports, 144 (2007) (in Polish).

[2] L. Pajdowski General chemistry, Polish Scientific Publishing House (1985) (in Polish).

[3] T. Kalak Influence of surface properties of washing powder solutions on their washing abilities, doctoral dissertation, University of Economy (2012) (in Polish).

[4] Normative Document ND 001/07 Cleaning agents intended for external and internal washing of rolling stock, Railway Scientific and Technical Center/Railway Institute (2007) (in Polish).

[5] ISO 1262:2004 Surface active agents. Determination of pH value of solutions and dispersions.

[6] PS-C-04238:1980 Rubber. Determination of hardness according to Shore method.

[7] ISO 37:2007 Rubber, vulcanized or thermoplastic. Determination of tensile stress – strain properties.

[8] ISO 2812-3:2012 Paints and varnishes. Determination of resistance to liquids. Part 3: Method using an absorbent medium.

[9] K. M. Pazdro Chemistry. Educational Outbuilding (1995) (in Polish).

Terotechnology 2017 Materials Research Forum LLC
Materials Research Proceedings **5** (2018) 78-83 doi: http://dx.doi.org/10.21741/9781945291814-14

Evolution of Microstructure in the Heat Affected Zone of S960MC GMAW Weld

JAMBOR Michał[1, a*], ULEWICZ Robert[2, b], NOVÝ František[1, c], BOKŮVKA Otakar[1, d], TRŠKO Libor[3, e], MIČIAN Miloš[4, f], HARMANIAK Daniel[4, g]

[1]University of Zilina, Department of Material Engineering, Univerzitná 8215/1, Žilina, Slovak Republic

[2]Politechnika Czestochowska, Department of Production Engineering, Dabrowskiego 69 Czestochowa, Poland

[3]University of Zilina, Research centre of University of Zilina, Univerzitná 8215/1, Žilina, Slovak Republic

[4]University of Zilina, Department of Technological Engineering, Univerzitná 8215/1, Žilina, Slovak Republic

[a*]michal.jambor@fstroj.uniza.sk, [b]robert.ulewicz@wz.pcz.pl, [c]frantisek.novy@fstroj.uniza.sk, [d]otakar.bokuvka@fstroj.uniza.sk, [e]libor.trsko@rc.uniza.sk, [f]milos.mician@fstroj.uniza.sk [g]daniel.harmaniak@fstroj.uniza.sk

Keywords: High Strength Steel, S960MC, Welding, Heat Affected Zone

Abstract. High strength steels were introduced to the welded construction at the beginning of the 21st century. Phase transformation occurring in the heat affected zone of the weld significantly affects the resulting properties of the whole weld. In this study, the Gas Metal Arc Weld (GMAW) of S960MC was investigated with a special emphasis on the microstructural evolution throughout the Heat Affected Zones (HAZ). In the HAZ, three main different sub-zones were recorded. The changes of microstructures in these zones depend on the level of thermal exposure and are varying with distance from the weld metal zone. Microhardness measurements showed a decrease of microhardness in all the examined sub-zones of HAZ. The most significant decrease was recorded in the Intercritical HAZ (ICHAZ), thus it can be stated, together with the tensile tests observations, that the ICHAZ is the most critical part of the HAZ and the whole welded joint.

Introduction

In every aspect of engineering of transport vehicles, there are increased demands for the reduction of total weight for the improvement in fuel efficiency and payload capacity. High strength steels are attractive material for designers, as they are able to carry higher loadings, which allows for the reduction in thickness of used plates and therefore reduces total weight [1, 2]. High mechanical properties of these steels are obtained by combined thermo-mechanical processing to produce desirable microstructures [3, 4]. Chemical composition of high strength steels is carefully designed to allow reaching high mechanical properties while maintaining considerably low carbon content to ensure sufficient welding properties. And it is welding of high strength steels which is the main task for materialists and technologists to ensure appropriate properties of the welded parts. As mechanical properties in high strength steels are obtained by the thermomechanical processing, any heat introduced to the metal during the welding process can significantly affect the resulting microstructure and thus affect properties of the whole welded joint [5-7]. This is even more reasonable for high strength steels with the yield strength higher than 700 MPa. The S960MC steel belongs to this category. In this paper, microstructural evolution in the S960MC high strength steel welds is described and supported with the microhardness evaluation, with the aim to describe microstructural changes in all sub-zones in the heat affected zone.

Material and experiments

In the study, the thermomechanical processed S960MC steel was used. The chemical composition of the experimental material is shown in Table 1. The GMAW weld was prepared from the S960MC sheets 300×300 mm with 3mm thickness in the atmosphere of protective gas mixture M21 (82%Ar + 18%CO_2). As a filler metal, the G 89 5 M21 Mn4Ni2.5CrMo wire was used (chemical composition is shown in Table 1, and the mechanical properties are shown in Table 2).

Table 1. Chemical composition of the experimental material and filler metal.

	C	Si	Mn	P	S	V	Ti	Cu	Cr	Ni	Mo	Al	Zr
S960 MC	0.087	0.18	1.11	0.009	0.001	0.01	0.022	0.017	1.08	0.06	0.128	-	-
G 89 5 M21 Mn4Ni2.5Cr Mo	0.11	0.66	1.77	0.009	0.007	0.007	0.069	0.17	0.41	2.43	0.46	0.007	0.0019

Table 2. Mechanical properties of the S960MC base metal and the welds and the G 89 5 M21 filler metal (data were provided by the supplier).

	UTS [MPa]	**YS [MPa]**	**Elongation [%]**
S960MC base metal	1150	1034	11.5
S960MC weld metal	853	823	3.5
G 89 5 M21 Mn4Ni2.5CrMo	≥980	≥930	≥14

Single pass, but joint was performed, without any pre and post heat treatment and the applied heat input during the welding process equalled 4.3 kJ/cm. Transversal cuts were made from the weld from which the specimens for the microstructural observation were prepared, using the standard method for the preparation of metallographic specimens. Microstructural characterization was carried out using light microscopy which allows for distinguishing differences in the microstructure caused by heat affecting during the welding process. Microhardness measurements in all the individual sub-zones of the HAZ were carried out, with the used loading of 1kp. Five measurements were performed for the every identified sub-zone in the HAZ and the average values were calculated. To evaluate tensile properties of the weld and their comparison with those for the base metal, the tensile tests according to the EN ISO 6892-1 standard were carried out, and the obtained results are shown in Table 2.

Materials Research Proceedings **5** (2018) 78-83 doi: http://dx.doi.org/10.21741/9781945291814-14

Results and discussion

Microstructural observation shows significant changes in the HAZ of the examined welds. The microstructure of the base metal is shown in Fig.1, and consists of a mixture of tempered martensite and bainite. Such a microstructure exhibits high mechanical strength while maintaining good toughness and ductility. The weld metal consists of the martensite and bainite, but no special attention was given to a weld metal, as according to the literature, the HAZ is the most problematic area in the high strength steels weld. According to the microstructure observation throughout the HAZ, several structural different sub-zones were recorded. The results of the microhardness measurements of these sub-zones are shown in Table 3. The phase transformations in the HAZ depend on the thermal exposure, to which individual parts of the HAZ were subjected and on the the time of this thermal exposure. Closer to the weld metal and fusion zone, the area was exposed to higher temperatures, but also the cooling rate was higher. In the HAZ of the examined weld, the three main sub-zones were identified (naturally, transition areas were present between these clearly distinguished sub-zones).

Fig. 1 Microstructure of the base metal.

In the direction from the weld metal to the base metal, the Coarse Grain Heat Affected Zone (CGHAZ) was the first distinguished zone (Fig.2c). During the welding procedure, metal in this zone was heated far above the A_{c3} temperature and thus the whole initial structure was transformed to austenite whose grains subsequently rapidly grew. Considerably high cooling rates in this area result in the formation of metastable structures such as bainite and martensite during the cooling stage. The amount of the grain growth depends on the level of overheating, and closer to the weld metal, the grain growth will be more significant. The Fine Grain Heat Affected Zone (FGHAZ) is the second resolved sub-zone (Fig.2b). The material in this zone was heated slightly above the A_{C3} temperature, but total time spent above this temperature was very short. During this thermal cycle, the material was transformed to austenite, but due to the low level of overheating, short exposure time and considerably high cooling rates, no grain growth occured. On the contrary, the refinement of the austenitic structure occurs, together with its transformation to the mixture of bainite and martensite. The last observed sub-zone, called the Intercritical Heat Affected Zone (ICHAZ) lies between the HAZ and base metal (Fig.2a). This area is very narrow (slightly more than 500 µm) and during the welding process, the thermal exposure was in the range between A_{C1} and A_{C3} temperatures. The microstructure of the base metal is partially transformed to austenite, while the rest is very strongly tempered. After rapid cooling, the austenitic part of the structure will transform to the bainite/martensite, while the rest will stay intact (it means that no visible phase transformation occurs, but strong tempering will result in significant changes of mechanical properties).

Materials Research Proceedings **5** (2018) 78-83 doi: http://dx.doi.org/10.21741/9781945291814-14

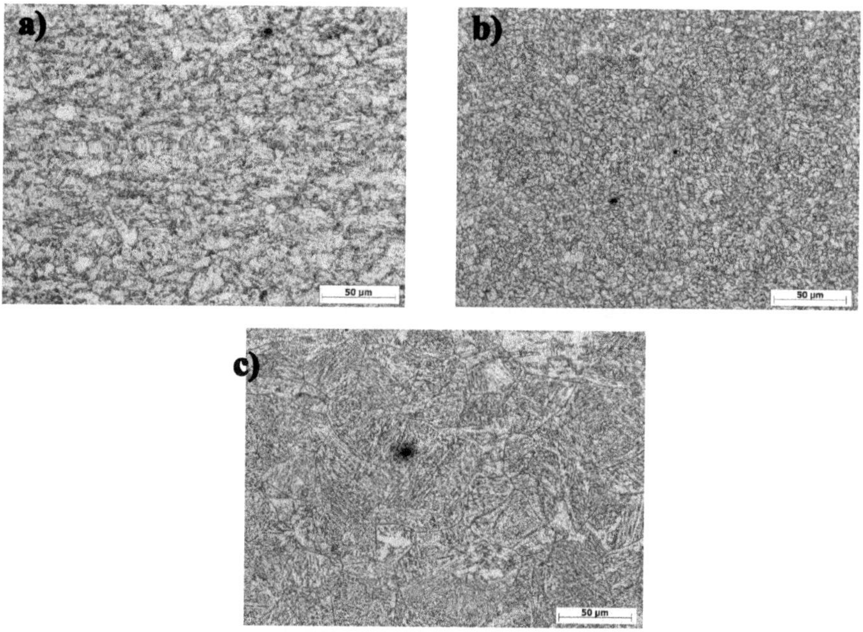

Fig. 2 Microstructure of the HAZ of the S960MC weld, a) CGHAZ, b) FGHAZ, c) ICHAZ.

The results of the tensile tests and the microhardness measurements showed several important findings. First, the tensile tests of the welded specimens showed significant decrease in the tensile strength, yield strength and also very significant reduction in the elongation of the specimens. The subsequent examination revealed that the rupture occurred in the ICHAZ in all cases. The microhardness of the base metal was almost 340 HV1, which corresponded with the microstructure of mixture of bainite and tempered martensite. In the CGHAZ, a significant decrease of the microhardness was recorded (270 HV1). This reduction of mechanical properties can be attributed to the tempering of the bainite/martensite structure and the excessive grain growth, as a result of the heat introduced to the material during the welding process. The FGHAZ also exhibited lower microhardness compared with the base metal (302 HV1), but this value is still higher than that of the CGHAZ. The decrease of the microhardness in the FGHAZ has similar reasons as that in the CGHAZ, except the grain growth (in the FGHAZ, the refinement of the grain structure was observed). The most significant reduction of microhardness was recorded for the ICHAZ. In this sub-zone of the HAZ, the value of only 228 HV1 was recorded. Similar observations were recorded by other authors [3, 5, 8, 9], the ICHAZ is the most critical part of the HAZ in high strength steels welds. In the literature, there is good agreement that the ICHAZ is the "weak point" of the weld from these steels and in most cases, the welds are broken in that area, but there is no good agreement in the description of microstructure in these areas and the explanation of low mechanical properties and ductility in this area. In the most of the available literature sources, the microstructure of the ICHAZ is described as a mixture of different types of martensite and bainite [5, 8]. In our study, the results of microhardness measurements suggest that microstructure in the ICHAZ cannot be considered as martensitic, due to ultra-low values of microhardness (228 HV1). The drop of mechanical properties in the ICHAZ observed in this study was larger than that observed by other authors [3, 5, 8, 9] (who used different welding methods), based on which it can be concluded that the GMAW is not an appropriate method for the welding of S960MC high strength steel. Low microhardness values in the ICHAZ suggest that strong tempering of the martensite/bainite mixture will result in the formation of difficult to describe type of ferrite-carbide mixture. This mixture cannot be considered as a martensitic nor bainitic structure, as the microhardness values are very low. The microstructure of the ICHAZ consists of a special kind of a disintegrated austenite structure (ferrite-carbide mixture), which is more similar to the unusual ferrite-pearlite mixture than to the martensite or bainite. Very low ductility is another significant property of this structure, which is a result of the reduction in elongation observed in this study, but also by other authors [3, 8]. For the determination of the exact nature of this microstructural mixture, additional tests are required, and this will be the subject of the following examinations. Even though high strength steels are used in engineering application from the beginning of the 21st century, there are still a lot of aspects, especially connected with the microstructure evolution in the HAZ, which are needed to be revealed.

Table. 1 Values of microhardness of individual sub-zones in HAZ

Sub-zone	**CGHAZ**	**FGHAZ**	**ICHAZ**	**Base metal**
Average value HV1	270.1	301.5	228.1	338.4

Conclusions

Based on the carried out experiments, the following can be concluded:

- The welding process significantly affects microstructures in the HAZ.
- In the HAZ, three main different sub-zones were recorded. The changes of microstructures in these zones depend on the level of thermal exposure and are varying with the distance from the weld metal zone.
- Observation of the three sub-zones in the HAZ was in consistency with the observations of other authors, and these were called the CGHAZ, FGHAZ and ICHAZ.
- Tensile tests showed a significant decrease of the tensile strength, yield strength and also tensile elongation. The fracture of all specimens occurred in the ICHAZ.
- Microhardness measurements showed a decrease of microhardness in all the examined sub-zones of the HAZ. The most significant decrease was recorded in the ICHAZ and thus, together with the tensile tests observations, it can be stated that the ICHAZ is the most critical part of the HAZ and the whole welded joint.
- Microstructure in the ICHAZ was a subject of increased interest in many studies, but the microstructure of this zone is still not completely clear. Based on the carried out experiments, the microstructure in the ICHAZ is some special form of ferrite-carbide mixture, but due to very low microhardness, it cannot be considered as a bainitic structure.
- Drop of microhardness values in the ICHAZ, which is more significant than that observed in other studies on a similar material suggests that the GMAW is not an appropriate method for the welding of the S960MC steel.

Acknowledgement

The research was supported by the Scientific Grant Agency of the Ministry of Education, Science, Research and Sport of the Slovak Republic under the VEGA contract no. 1/0951/17 and Slovak Research and Development Agency under the contract no. APVV-16-0276 - Development of robotic device for strengthening and nanocrystallization of welded joints surface layers with the ultrasonic surface attrition method.

References

[1] X. Qiang, X. Jiang, F. S. K. Bijlaard, H. Kolstein. Mechanical properties and design recommendations of very high strength steel S960 in fire. Engineering Structures. 112 (2016) 60-70. https://doi.org/10.1016/j.engstruct.2016.01.008

[2] M. S. Węglowski, M. Zeman. Prevention of cold cracking in ultra-high strength steel Weldox 1300. Archives of Civil and Mechanical Engineering. 14 (2014) 417-424. https://doi.org/10.1016/j.acme.2013.10.010

[3] W. Guo, L. Li, S. Dong, D. Crowther, A. Thompson. Comparison of microstructure and mechanical properties of ultra-narrow gap laser and gas-metal-arc welded S960 high strength steel. Optics and Lasers in Engineering. 91 (2017) 1-15. https://doi.org/10.1016/j.optlaseng.2016.11.011

[4] S. Błacha, M. S. Węglowski, S. Dymek, M. Kopyściański. Micorstructural and mechanical characterization of electron beam welded joints of high strength S960QL and WELDOX 1300 steel grades. Arch. Metall. Mater. 62 (2017), 2, 627-634.

[5] W. Guo, D. Crowther, J. A. Francis, A. Thompson, Z. Liu, L. Li. Microstructure and mechanical properties of laser welded S960 high strength steel. Materials and Design. 85 (2015) 534-548. https://doi.org/10.1016/j.matdes.2015.07.037

[6] M. Gosh, K.Kumar, R. S. Mishra. Analysis of microstructural evolution during friction stir welding of ultrahigh-strength steel. Scripta Materialia. 63 (2010) 851-854. https://doi.org/10.1016/j.scriptamat.2010.06.032

[7] D. Min, T. Xin-hua, L. Feng-gui, Y. Shun. Welding of quenched and tempered steels with high-spin arc narrow gap MAG system. International Journal of Advanced Manufacturing Technology. 55 (2011) 527-533. https://doi.org/10.1007/s00170-010-3052-1

[8] L. Yajiang, W. Juan, L. Peng. Fine structure in the inter-critical heat-affected zone of HQ130 super-high strength steel. Bull. Mater. Sci. 26/2 (2003) 273-278. https://doi.org/10.1007/BF02707803

[9] J. Youn-Suk, et al. Microstructural Evolution and Mechanical Properties of Resistance Spot Welded Ultra High Strength Steel Containing Boron. Materials Transactions. 52 (2011) 6, 1330-1333.

Materials Research Proceedings 5 (2018) 84-89 doi: http://dx.doi.org/10.21741/9781945291814-15

The Influence of Diamond-Like Coatings on the Properties of Titanium

PAWELEC Katarzyna[1,a,*], BARANOWICZ Paweł[2,b], WYSOKIŃSKA-MISZCZUK Joanna[3,c], MADEJ Monika[1,d]

[1] Kielce University of Technology, Faculty of Mechatronics and Mechanical Engineering, Tysiąclecia PP 7, 25-314 Kielce, Poland

[2] Medical University of Lublin, Karmelicka 7, 20-081 Lublin, Poland

[3] NZOZ Ekstradent, Juliusza Słowackiego 21/1U, 25-365 Kielce, Poland

[a]kpawelec@tu.kielce.pl, [b]pawel.baranowicz@ekstradent.pl, [c]periodontologia@tlen.pl, dmmadej@tu.kielce.pl

Keywords: Titanium, Diamond-Like Carbon (DLC) Coating, PACVD, Wear of Tribological Systems

Abstract: The paper compares the morphology and geometric structure of the surface and the results of tribological tests of polished titanium discs and PACVD diamond-like coatings deposited on them. Observations of surface morphology and determination of chemical composition were carried out using scanning microscopy and EDS analyzers. The geometric structure of the surface was investigated using an optical profilometer. Tribological properties were investigated by ball-on-disc testing under dry friction and under lubrication regime using Ringer's solution. The results obtained during the tests showed that the use of DLC coatings significantly improves the performance of titanium.

Introduction

Modern biomedical materials are to a large extent based on metal alloys, among which titanium occupies more and more space. The first applications of titanium in bone surgery date back to the forties of the last century. The choice of titanium is due to its good mechanical properties and corrosion resistance in the tissue environment. For this reason, it is widely used for various types of implants. Intraosseous implants allow obtaining satisfactory results in a very wide range of indications. Due to the physicochemical and mechanical properties of titanium and its alloys, they play an important role in implantology and dental prosthetics [1]. Titanium shows domination over other metallic materials owing to better biocompatibility in the environment of a living organism, and contributes to rapid osseointegration [2].

Titanium is characterized by high resistance to wear and high mechanical strength, showing the highest ratio of mechanical strength to its specific weight among all metal biomaterials [3, 4, 5]. In order to increase the service durability and to improve the quality of the surface, protective coatings are deposited.

Diamond-like carbon (DLC) coatings are one of the most durable varieties of protective layers. Their characteristics include a high modulus of elasticity, resistance to brittle fracture, high thermal conductivity, low coefficient of thermal expansion and chemical stability. These unique features make DLC coatings suitable for use in the surface engineering to control tribological damage as material with a low coefficient of friction and high resistance to friction wear [6]. The coatings are widely applied in medicine for the production of biomaterials [7, 8]. DLC coatings can be obtained by many techniques, including PACVD. In its essence, the PACVD process is a CVD process assisted by a glow discharge plasma, aimed at producing hard surface layers or layers exhibiting special surface and volume related properties (e.g., protective, anti-corrosion, tribological properties). In general, the PACVD technique uses the advantages of CVD, such as uniform deposition of layers on different materials with simultaneous elimination of their defects (high temperature at which the CVD process takes place). Applied treatments allowed obtaining surfaces with different topography and morphology [9, 10].

Materials and methods:

The tests were carried out on titanium discs with a diameter of 18 mm and a height of 6 mm. Before measurements and observations were made, the specimens were sanded and polished using a Bühler Automet 250 grinding and polishing machine. The surface preparation process was carried out in accordance with polishing techniques recommended by the manufacturer. Then, a PACVD coating was applied to some of the specimens.

Observation of surface morphology was carried out using a Phenom ProX scanning electron microscope, while a JEOL scanning microscope JEOL 7100F, equipped with an X-ray dispersion energy spectrometer was used to determine the chemical composition.

A Taylor Hobson coherent Talsurf CCI Lite co-op interferometer analyzed the geometric structure of the specimen surfaces before and after tribological tests.

Tribological ball-on-disc tests were performed using Tester T-01. A titanium specimen subjected to physicochemical treatment and an Al_2O_3 ball with a diameter of 10 mm constituted the friction pair. The motion resistance was determined under technically dry friction (TDF) and with the use of Ringer's solution (RS) as a lubricant. Ringer's solution was chosen because its chemical composition resembles that of the body fluid. The Ringer solution consisted of 8.6 kg/m^3 of NaCl, 0.30 kg/m^3 of KCl and 0.48 kg/m^3 of CaCl. The test load was P = 20 N over the sliding distance of 200 m. The tests were carried out in laboratory atmosphere at a relative humidity of 50±5% and a temperature of 23±1°C.

Results

Surface morphology and chemical composition analysis. Figure 1 shows the morphology of the surface of the polished specimen and of the specimen with a DLC coating at x4000 magnification.

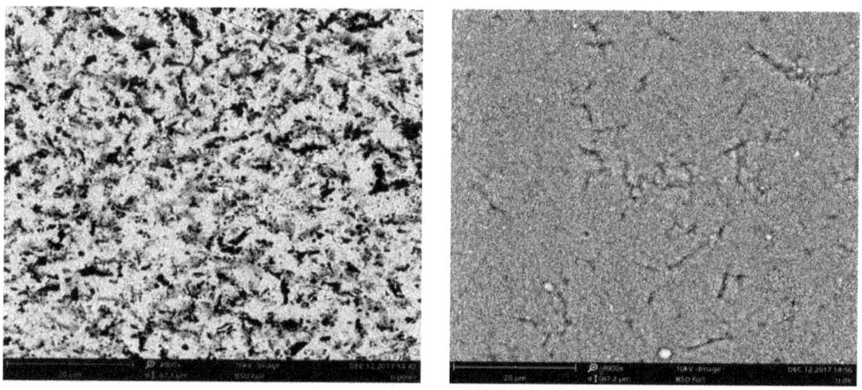

a) b)

Fig.1. SEM- morphology of titanium surfaces of discs a) polished b) with a:C-H coating

Materials Research Proceedings **5** (2018) 84-89 doi: http://dx.doi.org/10.21741/9781945291814-15

Observations made with the use of a SEM microscope showed that both the polished specimen and that with the a:C-H coating had a homogeneous morphology. The discs had a smooth surface with no visible traces of machining or defects. Figure 2 shows the chemical composition analysis in micro-areas.

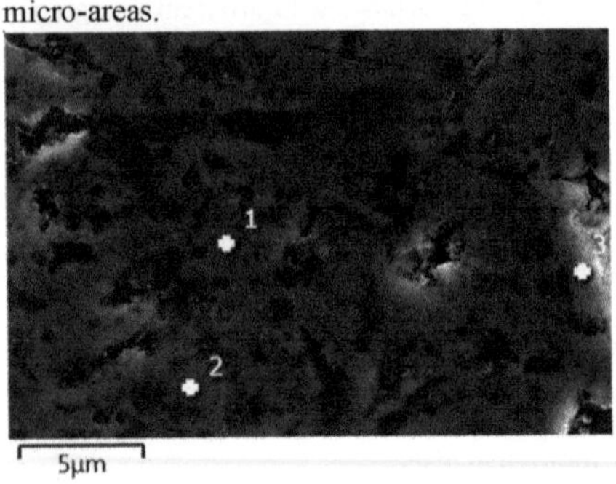

a)

Point	[Ti]	[Si]	Total
1	96.45	0.60	97.05
2	96.47	1.19	97.66
3	99.90	3.0	99.90

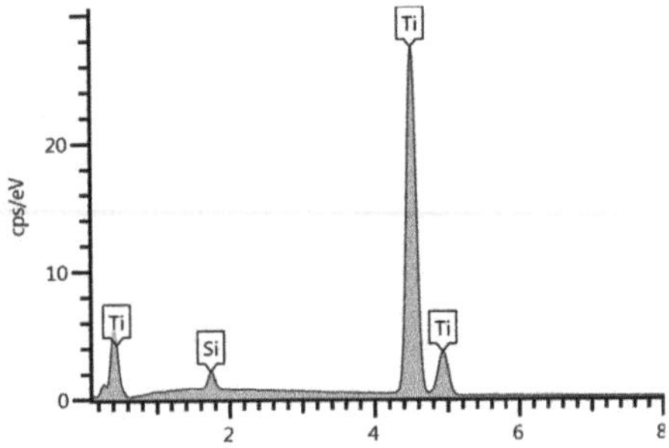

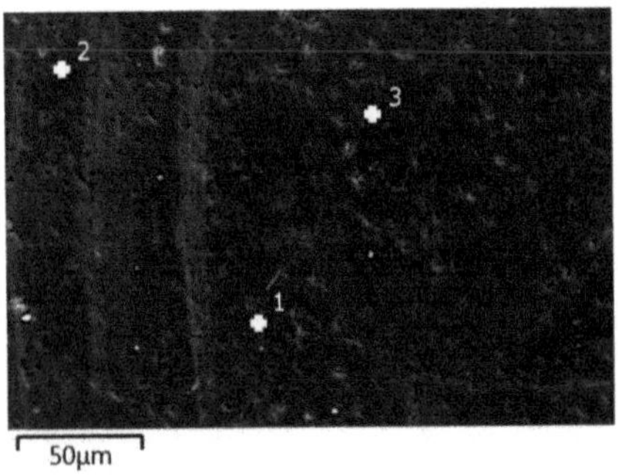

b)

Point	[Ti]	[C]	[W]	Total
1	0.19	80.35	19.46	100.00
2	0.24	80.76	19.01	100.00
3	0.09	81.71	18.19	100.00

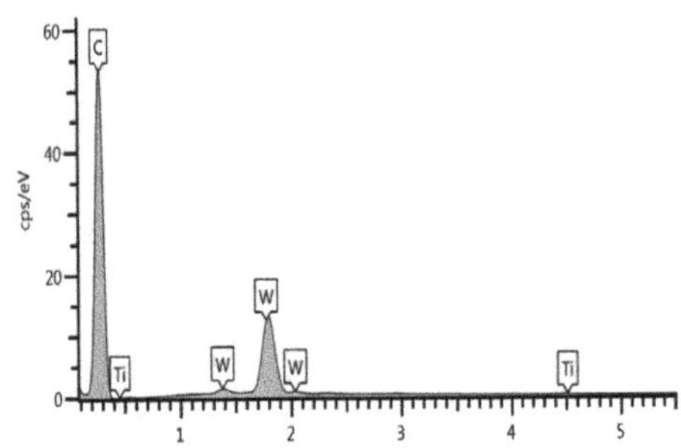

Fig. 2. Analysis of the chemical composition of titanium discs
a) polished, b) with a:C-H coating

Analysis of the chemical composition performed for the micro-areas of the polished specimen showed the presence of titanium and silicon. The presence of silicon in the polished specimen resulted from the SiC discs grinding and polishing processes. In the case of the DLC specimen, the analysis revealed a low content of titanium and a high content of carbon and tungsten.

Materials Research Proceedings **5** (2018) 84-89 doi: http://dx.doi.org/10.21741/9781945291814-15

Assessment of the geometric structure of the surface. Figures 3 and 4 show isometric images and ordinate distribution with the Abbot-Firestone curve.

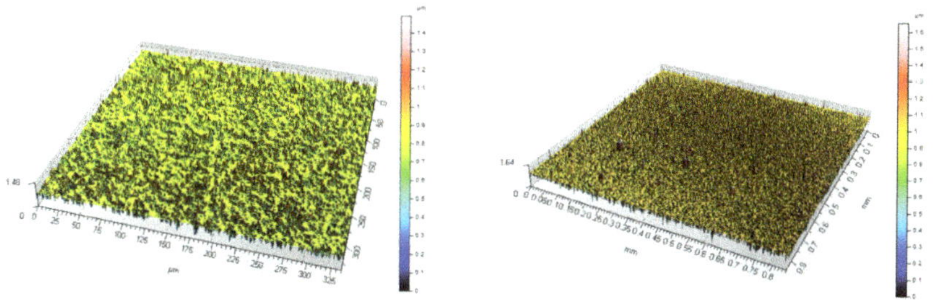

a) b)

Fig. 3. Isometric images of titanium discs a) polished, b) with a:C-H coating

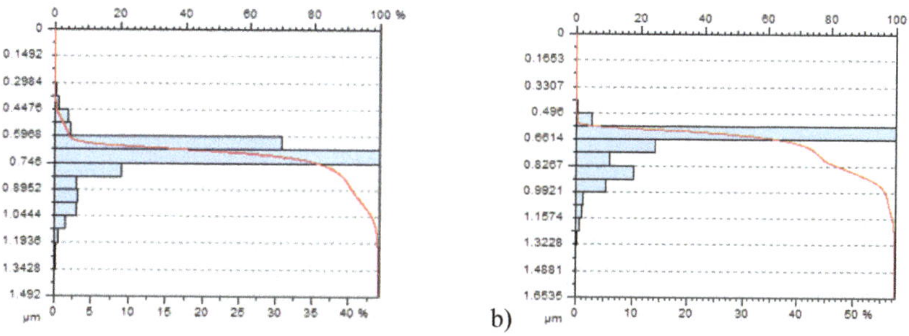

a) b)

Fig. 4. Distribution of ordinates with the bearing area curve of the specimen a) polished, b) with a:C-H coating

Tab. 1. Parameters of the surface geometry

Specimen	Parameters					
	[µm]					
	Sa	*Sp*	*Sv*	*Sz*	*Ssk*	*Sku*
polished	0.0730	0.7134	0.7786	1.4920	- 1.6956	8.6480
with a: C-H coating	0.1039	0.6939	0.9595	1.6535	-1.5709	7.1455

Three-dimensional images allowed the analysis of the geometrical structure of both specimens. Knowledge of surface geometry is very useful when assessing its quality. The test results revealed that the DLC coated disc has a more developed geometric surface structure. Additional information about the surface topography included the amplitude parameters shown in Table 1.

For the specimens with the DLC coating, the Sa, Sv and Sz values were higher than those for the polished specimen. This indicates greater surface roughness. The Ssk surface asymmetry coefficient (skewness) describes the symmetry of height distribution with respect to the mean plane [1]. The negative value of this coefficient indicates a surface with deep valleys in a smoother plateau. The analysis of the results presented in Table 1 shows negative skewness in both cases.

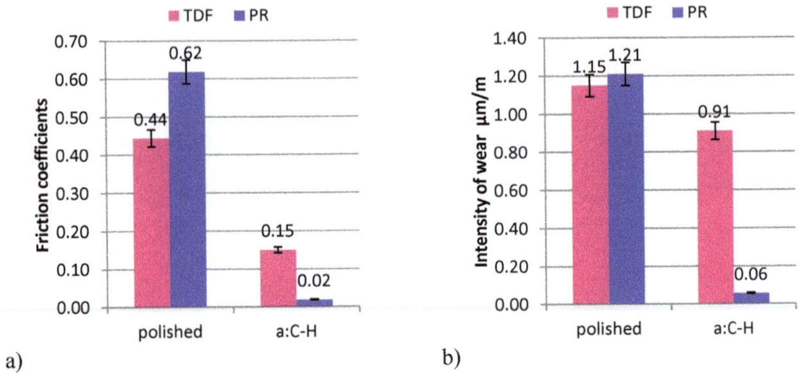

Fig.5. Coefficient of friction (a) and intensity of wear (b)

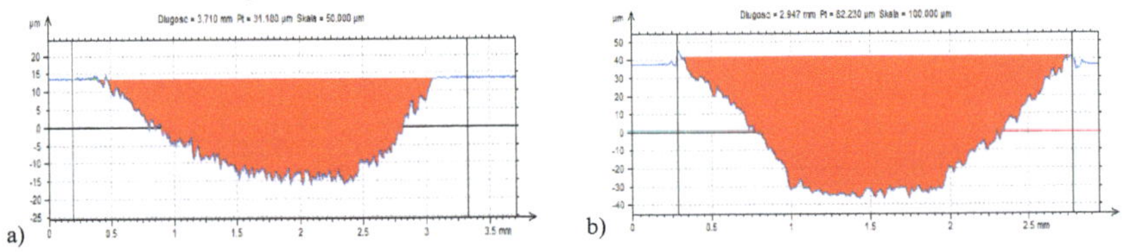

a) Max. depth: 30.107 µm Area: 51 550.090 µm² b) Max. depth: 78,843 µm Area:122 841,66 µm²

Fig. 6. Wear depth of the polished disc
a) technically dry friction, b) Ringer's solution lubricated friction

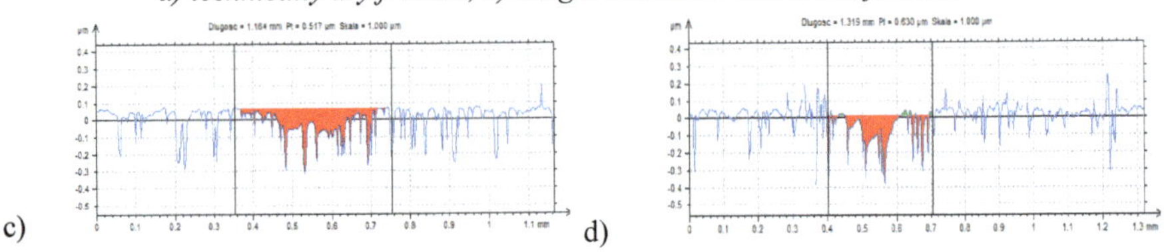

c) Max. depth: 0,390 µm Area:41,633 µm² d) Max. depth: 0,402 µm Area:29,695 µm²

Fig. 7. Wear depth of the disc with a:C-H coating
a) technically dry friction, b) Ringer's solution lubricated friction

Tribological tests. The results of the tests are presented as a comparison of the coefficient of friction (Fig. 5a) and intensity of wear (Fig. 5b) of the test elements depending on the types of friction pair materials and the lubricants used.

The lowest coefficients of friction were recorded for the diamond-like coating, both during technically dry friction and with Ringer's lubrication. The lowest wear intensity value was observed for the specimen with a:C-H coating during lubricated friction. After the tribological tests, wear depths were measured. Figures 6 and 7 summarize the measurement results.

The comparison of the maximum depths shows that the polished disc was a faster wear material in the Al_2O_3 friction pair. In the case of the specimen with the diamond-like coating, the smallest maximum wear depth was recorded during technically dry friction. However, when comparing the values of the hole surface area, the smallest wear was observed on the specimen with the a:C-H coating under the lubricated friction. The obtained results indicate that the use of the diamond-like carbon (DLC) coating ensured tribocorrosion protection of the titanium disc in the environment of the solution simulating body fluid.

Conclusions

The results of morphology, topography tests, chemical composition and geometric structure analysis were the bases for the following conclusions. The microscopic studies showed that the observed discs had different surface morphologies and geometric structures depending on the physico-chemical treatments used. Tribological tests revealed that the best tribological characteristics were obtained for surfaces with the diamond-like carbon (DLC) coating. Compared with the polished disc, the coefficient of friction recorded during technically dry friction was three times lower and 30 times lower during Ringer's lubricated friction. The comparing of wear scars showed that the DLC coated disc was the least worn material in the friction pair with the Al_2O_3 ball. The results obtained during the tests showed that the use of DLC coating improves tribological properties and significantly increases the service life of titanium.

References

[1] M. Madej, P. Baranowicz, Materials research and clinical studies of titanium dental implants, Mechanics in Medicine, Rzeszów, (2016) 131-143.

[2] Zasińska K., Piątkowska A.: Evaluation of abrasive wear of the Ti13Nb13Zr alloy implanted by nitrogen ions for friction components of the hip joint endoprostheses. Tribology 6 (2015)

[3] T. Okabe, H. Hero, The use of titanium in dentistry. Cells Mater. 5 (1995) 211–230. English.

[4] H.F Kappert. Titan als Werkstoff für die zahnärztliche Prothetik und Implantologie. Dtsch. Zahnärztl. Z, German (1994) 573–583.

[5] N. Radek, E. Wajs, M. Luchka, The WC-Co electrospark alloying coatings modified by laser treatment, Powder Metallurgy and Metal Ceramics 47 (2008)197-201. https://doi.org/10.1007/s11106-008-9005-7

[6] M. Madej, D. Ozimina, The tribological properties of DLC coatings, Trybology 3 (2012) 95

[7] M. Moseler, P. Gumbsch, C. Casiraghi, A.C Ferrari., J. Robertson, The Ultra smoothness of Diamond-like carbon surfaces, Science (2005) 1545–1548. https://doi.org/10.1126/science.1114577

[8] R. Narayan, Nanostructured diamond like carbon thin films for medical applications, Materials Science and Engineering C, (2005) 405–416. https://doi.org/10.1016/j.msec.2005.01.026

[9] Information on http:// www.plazma.certigo.com.pl/main.htm

[10] M. Madej, D. Ozimina, K. Kurzydłowski, T. Płociński, P. Wieciński, P. Baranowicz, Diamond-like carbon coatings in biotribological applications, Kovove Mater 54 (2016) 185-191. https://doi.org/10.4149/km_2016_3_185

Terotechnology 2017 Materials Research Forum LLC
Materials Research Proceedings **5** (2018) 90-95 doi: http://dx.doi.org/10.21741/9781945291814-16

Fire Tests of Non-Metallic Materials for Walls and Ceilings in Rolling Stock

RADZISZEWSKA-WOLIŃSKA Jolanta Maria [a*], MILCZAREK Danuta [b]

Instytut Kolejnictwa, Materials & Structure Laboratory, Chlopickiego 50 Street, 04-275 Warsaw, Poland

[*a]jradziszewska-wolinska@ikolej.pl, [b]dmilczarek@ikolej.pl

Keywords: Fire Tests of Railway Materials, Critical Heat Flux at Extinguishment, Lateral Spread of Flame on Products, Cone Calorimeter, Maximum Average Rate of Heat Emission

Abstract. The article discusses fire parameters of composites made on the basis of polyester resin, proposed for use in rail transport. The modifications and their effect on the flammable properties of these materials have been described. It has been shown that achieving the intended goals such as weight reduction or improvement of the aesthetics of the surface significantly changes parameters related to heat emission and flame propagation. In connection with the above, the scope of the introduced changes requires conducting control tests in the field of fire safety.

Introduction

The dynamic development of rail transport forces the rolling stock manufacturers to apply more and more modern technical and material solutions. It is associated with the need to ensure safety, including fire safety. This great emphasis on safety in means of railway transport caused intense development of requirements and methods in this area, and, as a consequence, contributed to the development of materials and products. There was a need to develop new and modify existing materials used in rail vehicles, which reduce the risk of people being threatened and improve the overall safety. An important effect of the existing fire (posing a deadly threat to passengers and impeding evacuation) is the spread of flame on the surface and the rate of heat released as a result of the combustion of materials constituting the lining of walls and ceilings of railway vehicles. Therefore, these materials, in accordance with the requirements of EN 45545-2 [1], must be subjected, inter alia, to tests according to ISO 5658-2 [2] and ISO 5660-1 [3]. These are fire parameters that characterize the material's resistance to external sources of fire.

The most commonly used materials are composites based on unsaturated polyester resins, which are characterized by high resistance to atmospheric and chemical factors, ease of processing and good mechanical properties. On the other hand, the ignition of resins by introducing appropriate flame retardants improves their fire-resistant properties [4-9]. These features are important from the point of view of their use as construction materials for walls and ceilings of rail vehicles [10-11].

Testes of selected fire properties

Laminates made on the basis of fiberglass and polyester resins have been tested. The composition of individual materials is presented in Table 1. However, the research was carried out in accordance with the methodologies described below.

Table 1. Determination and composition of samples of composites based on polyester resin

Sample	Resine	Modifications of the laminate
GRP(D1)	**resin: Giralithe Ditra 2109-10XP** glass mat **gelocoat: Nuvopol Gelcoat 37-05 TGP**	**topcoat**
GRP(D2)		**topcoat**
GRP(D3)		**topcoat**
GRP(E1)	**resin: Endyne H86181 TF** glass mat **gelocoat: Polytor 2335**	**none**
GRP(E2)		**foam PUR 20 mm**
GRP(E3)		**foam PUR 15 mm, topcoat**
GRP(E4)		**foam PUR 20 mm, topcoat**
GRP(B1)	**resin: Büfa-Firestop 8175-W-1** glass mat **gelocoat: Büfa-Firestop S285**	**none**
GRP(B)		**none**
GRP(B2)		**none**
GRP(B3)		**topcoat**
GRP(B4)		**topcoat**
GRP(B5)		**topcoat**
GRP(B6)		**coremat 5 mm, topcoat**
GRP(B7)		**foam PUR 5 mm, topcoat**

Fire test according to ISO 5658-2 [2]. The standard ISO 5658-2 [2] is based on the method of the IMO (International Maritime Organization) published as IMO resolution A.653 (16) [12]. It specifies a method of test for measuring the lateral spread of flame along the surface of product oriented in a vertical position as shown in Fig. 1. During the test, the *CFE* (Critical Heat Flux at *Extinguishment)* parameter is determined. The lower is the value of *CFE*, the greater is the fire hazard.

Fig. 1 The sample during test in IK according to ISO 5658-2 [2]

Fire test according to ISO 5660-1 [3]. Standard ISO 5660-1 [2] specifies a method for assessing heat release rate of specimen exposed in the horizontal orientation to controlled levels of irradiance with an external igniter using a cone calorimeter (Fig.2). The heat release rate is determined by the measurement of oxygen consumption based on the measure of oxygen concentration and the flow rate in the combustion products stream.

Materials Research Proceedings **5** (2018) 90-95 doi: http://dx.doi.org/10.21741/9781945291814-16

Fig. 2 ISO 5660-1[3] stand and sample during the test in IK

Generally, the net heat of combustion is proportional to the amount of oxygen used to burn a specimen. That is approximately 13.1×10^3 kJ of heat release per kilogram of oxygen consumed. During the test, the *MARHE* parameter (Maximum Average Rate of Heat Emission) is determined. The higher its value, the greater the fire hazard. An example of a graph of heat release rate with *MARHE* (orange line) for GPR is presented in Fig. 3.

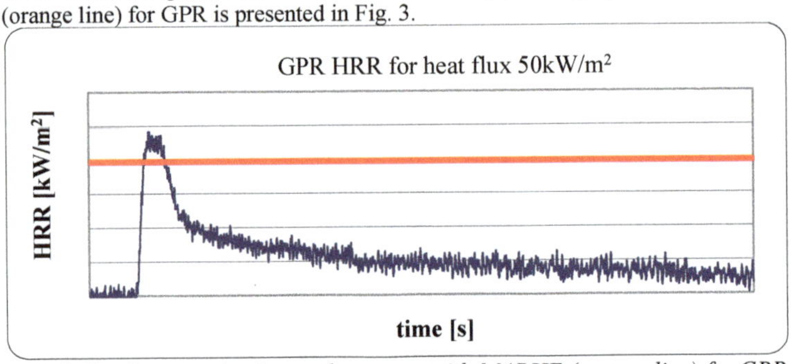

Fig. 3 Graph presenting heat release rate with MARHE (orange line) for GPR

Studies of fire properties

The results of the tests of the flame spreading on the surface and the heat release rate during combustion of the tested laminates are presented in Tables 3-5.

Table 3. Fire properties of polyester-glass laminates made on the basis of Giralithe Ditra 2109-10XP resin

Sample	**thk [mm]**	t_{ig} [s]	**CFE**[kW/m^2]	MARHE [kW/m^2]	HRR$_{max}$ [kW/m^2]
GRP(D1)	3	31	20.7	79.3	92.5
GRP(D2)	5	33	21.8	73.0	82.65
GRP(D3)	9	44	24.3	50.2	5.0

Table 4. Fire properties of polyester-glass laminates made on the basis of Endyne H86181 TF resin

Sample	thk [mm]	t_{ig} [s]	CFE [kW/m²]	MARHE [kW/m²]	HRR_{max} [kW/m²]
GRP(E1)	3.5	45	36.2	51.0	107.0
GRP(E2)	30.0	49	23.9	79.0	127.0
GRP(E3)	30.0	51	23.9	83.2	124.2
GRP(E4)	30.0	206	25.8	81.4	126.9

Table 5. Fire properties of polyester-glass laminates made on the basis of Büfa-Firestop 8175-W-1resin

Sample	thk mm]	t_{ig} [s]	CFE [kW/m²]	MARHE [kW/m²]	HRR_{max}[kW/m²]
GRP(B1)	3.0	113	27.5	91.9	194.1
GRP(B)	5.0	160	31.2	84.7	185.0
GRP(B2)	6.5	205	34.9	69.5	174.9
GRP(B3)	3.0	20	21.3	74.6	125.0
GRP(B4)	5.0	29	22.9	77.0	125.0
GRP(B5)	6.0	36	26.2	81.6	144.4
GRP(B6)	10.0	25	28.0	68.3	109.2
GRP(B7)	10.0	31	23.9	69.9	165.0

The influence of the thickness of GRP laminates on their fire properties. The analysis was carried out on laminate samples made on the basis of Giralithe Ditra 2109-10XP resin. The obtained results are shown in Fig. 5.

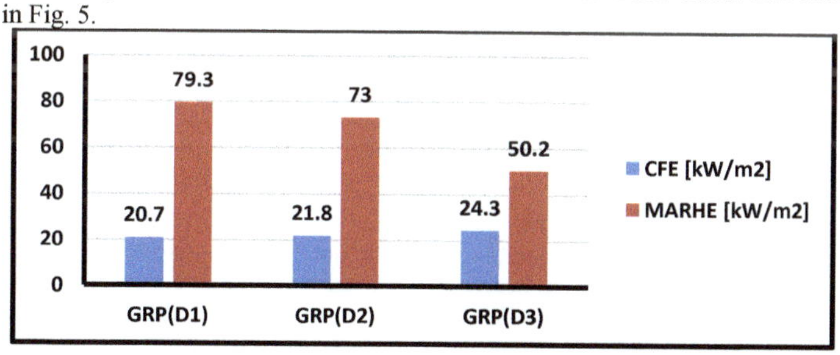

Fig. 5. CFE and MARHE values of laminate (Giralithe Ditra 2109-10XP resin) with different thicknesses

The GRP sample (D1), which was the thinnest of the respondents, revealed the fastest ignition under the influence of thermal radiation and obtained the largest range of flame, which is associated with the smallest critical radiation stream (*CFE*). However, the GRP sample (D3), which was the thickest of the studied, under the influence of thermal radiation ignited after the longest time and obtained the smallest range of the flame, which is associated with the largest critical radiation flux

(*CFE*). A similar relationship was also observed when determining the maximum average heat release rate. The GRP sample (D1) during the test on a cone calorimeter was characterized by considerable flammability. The maximum value of *MARHE* for this sample was 79.23 kW/m^2 and was greater by more than 39% from the *MARHE* GRP sample (D3).

Influence of modification on fire properties of GRP laminates

The modification of the tested polyester-glass laminates consisted of the following:

- applying a lacquer coat to the gelcoat in order to improve the aesthetics of the surface,
- adding polyurethane foam or cork to increase the physico-mechanical properties and reduce weight with the need to use samples with thicknesses> 10 mm.

The obtained results are shown in Figs. 6 and 7.

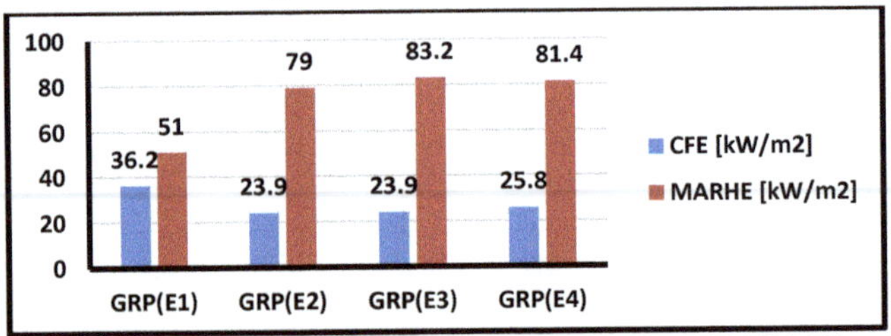

Fig. 6. Impact of thickness of laminate samples (Endyne H86181 TF resin) on CFE and MARHE values

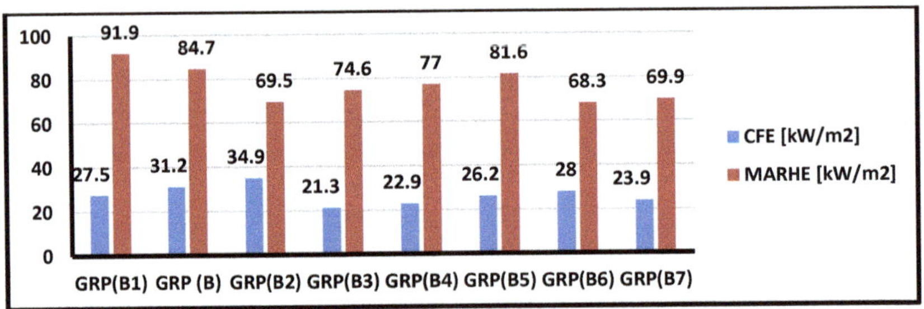

Fig. 7. Effect of thickness and modification of laminate samples (Büfa-Firestop 8175-W-1 resin) on CFE and MARHE

In the case of polyester-glass laminates based on the Endyne H86181 TF resin, a deterioration of the *CFE* parameter with respect to the initial sample was noticed. This may be due to the application of a lacquer coat to the gelcoat, which accelerates the ignition time and increases the range of the flame. However, the increase in the *MARHE* parameter is due to the addition of a 25 mm layer of polyurethane foam, which burns with the emission of a large amount of heat.

A similar relationship was also observed in the case of Büfa-Firestop 8175-W-1 resin laminates (Fig.7). Both modification methods (application of a varnish coat, laminating foam) caused a deterioration of the CFE parameter. It was associated with an additional layer of material that was easier to ignite, so that its critical heat flow supporting the burning was smaller. A reverse dependence occurred for the *MARHE* parameter. In the case of heat emission for both modifications, this parameter was improved. In the case of varnish coatings this may be due to cutting off the air directly to the surface of the laminate. The additional coating ignites and isolates the remaining layers of material. This causes a slower combustion of the whole sample, which extends the burning time of the material but reduces its heat release intensity (despite the addition of an additional layer in the form of 5 mm foam or 5 mm cork).

Summary

The tests of polyester-glass laminates based on various polyester-glass resins has confirmed the effect of thickness on the critical heat flux (*CFE*) and the maximum average rate of heat release (*MARHE*). The increase in thickness of the tested samples improved both the value of the *CFE* and *MARHE* parameters.

Studies have also shown that:

- laminate modifications aimed at improving the aesthetics of the surface by applying a lacquer coating caused a deterioration of the fire properties in the area of critical heat flux (*CFE*). The *MARHE* parameter was improved.
- modifications of laminates, aimed at improving their physicochemical properties and weight reduction by adding an inner layer of foam, affected the deterioration of the *MARHE* parameter. However, the *CFE* parameter did not change.

In connection with the above, the scope of introduced changes requires each time a control test in the field of fire safety.

References

[1] EN 45545-2:2013 Railway applications – Fire protection on railway vehicles – Part 2: Requirements for fire behaviour of materials and components

[2] ISO 5658-2:2006 Reaction to fire tests - Spread of flame. Part 2: Lateral spread on building and transport products in vertical configuration.

[3] ISO 5660-1:2015 Plastics – Reaction-to-fire tests-Heat release, smoke production and mass loss rate – Part 1: Heat release rate (cone calorimeter method) and smoke production rate (dynamic measurement).

[4] S. Boryniec, W. Przygocki , Procesy spalania polimerów. Cz III Opóźnianie spalania materiałów polimerowych, Polimery, 44 (1999) 656-665 (in Polish).

[5] E. Kocko-Walczak, Uniepalnianie nienasyconych żywic poliestrowych . Dotychczasowe osiągnięcia oraz nowe propozycje, Przemysł Chemiczny, 80 (2001) 576-579 (in Polish).

[6] S. Bourbigot, S. Duquesne, Fire retardant polymers, recent developments and opportunities, Journal of Materials Chemistry, 17 (2007) 2283-2300. https://doi.org/10.1039/b702511d

[7] E. Kocko-Walczak, Nowe bezhalogenowe antypireny-uniepalnianie nienasyconych żywic poliestrowych z zastosowaniem związku boru, Polimery 53 (2008) 126-132 (in Polish).

[8] D. Riegert, Sposoby modyfikowania właściwości palnych tworzyw sztucznych, Bezpieczeństwo i Technika Pożarnicza, 30 (2013) 51-57 (in Polish).

[9] W. Zatorski, K. Sałasińska, Analiza palności nienasyconych żywic poliestrowych modyfikowanych nanocząsteczkami, Polimery 61 (2016) 815-823 (in Polish). https://doi.org/10.14314/polimery.2016.815

[10] J.M. Radziszewska-Wolińska, D. Milczarek, Uniepalnienie materiałów niemetalowych a ich właściwości funkcjonalne, TTS 11-12 (2012) 56-59 (in Polish).

[11] D. Hohenwarter, Experience Gained from Fire Testes According to EN 45545-2 and DIN 5510-2 for Testing of Seats, Railway Reports, 60 (2016) 27-38.

[12] IMO resolution A653 (16), International Code for Application of Fire Test Procedures, 1998 Edition.

Stay-in-Place Formworks Applied in Civil Engineering – Requirements and Scope of the Control

BROZDA Kinga[1,a *], SELEJDAK Jacek[2,b]

[1]Czestochowa University of Technology, Faculty of Civil Engineering, Department of Concrete Structures and Geotechnics, Akademicka 3, PL- 42200 Czestochowa, Poland

[2]Czestochowa University of Technology, Faculty of Civil Engineering, Department of Concrete Structures and Geotechnics, Akademicka 3, PL- 42200 Czestochowa, Poland

[a*]kbrozda@bud.pcz.czest.pl, [b]jselejdak@bud.pcz.czest.pl

Keywords: Stay-in-Place Formwork, Functions of Building Formwork, Quality Control, General Requirements, Corrective Procedure

Abstract. Formworks are an inseparable element of the building and engineering structures shaping, which are implemented in technology of the monolithic concrete. This article attempts to collect information related to construction and material, as well as utility conditions which formworks should meet. Particular attention was paid to issues related to the use of SIP (stay-in-place) formwork systems. General requirements and functions that should be met by formworks (especially SIP formworks) are discussed, with an indication of loads acting on those formworks. In addition, the scope of control carried out as part of the execution supervision of formworks was analyzed and necessary actions to be taken in the event of a detected non-compliance were presented.

Introduction

It is estimated that 60-70% of all currently constructed building and engineering structures are made in concrete structure technology, in both prefabricated and monolithic constructions [1]. The undeniable advantages of this technology include, among others, the following [2, 3]: freedom of architectural forming, the possibility of using convenient static systems of individual structural elements, the possibility of recycling of selected elements after the end of the construction period, increased durability and resistance of the structure exposed to seismic movements or impact of the dynamics.

The shaping of almost every structural element made in monolithic technology takes place on the building site. This creates the necessity of using special forming devices, the so-called formworks [4], tasked with giving the right shape to a concrete mix and supporting a formed element before it reaches adequate strength. The stay-in-place (SIP) formwork is a separate category because it is an element that cannot be removed. This type of formwork is permanently installed and after the concrete work is an indispensable part of the object. During implementation, the SIP formwork acts as a forming and supporting element, while during operation – a construction, insulation (including insulation against corrosion) or construction-insulation element [5-8].

General requirements

According to the norm [9], structural formwork is used to give a concrete mixture the right form, consistent with the assumed architectural and construction design. Its task is to ensure proper support of a reinforcement during concreting and also to a concrete mixture until the concrete hardening process is completed and the desired strength is obtained. One of the most important requirements for a formwork is the possibility of its execution in accordance with the design model and applicable standards [10-13]. Formworks are a very special type of a temporary construction, where the pressure coming from the concrete mix is the leading parameter associated with the structural design [10, 14].

In addition, a SIP formwork after the curing process of the concrete, during operation of the structure can act as an additional external reinforcement, integrated with the entire structure, transferring other external loads associated with the work of this structure. SIP formworks can also give a construction structure other special features, such as thermal or damp insulation [2, 3]. Basic conditions that should be met by SIP formworks are to ensure proper connection of the formwork with the monolithic part of the structure and allow a correct execution of reinforcement work. Moreover, at the moment of commencement of concrete works, formwork bearing elements should have proper strength and stiffness to enable a safe transfer of loads from the weight or pressure of the concrete mix [15, 16].

There are three main functions that building formwork should fulfill [15, 17]:

- ensuring the safety of the structure - it is necessary that a formwork meets all load capacity conditions, while maintaining appropriate safety factors,
- maintaining appropriate technical quality - designing and execution of a formwork should ensure proper stiffness, positioning and geometry of a formwork construction,
- economy - minimizing the time and costs associated with the implementation of a construction structure is directly related to the effective execution of a formwork.

The last factor is as important as the other two because the cost of a formwork can reach 35% to 60% of the total cost of constructing concrete [17]. The economics of formwork structures is also associated with the type of material from which the formwork is made. It is advisable to use materials with high strength and resistance [2].

Quality Control of formworks

Supervision and control are carried out in order to check the compliance of construction works with the design and technical specification. It refers not only to the verification of works, but also to the properties of the used products and materials [18, 19]. This control may concern both the whole structure and its individual elements or some technologies used in the structures. The requirement regarding the scope of control is determined in relation to the selected control class (Table 1). The expected control class should be specified in the design specification [3].

Depending on the control class, the scope of contracting supervision includes the following [3]:

- 1st class of control – a visual inspection and testing of randomly selected elements,
- 2nd class of control – a visual inspection and systematic inspection of a large formwork carried out before concreting,
- 3rd class of control – a visual inspection and detailed inspection of all works related to the formwork, affecting the achievement of adequate load-bearing capacity and durability of the formed structure (also made before the start of concrete works).

According to the recommendations in the standard [10], the control of a formwork should be done not only before concreting, but also after laying the concrete mix. The purpose of a control performed after concreting is to check strength parameters of the concrete, while the following things are subject to a control before concreting [15]:

- geometry of a formwork,
- leak tightness of a formwork,
- stability of a formwork and props (including the substructure of props),
- elimination of pollution such as dust, ice, snow etc.,
- elimination of water from the formwork surface,
- proper preparation of the formwork surface.

If the control reveals some non-compliance, proper measures should be taken as soon as possible to ensure that the construction fulfills functions consistent with the project. It is recommended to carry out a corrective procedure, where the analysis should be made in the strictly specified order [3]. The repair procedure is presented in Fig. 1.

Materials Research Proceedings **5** (2018) 96-100
doi: http://dx.doi.org/10.21741/9781945291814-17

Table 1. Control classes to which structural elements and building structures are subject [10].

Type of item	1st class of control	2nd class of control	3rd class of control
Structural elements	single walls single columns single foundations as foundational footings RC (reinforced concrete) plates and beams with a span of *less than 10 m*	slender walls slender column pile caps arches with a span of *less than 10 m* RC (reinforced concrete) plates and beams with a span of *more than 10 m*	RC arches and vaults arches with a span of *more than 10 m* complex foundation structures elements that are affected by high values of compressive forces
Building structures	buildings with *less than two floors*	buildings with *more than two floors* (no more than 10 floors) typical bridge structures	high-rise buildings special bridge structures nuclear reactor buildings large dams and hydrotechnical structures other security housing

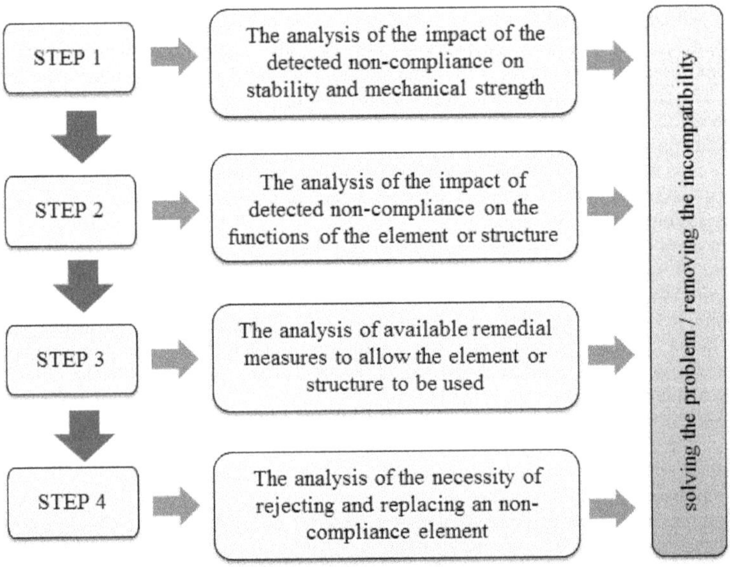

Fig. 1. The schema of the repair procedure [3].

Works related to the execution of the formwork of a structure are often at heights. Then it is extremely important to protect employees against a life-threatening fall. Ensuring a proper safety and health plan and compliance with applicable health and safety regulations is a superior responsibility of the safety and health plan [20]. One of the most common technical and organizational mistakes include downplaying the necessity of using safety equipment for a formwork [21]. If any violation is detected, all measures must be taken immediately to remedy the irregularity.

Summary

Most of the currently implemented construction projects are made in the monolithic concrete technology, which often requires the use of individually designed construction formworks. Due to their functions, such as ensuring the correct shape of a concrete element, transferring the accompanying impacts and establishing an integral part of the structure during operation, the SIP formworks must meet a number of specific requirements. Above all, they should ensure safety of people and constructions, maintain appropriate technical quality and be used as effectively as possible, which translates into the economy of the entire investment. In addition, each formwork (especially a SIP formwork) is subject to strict supervision and compliance control with the design and technical specification. In the case of a non-compliance and excessive intolerance, it is necessary to carry out an appropriate repair procedure, ensuring that the structure performs functions consistent with the project.

References

[1] P. Ignatowski, Realizacja żelbetowych konstrukcji nowoczesnych budynków użyteczności publicznej, Materiały Budowlane 6 (2008) 85-87.

[2] K. Brózda, J. Selejdak, Technologie deskowań traconych stosowane w ramach technologii budownictwa monolitycznego, in: J. Pietraszek, M. Mazur (Eds.), Zasoby wiedzy i technologii w kreowaniu wartości przedsiębiorstwa, Oficyna Wydawnicza Stowarzyszenia Menedżerów Jakości i Produkcji, Częstochowa, 2017, pp. 40-50.

[3] Z. Orłowski, Podstawy technologii betonowego budownictwa monolitycznego, Wydawnictwo Naukowe PWN, Warszawa, 2013.

[4] J. Jasiczak, Technologie budowlane II, Politechnika Poznańska, Wydział Budownictwa, Architektury i Inżynierii Środowiska, Instytut Konstrukcji Budowlanych, Poznań, 2003.

[5] P. Koteš and J. Vičan, Using stay-in-place GFRP formwork and concrete slab as hybrid composite structure, “7th International Conference on FRP Composites in Civil Engineering”, Vancouver, 2014.

[6] M.S. Nelson, A.Z. Fam, J.P. Busel, C.E. Bakis, A. Nanni, L.C. Bank, M. Henderson and J. Hanus, FRP Stay-in-Place Structural Forms for Concrete Bridge Decks: A State-of-the-Art Review, Aci Structural Journal 111 (2014) 1069-1080. https://doi.org/10.14359/51686810

[7] T. Lipinski, Corrosion effect of 20 % NaCl solution on basic carbon structural S235JR steel, 16th International Scientific Conference: Engineering for Rural Development, Book Series: Engineering for Rural Development (2017) 1069-1074. https://doi.org/10.22616/ERDev2017.16.N225

[8] I. Pliszka and N. Radek, Corrosion Resistance of WC-Cu Coatings Produced by Electrospark Deposition, Procedia Engineering 192 (2017) 707-712. https://doi.org/10.1016/j.proeng.2017.06.122

Materials Research Proceedings 5 (2018) 96-100 doi: http://dx.doi.org/10.21741/9781945291814-17

[9] PN-90/M-47850 Deskowania dla budownictwa monolitycznego. Deskowania uniwersalne. Terminologia, podział i główne elementy składowe.

[10] PN-EN 13670:2011. Wykonywanie konstrukcji z betonu.

[11] PN-EN 1065:2001. Regulowane teleskopowe podpory stalowe. Charakterystyka, konstrukcja i ocena na podstawie obliczeń i badań.

[12] PN-M-47900-2:1996. Rusztowania stojące metalowe robocze. Rusztowania stojakowe z rur.

[13] PN-M-47900-3:1996. Rusztowania stojące metalowe robocze. Rusztowania ramowe.

[14] DIN 18218. Frischbetondruck auf lotrechte Schalungen.

[15] A. Piekarczyk, Ł. Drobiec, R. Jasiński, Kontrola robót betonowych i żelbetowych w trakcie wykonywania oraz kontrola odbiorcza, XXV Ogólnopolskie Warsztaty Pracy Projektanta Konstrukcji, Szczyrk, 2010.

[16] A. Szczotok, J. Nawrocki and J. Pietraszek, The Impact of the Thickness of the Ceramic Shell Mould on the ($\gamma + \gamma'$) Eutectic in the IN713C Superalloy Airfoil Blade Casting, Archives of Metallurgy and Materials, 62(2) (2017) 587-593. https://doi.org/10.1515/amm-2017-0087

[17] D.W. Johnston, Design and Construction of Concrete Formwork, in: E. G. Nawy (Ed.) Concrete Construction Engineering Handbook, CRC Press, Boca Raton, 2008. https://doi.org/10.1201/9781420007657.ch7

[18] R. Ulewicz, Quality management system operation in the woodworking industry, Path Forward for Wood Products: A Global perspective, proceedings of scientific papers (2016) 51-56.

[19] J. Selejdak, R. Ulewicz, M. Ingaldi, The evaluation of the use of a device for producing metal elements applied in civil engineering, 23rd International Conference on Metallurgy and Materials (2014) 1882-1888.

[20] M. Szruba, Deskowania i rusztowania, Nowoczesne Budownictwo Inżynieryjne 3-4 (2016) 64-67.

[21] P. Ignatowski, Deskowania - najczęstsze błędy przy ich doborze, Inżynier budownictwa 10 (2011) 82-86.

Materials Research Proceedings **5** (2018) 101-104 doi: http://dx.doi.org/10.21741/9781945291814-18

Selection of Laser Ablation Parameters on the Basis of Accuracy of Performing a Single Texture Element

SĘK Piotr

Kielce University of Technology, Faculty of Mechatronics and Mechanical Engineering, Laser Materials Processing Research Center, Al. 1000-lecia P.P. 7, 25-314 Kielce, Poland

psek@tu.kielce.pl

Keywords: Laser Ablation, Laser Surface Texturing - LST, Picosecond Laser, Fluence

Abstract. Laser surface texturing is currently the most developed technique of producing microcavities in a completely repetitive way on the surface of machine elements. The article shows the method of a selection of laser micromachining parameters such as beam trajectory, power and radiation frequency to obtain the right quality of the canopy shape microcavities.

Introduction

Laser surface texturing is used in modern machines for various purposes, including lubrication intensification, intensification of heat exchange, stimulation of microflows and increased chemical activity of the surface. The subject of the study focuses on the selection of laser micromachining parameters with picosecond pulses in order to achieve the highest possible efficiency and accuracy of the implementation of a single element of texture. The development of technology places very high demands on the reliability and durability of machine components [1, 2]. One of the ways to meet these requirements is the action consisting in shaping microgeometry of selected surfaces of machine elements. Laser ablation is the removal of material from the substrate by direct absorption of laser energy. This is the most efficient technology for surface development [3, 4]. The ablation starts above the threshold fluence (energy amount per unit area), which depends on the absorption mechanism, properties of particular material, microstructure, morphology, the presence of defects and laser parameters such as wavelength and duration of the pulse.

Microsecond and nanosecond pulses are called long, picosecond and femtosecond pulses are called short or ultra-short. The mechanism of long pulse ablation is called photothermal (due to the HAZ) and ultra-short ablation is called photochemical or cold ablation [5, 6].

Laser surface texturing exceeds chemical, electrical and mechanical methods, because it allows localized changes with a high degree of control over the shape and size of the resulting elements and enables greater size diversification. Different textures can be produced in a strictly defined way by controlling process parameters such as beam intensity, spatial and temporal profile, wavelength and environmental selection. Basic dimensions of surface features are usually determined by the shape and size of the beam. However, secondary characteristics in the microscale and even at the nanoscale may be formed around the LST irradiance area due to the variety of mechanisms involved, including post-ablation melting, re-solidification, or spattering of surface fluid due to recoil pressure [7, 8].

Experiment

The research aiming at achieving the highest possible accuracy in the performance of a single texture element was carried out using a laser micromachining station equipped with a TruMICRO 5000 5325c picosecond laser with an average power of 5W max, pulse energy up to 12.6μJ and M^2 radius quality <1.3. The maximum pulse frequency equals 400kHz and can be modified by entering the pulse divider assuming values of 200kHz, 133.33kHz, 100kHz, 80kHz, 66.66kHz, etc. The radiation wave length of the TruMICRO 5325c laser is 343 nm. The TruMICRO 5325c laser has been equipped with a SCANLAB "galvo" scanning head with a square machining area of 90 mm side and scanning speed (laser beam through material feed) in the range of 100 - 5000 mm/s and with a lens with a focal length of 160mm. The micro-machining of 100Cr6 bearing steel was the subject of interest.

In order to determine the optimal method of producing microcavities in the form of a canopy, three possible trajectories of the laser beam offered by the laser software were examined. Each of the beam guidance methods produced different results. In the case of a spiral trajectory, the machine software allows you to determine the number of coils and their mutual distance. Unfortunately, it is not possible to program the closure of the spiral at its center point, which resulted in insufficient removal of the material in the middle of the microforming. The outer spiral coil propagates in relation to the central point at a given angle, which resulted in poor representation of the preset shape in the workpiece - the microcavity produced was not completely circular. In the case of a trajectory of parallel lines, a microcavity of a suitable shape was created, but its edges and slopes were very ragged. This is caused by a software limitation that does not allow the laser to get the desired power from the beginning of the line. The power increase to a given value occurs exponentially over a period of time and is called a ramp. The best results were obtained using the trajectories of concentric circles. Microcavities produced with this trajectory possessed the desired geometry with smooth undamaged edges and a smooth bottom. Further investigations were carried out using this trajectory of the beam.

During the production of samples for the tests, the range of average laser power from 5W to 1.5W and the frequency of laser pulses 400 kHz, 200 kHz, 133.33 kHz, 100 kHz, 80 kHz, 66.66 kHz were examined. For average power of less than 1.5W and frequencies lower than 66.66k Hz, the fluence F was less than 1.33 J/cm^2; no significant influence of the treatment on the material and loss of ablation mass was observed. Thus, it can be considered that 1.33 J/cm^2 is the threshold fluence for the 100Cr6 steel. All the ninety test samples were made with a constant scanning speed of 100 mm/s. The set diameter of the microcavity equalled 100 μm. The surfaces of the samples were prepared before laser processing on a polishing machine (Ra=0.392÷0.694 μm, Rz=1.336÷1.919 μm).

To determine the geometry of the created microcavities, a Talysurf CCI optical profiler by Taylor Hobson was used (using a 2.2mm coherent correlation-based interferometry and a 0.01nm resolution lens with an x50 magnification to measure the area of 330 μm x 330 μm and a slope of up to 27.7°). Three measurements were made for each sample, concentrating on the size of the deviation of the obtained diameter from the set one - Δø.

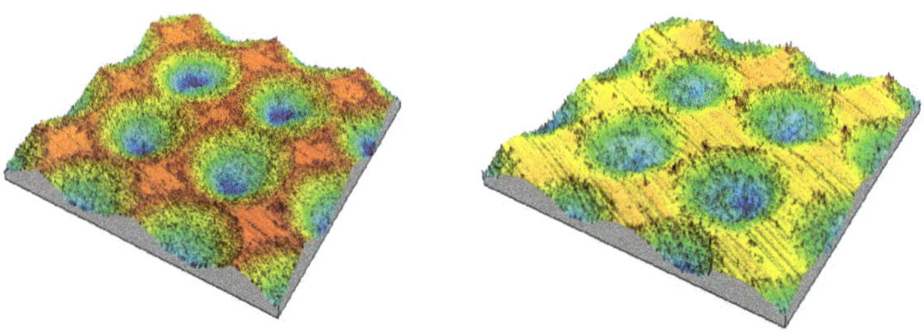

Fig. 1. Isometric views of samples made with fluences of: left - 3, 75, right 2,31 J/cm^2

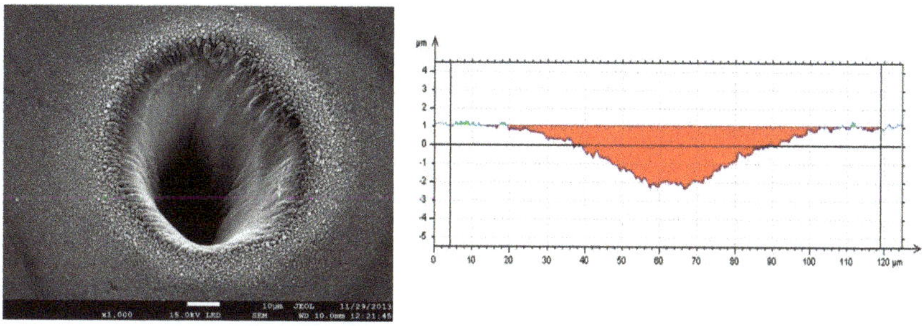

Fig. 2. View of sample made with the fluence of 2,31 J/cm²:
left – microscopic view, right – geometry measurement

Analysis of results

In order to determine the effect of fluence on the geometry of a single microclimate, the analysis of the dependence of quantitative variables was used. The analysis was performed in the SAS System version 9.4, determining the mathematical dependence of the factors considered. The evaluation of the quadratic parameters of the regression function, determined by the least squares method, leads to the following equation describing the relationship between F and Δø:

$$\Delta ø = 0{,}282F^2 + 2{,}273 \qquad (1)$$

Figure 1 presents an estimation of the square regression function in a graphical form, estimated by the least squares method. The cut off of the apex of the parabola, slightly smaller than 3, determines the value of F, for which the smallest Δø is observed.

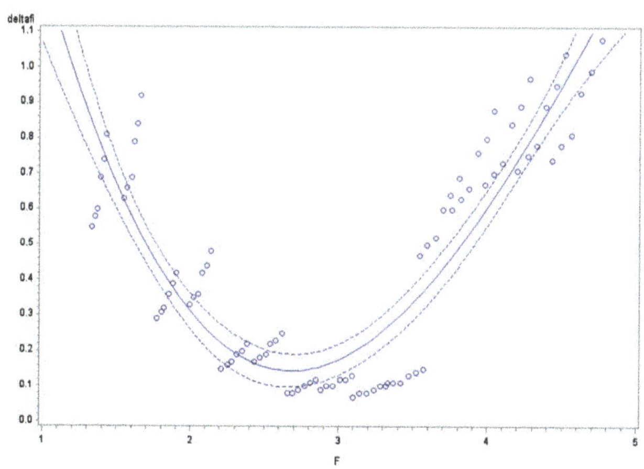

Fig. 1. Dependence of F and ΔøSummary

The analysis of the dependence of the quality of the implementation of a single texture element on the fluence of laser radiation showed a strong influence of the laser processing parameters on the shape and quality of the microfusion. The highest accuracy of microcubber performance was obtained for fluence from the range of 2.62 to 3.57. For the fluence values below 2.62 J/cm^2, we observe insufficient quality of a single texture element caused by the insufficient amount of energy delivered to the work surface. Micro cavities made at these fluence values were not precisely made, and their diameter was much smaller than the set one. For machining with fluence greater than 3.57 J/cm^2, we observe microcavities of irregular shape resulting from remelting material. The diameter of the microcavities made with the fluence values higher than J/cm^2 was greater than the set value.

References

[1] Chichkov B.N., Momma C., Nolte S., von Alvensleben F., Tünnermann A., Femtosecond, picosecond and nanosecond laser ablation of solid, Applied Physics A: Material Science & Processing 63: 109, 1996. https://doi.org/10.1007/BF01567637

[2] Chrisey D.B., Hubler G.K., Pulsed Laser Deposition of Thin Films,Wiley-Interscience, New York, 1994.

[3] N. Radek, E. Wajs, M. Luchka, The WC-Co electrospark alloying coatings modified by laser treatment, Powder Metallurgy and Metal Ceramics 47 (2008)197-201. https://doi.org/10.1007/s11106-008-9005-7

[4] J. Pietraszek, N. Radek, K. Bartkowiak, Advanced statistical refinement of surface layer's discretization in the case of electro-spark deposited carbide-ceramic coatings modified by a laser beam, Solid State Phenom. 197 (2013) 198-202. https://doi.org/10.4028/www.scientific.net/SSP.197.198

[5] N. Radek, J. Pietraszek, B. Antoszewski, The average friction coefficient of laser textured surfaces of silicon carbide identified by RSM methodology, Advanced Materials Research, 874 (2014) 29-34. https://doi.org/10.4028/www.scientific.net/AMR.874.29

[6] B. Antoszewski, P. Sęk, Infulence of laser beam intensity on geometry parameters of a single surface texture element, Archives of metallurgy and materials, Tome: 60, 3B (2015)117-121.

[7] Gregson V., Laser Material Processing, Holland Publishing Company, Holland, 1984 Valette S., LeHarzic R., Huot N., Audouard E., Fortunier R., 2D calculations of the thermal effects due to femtosecond laser-metal interaction, Applied Surface Science, 247: 238, 2005.

[8] Ion J.C., Laser Processing of Engineering Materials: Principles, Procedure and Industrial Applications, Elsevier Butterworth-Heinemann, Oxford, 2005.

Materials Research Proceedings **5** (2018) 105-110 doi: http://dx.doi.org/10.21741/9781945291814-19

Influence of Reinforcement on the Strength of the Concrete Sleepers

ANTOLIK Lukasz [1*], ZAWISZA Andrzej [2]

[1, 2]Instytut Kolejnictwa, Materials and Structure Laboratory, Chlopickiego Street 50, 04-275 Warsaw, Poland

[1]lantolik@ikolej.pl, [2]azawisza@ikolej.pl

Keywords: Prestressed Monoblock Sleeper, Prestressing Wire, Structural Composite

Abstract. Prestressed concrete sleepers are the most commonly used composite in the track construction in Polish railways. Different types of prestressed concrete sleepers (e.g. PS-83, PS-93 or PS-94) are used depending on the permissible load and maximum speed. The Polish largest company that manages railway infrastructure requires to use sleepers produced in accordance with the accepted technology and using approved materials. Because there are producers who offer complementary technology with alternative methods and types of reinforcement, in this article the authors take into consideration the theme of comparing strength of the sleepers in the laboratory and operating conditions. A PS-94 type sleeper, reinforced in two different ways and widely used in the construction of railroads in Poland, is an example of this problem.

Introduction

Reinforced concrete sleepers have been used on Polish tracks for decades since the time concrete technology and their reinforcing were developed. Technological progress has also resulted in the development of new constructions used in Polish railways. While at the end of the twentieth century, reinforced concrete PS-83 type sleepers and their derivatives were used in Poland, currently PS-83, PS-93 and PS-94 type constructions are accepted in different variations. The range of using concrete sleepers is regulated by instructions and mainly depends on the track category. The most commonly used is the PS-94 sleeper (Fig. 1), whose equivalent in German railways is the B70 type sleeper. The main Polish infrastructure manager after many years of operating tests decided to use only reinforcement consisting of smooth drawn bars of high-strength steel, whose minimum tensile strength is 1670 MPa and reinforced anchoring with retaining discs mounted on the sleepers' side surfaces.

The reason of such a decision was the information about cracks of sleepers reinforced with bars without anchoring plates (Fig. 2), which resulted in the lack of trust of the manager to this technology.

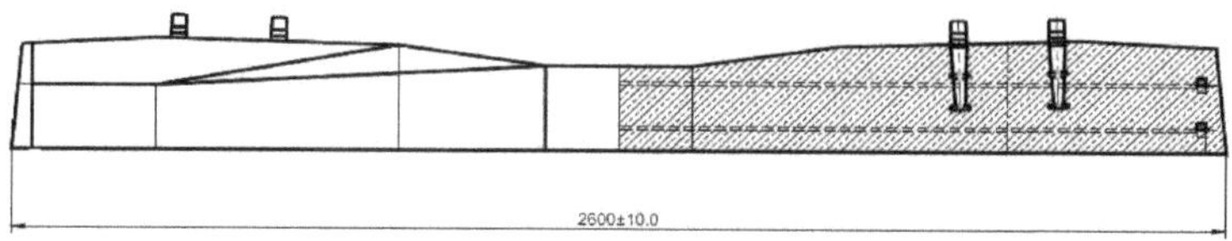

Fig 1. Half-view and half-section of the PS-94 type sleeper (source [1]).

Although it has been proven that reinforcing technology assures high strength and durability of ready structural composites, it was attempted to make the PS-94 type sleeper reinforced with eight burnished wire braids of 3 mm diameter and tensile strength of 1850 MPa. Such a type of reinforcement is also pre-tensioned before filling up the mould, which after mortar curing and releasing of tension compresses the sleeper and makes the required strength.

Fig. 2. Longitudinal cracks of the B70 type sleeper during operation (left) and new sleepers (right), (Source [2]).

According to the manufacturer's information, this type of reinforcement has many times more surface area of connection with concrete than in the case of the retaining discs use.
In the following chapters the results of tests and final conclusions will be presented.

Laboratory testing

Comparative tests of the PS-94 type sleepers have been made at the Materials and Structure Laboratory of the Railway Research Institute (Instytut Kolejnictwa) according to the procedure described in the PN-EN 13230-2:2016 [3] standard.

During the static test at a rail seat section, (Fig. 3) the sleeper was initially loaded up to the value of force equal to Fr_0 = 161 kN. From that moment, the loading was gradually increased by the value of 10 kN. The interval between the consecutive increases of loading was between 30 to 60 sec. After each increment of loading the presence of cracks in the rail seat zone was verified. If the cracks occurred, their width was measured at 15 mm from the bottom edge of the element. After the recording of the crack width with the opening of 0.05 mm, the bearing capacity of the sleeper was tested. During the whole test, the speed of the load increment was 120 kN/min

The sleepers tests at the center section (Fig. 4) have been done with initial loadings equal to Fc_o=27,6kN for the normal position and Fc_{on} = 37,7 kN for the negative position. The load steps were equal to 5 kN up to the level of visible crack and next up to breaking of the sleeper.

The purpose of testing of the sleeper under dynamic loading (Fig. 5) was to verify crack resistance of the element under the exploitation loading conditions (railway rolling stock induces dynamic forces). The sleeper was loaded with sequences of 5 000 cycles, whereas at each following stage of loading the amplitude was increased by 20 kN (the loading amplitude was increased). During the test, the level of cracking loading (Fr_r), crack opening of 0.05 mm ($Fr_{0.05}$) and destruction load (Fr_B) were registered. The frequency was constant throughout the whole test and equal to 4 Hz [4].

The purpose of the long term fatigue loading test (Fig. 6) was to simulate the real loading characteristics of sleepers caused by the railway traffic. The tested element was initially loaded up to the occurring of the crack. Next, it was subjected to the 2 000 000 cycles of fatigue loading in the range from 50 kN to 161 kN. The frequency of loading throughout the whole test was 4 Hz. After those 2 000 000 cycles the sleeper was loaded up to the destruction [4].

Materials Research Proceedings **5** (2018) 105-110 doi: http://dx.doi.org/10.21741/9781945291814-19

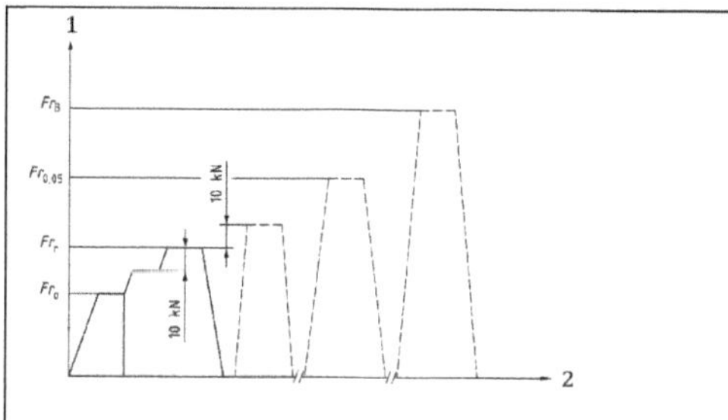

Fig. 3. Scheme of the static loading at the rail seat section.

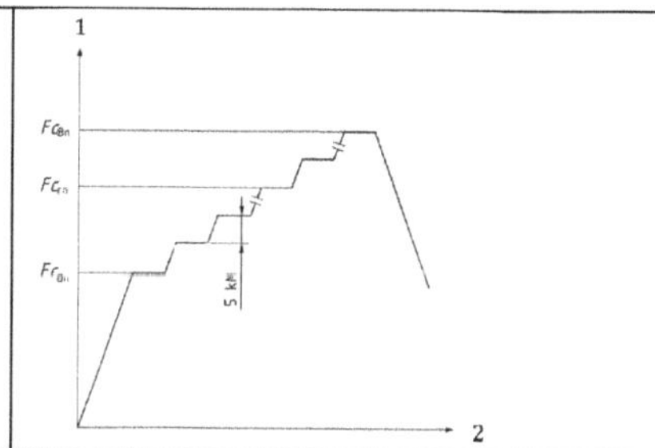

Fig. 4. Scheme of the static loading at the rail center section.

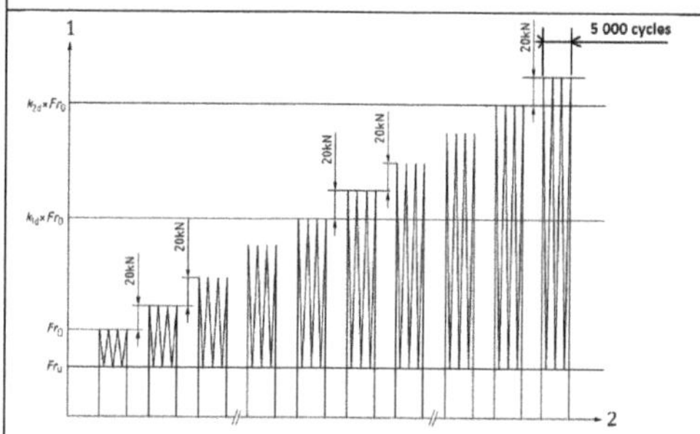

Fig. 5. Scheme of the dynamic loading at the rail seat section.

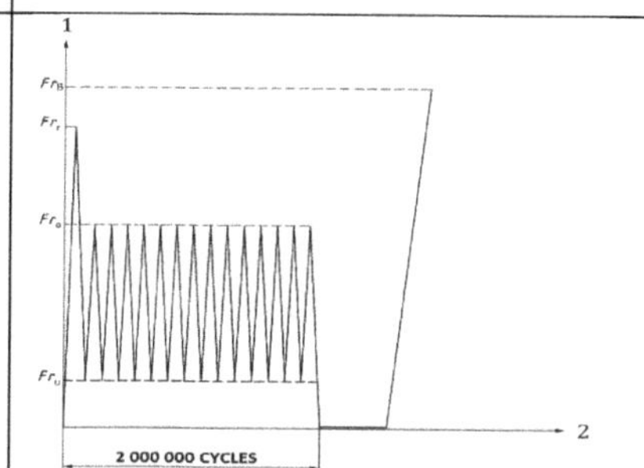

Fig. 6. Scheme of the fatigue loading at the rail seat section.

Tests

The tests were performed under repeatability conditions, using the same test stand and special instrumentation.

The sample of the destroyed PS-94 type sleeper reinforced with burnished wire braids is presented in Fig. 7.

The example of the destroyed sleeper reinforced with smooth bar with retaining discs is presented in Fig 8.

All the tested sleepers meet the minimum time of seasoning of 28 days.

The results of the endurance tests were presented in the graphs (Fig. 9-11).

All the results of the tests, beyond the maximum destructive force in the static test at the rail seat section, confirm the advantage of the sleeper reinforced with the braided wire.

From the point of the in-service properties, the value of the force which causes the first scratch on the bended surface is important. In this case, all the tests have proven the advantage of the solution with the braided wire.

The relative increase of the force that causes the first scratch of the sleeper reinforced with the braided wire are as follows:

- 12% - static tests at the rail seat section,
- 5% - dynamic tests at the rail seat section,
- 13% - static tests at the center seat section for the normal position,
- 17% - static tests at the center section for the negative position.

Fig. 7. Fracture of the PS-94 sleeper reinforced with a braided wire.

Fig. 8. Fracture of the PS-94 sleeper reinforced with a smooth bar.

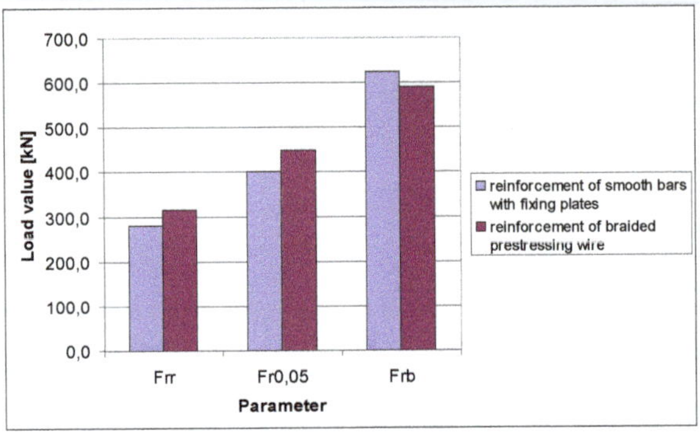

Fig. 9. Result of the static test at the rail seat section.

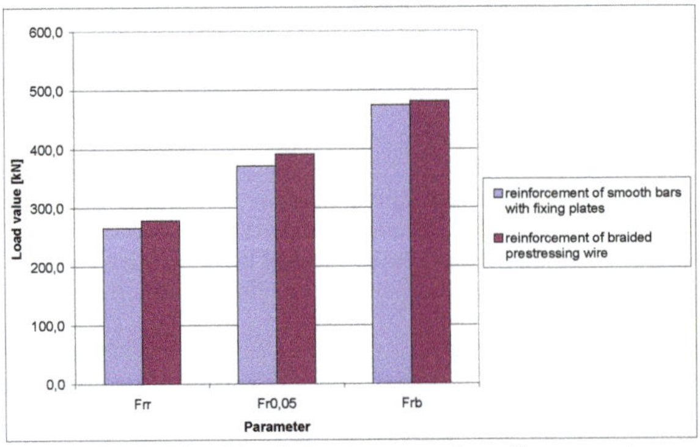

Fig. 10. Result of the dynamic test at the rail seat section.

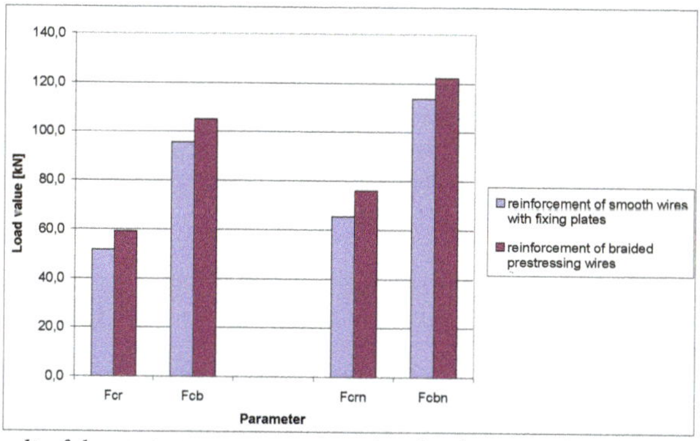

Fig. 11. Result of the static test at center section for the normal and negative position.

Photos (Fig. 7 and 8) show the way of the reinforcement cracking. After cracks examination of the smooth bars and braided wires it can be concluded that both reinforcements broke as a result of tension. In connection with low adhesion of smooth wires for concrete, pulling out of bars before their failure is visible. Such effect is not visible in the case of sleepers reinforced with braided wires due to a greater contact surface of bars and concrete and the shape of wire that causes large adhesive forces between concrete and steel as well as the large retain surface of the entire length of the braided wire.

In-service testing

In service tests have been carried out on the sample in accordance with the requirements and instructions of the company managing infrastructure.
The section of tests consisted of 500 sleepers fastened with rails by the fastening SB type system.
A part of the experimental section is shown in the picture below (Fig. 12).

Fig. 12. A section of the rail road with sleepers that have reinforcement with braided prestressing wires (source: [5])

The testing track has been built in the curved track with the radius of 2250 m and a cant track of 50 mm, using 60E1 type rails. Passenger trains velocity did not excess 120 km/h (75 mph) and velocity of the freight trains 70 km/h (43 mph).

After two years of operation and summary load of 20 Tg, there were no data about any technological problems with the tested sleepers and any factors that would indicate the reduction of impact strength in comparison with the sleepers reinforced with a smooth bar with retaining discs. There were no signs of weathering of concrete.

Summary

Reinforced concrete sleepers are very reliable parts of the railway track that keep right the track gauge. Their manufacturers use mainly approved technology based on reinforcement with smooth bars with retaining discs.

As a result of the laboratory tests which were carried out, it can be concluded that the PS-94 sleepers reinforced with the braided wire are at least as strong as the commonly used the PS-94 sleepers reinforced with technology approved by the company of tracks manager.

Operational tests carried out in the field in the range of 20 Tg summary load confirmed high durability of the products and exploitation without any problems. There were no cracks that characterize composites without retaining discs and in many cases, appear at the beginning of use.

The infrastructure manager has the right to additional requirements for safety railways transport and in connection with potential costs of using alternative technologies.

References

[1] Technical Conditions of Production and Collecting Reinforced Concrete Sleepers and Bearers No WTWiO-ILK3a/5187/01/05.

[2] F. Rezaie, M.R. Shiri, S.M. Farnam, Experimental and numerical studies of longitudinal crack control for pre-stressed concrete sleepers. Engineering Failure Analysis 26 (2012) 21–30. https://doi.org/10.1016/j.engfailanal.2012.07.001

[3] PN-EN 13230-2:2016 Railway application – Track – Concrete sleepers and bearers – Part 2: Prestressed monoblock sleepers.

[4] W. Derkowski, B. Słyś, M. Szmit. Effect of strands anchorage system in railway sleepers on behaviour of its rail seat zone. Architecture, Civil Engineering, Environment 1 (2015) 61-67.

[5] Information on http://precon.com.pl/produkty/podklady-kolejowe-strunobetonowe/

Terotechnology 2017 Materials Research Forum LLC
Materials Research Proceedings **5** (2018) 111-116 doi: http://dx.doi.org/10.21741/9781945291814-20

The Evaluation of Structure and Destruction of Ribbed Bars as a Result of Processing and Heat Treatment

INGALDI Manuela

Czestochowa University of Technology, Faculty of Management, Department of Production Engineering and Safety, al. Armii Krajowej 19b, 42-200 Czestochowa, Poland

manuela@gazeta.pl

Keywords: Ribbed Bars, Steel Microstructure, Ferrite, Widmanstätten Structure

Abstract. In this paper the images of steel microstructures (on elements taken from ribbed bars were rolled using QTB technology) have been presented. The observation has been carried out on two types of steel which are most commonly used for ribbed bars, 34GS, BSt500S and 16, 25 mm in diameters. The observations on the structure of materials have been performed on samples perpendicular and parallel to the rolling direction (after surface nital etching). The analysis of a macro- and microstructure gave the possibility to approximate the causes of non-conformity of finished products, as a major problem in ribbed bars production and processing.

Introduction

Together with emerging investments and building highways, modern construction shows a great demand, among others, for ribbed bars for concrete reinforcement. Constructions of these bars connected with concrete allow for significant savings of steel, aggregates, cement, wood and shortening of construction time [1]. Ribbed bars in reinforced concrete structures guarantee better adhesion of reinforcement to concrete due to a special shape in which humps (ribs) are applied to the round core. Recipients of the ribbed bars demand that they have high strength and ductility, corrosion resistance, good weldability and high fatigue strength [2, 3].

Ribbed bars, like other metallurgical products, must meet relevant requirements resulting from standards and specified by customers in their orders. The following dimensions of ribbed bars are specified in the standards: nominal diameter and their permissible deviations, mass range of 1 m of a ribbed bar, core diameter, rib height, width and length of ribs longitudinal. For individual steel grades, their chemical composition is also defined. Requirements regarding mechanical properties, plasticity and weldability are also specified [4]. These properties are obtained due to their properly carried out production process [5-6].

The purpose of the article is to assess the structure of ribbed bars in terms of their production process. The samples for the analysis were obtained from one of the rolling mills in Poland.

Experimental

The strength and technological properties of steel are primarily related to its microstructure which depends on heat treatment. Due to physicochemical phenomena occurring while heating and cooling steel, it is possible to create the desired structural components, and thus, to give specific strength properties [7, 11].

To illustrate the diversity of the research material, a microstructure of selected steel grades was analyzed. The observation has been carried out on two types of steel which are most commonly used for ribbed bars, 34GS, BSt500S and 16, 25 mm in diameters. These bars were rolled with the use of QTB technology.

The observations on the structure of materials have been performed on samples perpendicular and parallel to the rolling direction (after surface nital etching). The metallographic analysis was carried

Materials Research Proceedings **5** (2018) 111-116
doi: http://dx.doi.org/10.21741/9781945291814-20

out using an Axiovert 25 optical microscope cooperating with a Mavica MVC-FD87 digital camera from Sony.

The size of ferrite grains was evaluated on the basis of the PN-EN ISO 643: 2005 standard [8] and determined by comparison of the scale model with a sample zoomed 100x. With the use of standardized scale models, the grain size standard was indicated, which in the best way approximated the size of the sample grain in the three analyzed areas of the sample. The average grain diameter $\bar{d}$ was taken form Table C.1. and included in the standard in which for individual grain size numbers from the scale model, the average grain diameter $\bar{d}$ [mm] corresponds to it.

In case of the metallographic sections which are parallel to the rolling direction, a banded microstructure was evaluated according to the PN-63/H-04504 standard [9], i.e. on the basis of a comparison of the scale models with two worst structures found on a given microsection, separately in each of the bar thickness zones.

Widmannstätten ferrite [10] was evaluated according to the PN–63/H–04504 standard [9] by comparison with the scale models in at least three fields of view from ten observed fields.

Results

Figure 1 presents the characteristics of the microstructure which occurs in the ribbed bars with a diameter of 16 mm from steel 34GS.

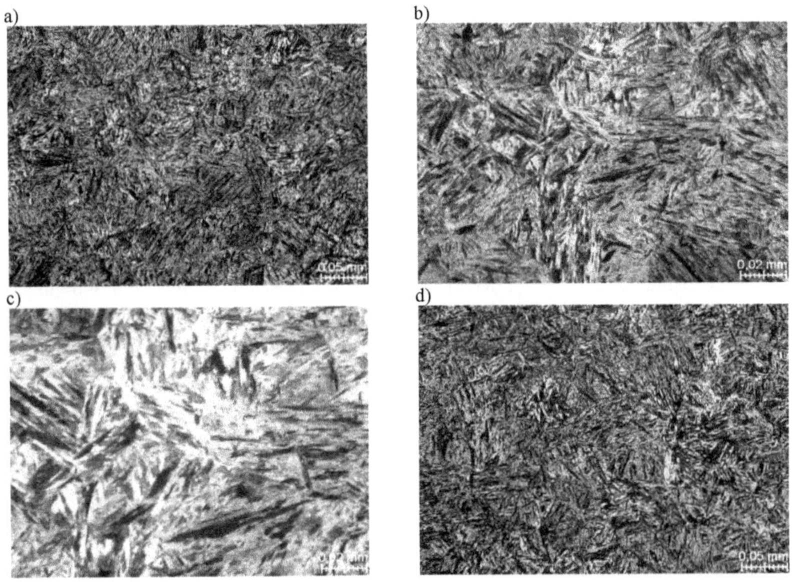

Materials Research Proceedings 5 (2018) 111-116 doi: http://dx.doi.org/10.21741/9781945291814-20

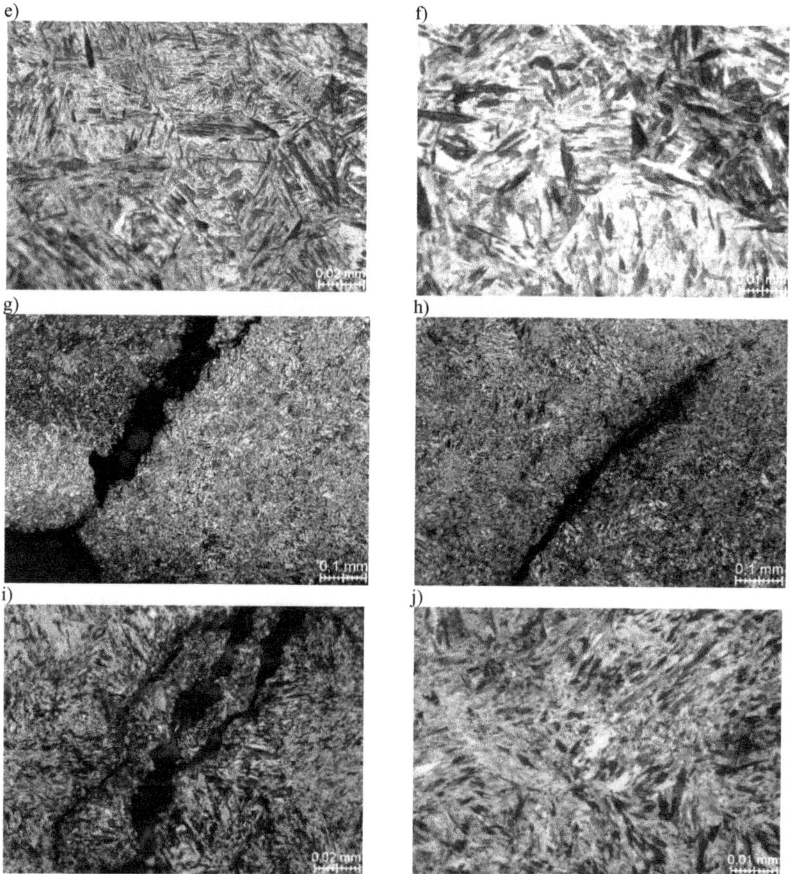

Fig. 1. Microstructure of the ribbed bar with a diameter of 16 mm from steel 34GS. After surface nital etching [4]

a) in the middle of the bar, zoom 200x, parallel metallographic section; b) in the middle of the bar, zoom 500x, parallel metallographic section; c) in the middle of the bar, zoom 1000x, parallel metallographic section; d) near the surface of the bar, zoom 200x, parallel metallographic section; e) near the surface of the bar, zoom 500x, parallel metallographic section; f) near the surface of the bar, zoom 1000x, parallel metallographic section; g) near the surface of the bar, zoom 100x, perpendicular metallographic section, crack near the surface of the bar – inlet; h) near the surface of the bar, zoom 100x, perpendicular metallographic section – end of the crack; i) near the surface of the bar, zoom 500x perpendicular metallographic section – crack of the bar; j) near the surface of the bar, zoom 1000x, perpendicular metallographic section

In the sample made of steel 34GS, the presence of bainitic-martensitic structures is observed (Fig. 1a-1g). In areas corresponding to the center of the bar thickness, the extent of the bainitic structure in the bainitic-martensitic structure has been noticed (Fig. 1c, 1j).

In the ribbed bar made of steel 34GS, cracks occurred starting from the surface of the bar until the end of the crack almost in the middle, i.e. a depth of about 10 mm (Fig. 1g-1i). No decarburization at the crack surface (Fig. 1g-1i) and the nature of its course (Fig. 1i) allow for making a conclusion that it is a crack which occurred during cooling of the bar from the austenitizing temperature. The visible particles (Fig. 1g, 1i) are most likely pressed into the crack during the preparation of the metallographic section from the abrasive papers. The cracking in the sample is shown in Figure 2.

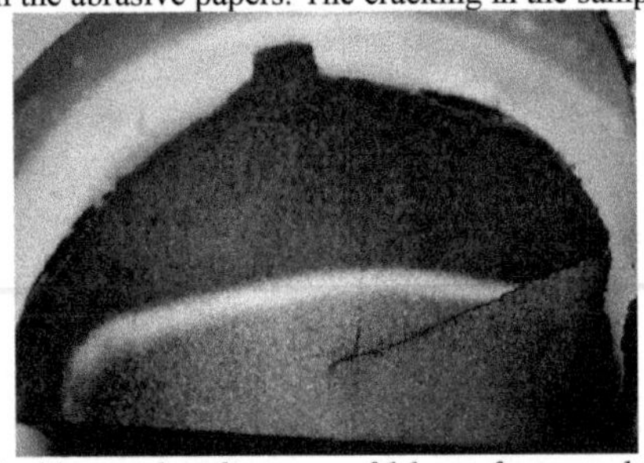

Fig. 2. Crack of the ribbed bar with a diameter of 16 mm from steel 34GS. After surface nital etching. A photo of the whole sample *[4]*

Figure 3 presents the characteristics of the microstructure which occurs in the ribbed bars with a diameter of 25 mm from steel BSt500S.

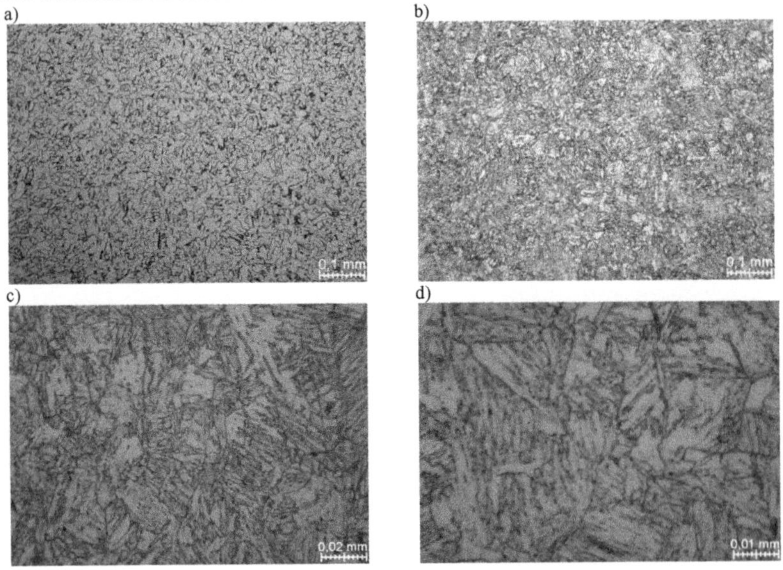

Materials Research Proceedings **5** (2018) 111-116 doi: http://dx.doi.org/10.21741/9781945291814-20

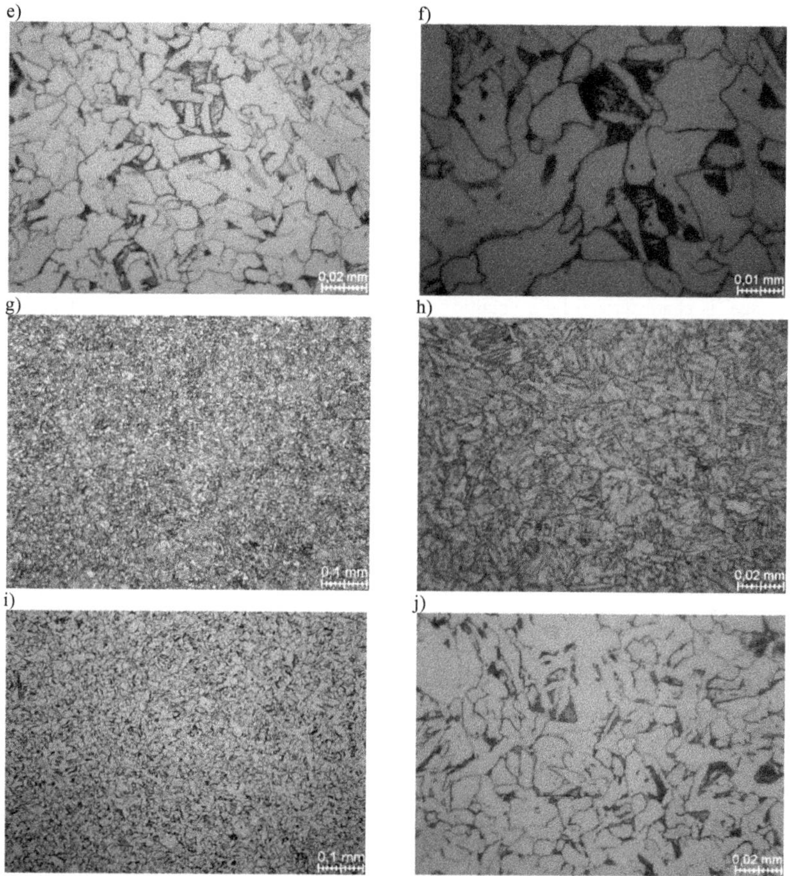

Fig. 3. Microstructure of the ribbed bar with a diameter of 25 mm from steel BSt500S. After surface nital etching [4]

a) in the middle of the bar, zoom 100x, parallel metallographic section; b) near the surface of the bar, zoom 100x, parallel metallographic section; c) near the surface of the bar, zoom 500x parallel metallographic section; d) near the surface of the bar, zoom 1000x, parallel metallographic section; e) in the middle of the bar, zoom500x, parallel metallographic section; f) in the middle of the bar, zoom 1000x, parallel metallographic section; g) near the surface of the bar, zoom 100x, perpendicular metallographic section; h) near the surface of the bar, zoom 500x, perpendicular metallographic section; i) in the middle of the bar, zoom 100x, perpendicular metallographic section; j) in the middle of the bar, zoom 500x, perpendicular metallographic section

In the sample made of steel BSt500S at a distance of about 5 mm from the surface, occurrence of bainitic structures and ferrite grains with carbide precipitates have been observed (Fig. 3b-3d). In the middle of the bar there is a ferritic-pearlitic structure (Fig. 3a) with locally occurring bainite areas

(Fig. 3e-3f). The ferrite grain size (Fig. 3a, 3i) was assigned to standards 8.5 and 9 of the scale models, which corresponds to the average grain diameter of $\bar{d} = 0{,}0186 \div 0{,}0156$ mm. The banded microstructure corresponded to model 0 on the A scale, which means no banding occurrence. The increase in the occurrence of Widmannstätten ferrite (Fig. 3i) corresponded to model 2 on the A scale.

Summary

On the basis of microstructure image observations, the effect of heat treatment and use of QTB technology on the structure of 34GS, BSt500S steels which are most commonly used for ribbed wire has been determined. It has been noticed that after the production process, 34GS steel is characterized by a bainitic-martensitic structure, where in the areas corresponding to the center of the bar thickness, the bainitic structure in the bainitic-martensitic structure was found. On the other hand, BSt500S steel is characterized by a bainitic structure with the area of occurrence of ferrite grains with carbide precipitates, where in the areas corresponding to the center of the bar thickness, the ferritic-pearlitic structure with locally occurring bainite areas was found. The occurrence of Widmannstätten ferrite was also confirmed.

References

[1] B. Golis, Z. Błażejowski, J.W. Pilarczyk, Z. Muskalski, Stalowe druty, pręty i wyroby z drutów do zbrojenia betonu, Series: Metalurgia nr 21, first ed., Wydawnictwo WIPMiFS Politechniki Częstochowskiej, Czestochowa, 2002.

[2] A. Almsallam, Effect of degree of corrosion on the properties of reinforcing steel bars, Constructions and Building Materials 15 (2001), 361-368. https://doi.org/10.1016/S0950-0618(01)00009-5

[3] V. Danchenko, H. Dyja, L. Lesik, L. Mashkin, A. Milenin, Technologia i modelowanie procesów walcowania w wykrojach. Series: Metalurgia nr 28, first ed., Wydawnictwo WIPMiFS Politechniki Częstochowskiej, Czestochowa, 2002.

[4] M. Konstanciak, Analiza strat w produkcji prętów żebrowanych, Praca doktorska, WIPiMFS, Częstochowa Politechnika Czestochowska, 2010.

[5] P. Sygut, Theoretical analysis of unit friction force working on the metal contact surface with the roll change during feedstock with non-uniform temperature distribution rolling process, Archives of Metallurgy And Materials, 61/2(2016), 927-932. https://doi.org/10.1515/amm-2016-0157

[6] P. Sygut, D. Klimecka-Tatar, S. Borkowski, Theoretical analysis of the influence of longitudinal stress changes on band dimensions during continuous rolling process, Archives of Metallurgy And Materials, 61/1(2016), 183-188. https://doi.org/10.1515/amm-2016-0032

[7] N. Toropov, D. Bajare, G. Shakhmenko, A. Korjakins, J. Justs, Effect of thermal treatment on properties of high strength concrete, Civil Engineering '13, PT I, Civil Engineering-Jelgava, 4/1 (2013), 129-133.

[8] PN-EN ISO 643:2005.

[9] PN-63/H-04504.

[10] R.P. Todorov, K.G. Khristov, Widmanstatten structure of carbon steels, Metal Science and Heat Treatment, 46/1-2 (2004), 49-53. https://doi.org/10.1023/B:MSAT.0000029601.58461.bd

[11] P. Szataniak, F. Novy, R. Ulewicz, HSLA steels - comparison of cutting techniques, 23rd International Conference on Metallurgy and Materials, METAL (2014), 778-783.

Materials Research Proceedings **5** (2018) 117-122 doi: http://dx.doi.org/10.21741/9781945291814-21

The Impact of Performance Improvement Achieved by the Closing Machine up to the Level of World-Class OEE on the Results of the Production Process

KNOP Krzysztof

Department of Production Engineering and Safety, Faculty of Management, Czestochowa University of Technology, Armii Krajowej 19B, 42-200 Czestochowa, Poland

krzysztof.knop@wz.pcz.pl

Keywords: Closing Machine, Performance, OEE, World-Class OEE, Profits

Abstract. The article presents the results of a simulation analysis related to the improvement of the efficiency (performance level) of a closing machine from the current, calculated level to the World-Class OEE level and the impact of this improvement on the results of the production process. The OEE indicator and its components for the analysed machine were calculated in the period of 1 year of its operation. Based on the calculations made, the level of improvement of OEE indicators was adopted in order to reach world-class levels. The analysis has shown what specific benefits in the form of the number of manufactured products and financial profits can be derived by improving the efficiency of using the analysed machine. It was also determined what improvement actions should be taken to get closer to the expected, world level of OEE indicators.

Introduction

The key to success in today's competitive world is the optimization of the use of existing production resources (assets), in the area of "4 x P", i.e. people, processes, plant equipment and predictive maintenance, which leads to improved efficiency [1]. Optimizing the functionality and time of uninterrupted operation of devices is an important issue related to increasing the efficiency of the company and its competitiveness on the market in any industry [2]. The efficiency of machines and devices is one of the most important determinant of product quality in many industries [3, 4]. In every company, many machines could certainly function more efficiently. Research shows that most devices produce only half of what they could produce, and the total use of resources is around 30-50% [5]. The reduction of costs by the proverbial "zloty" brings a greater profit to the company than the increase in sales by the same amount. Through the optimal use of production resources, production companies increase their profits from each zloty invested [6].

The introduction of TPM – Total Productive Maintenance program in the company is one of the ways to increase efficiency of devices. TPM is a system that allows world-leading companies to use devices in the most effective way [7]. Progress in TPM is measured, among others, by calculating OEE (Overall Equipment Effectiveness), which is an indicator combining the availability of machines, the performance of their work and the quality of the manufacturing process [4]. Frequently used measures in the TPM program also include the calculation of the Mean Time Between Failure (MTBF) and the Mean Time to Repair (MTTR) [8]. However, the OEE coefficient [5] is the key factor in TPM. Using this indicator is very important for production plants. Why? Information about the percentage of OEE gives close and true knowledge about the efficiency of production and the increase of this indicator by even a few percent brings a company tremendous financial benefits [2]. The project of introducing TPM in the company is considered successful when it reaches OEE = 85%. This means time availability higher than 90%, operational efficiency not less than 95% and quality index at the level of 99% [6]. 60% is a satisfactory level of OEE.

The article simulates the benefits of improving the efficiency of the analysed machine from its current, calculated level to the World-Class OEE level. Problems related to low machine work efficiency were identified and actions were proposed to improve this efficiency to the indicated level.

Research object and methodology

An automatic closing machine of the TO 43 +/- 89 mm type of an American manufacturer from 2002 is the object of the research. The closing machine is used in the confectioning line of liquid and semi-liquid products (such as concentrates, purees) in one of the food industry companies located in the northern part of the Silesian province. The view of the analysed machine and its dimensions is shown in Fig. 1.

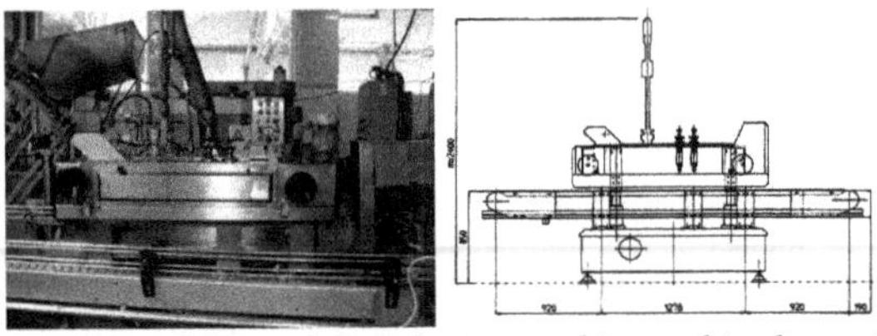

Fig. 1. View of the analysed closing machine and its dimensions.

The automatic closing machine is one of the machines in the dosing and closing line which, together with the pasteurization line and the labelling and packaging line, constitute a complete packaging line for concentrate and puree products in the company under study. All the machines in the line are connected with each other by means of plate conveyors. The automatic closing machine is designed to close the TO jars in a steam-vacuum system called the Whitecap system. The essence of the steam-vacuum sealing of the TO jars in the Whitecap system is evaporation of air through steam from the free space above the product just before and during the application and closing of the cap. The device works on the principle of a straight line system of continuous passing of packaging through the following positions: taking lids, steam injection and closing. The closure of the jar consists in turning the lid by 1/4 of a turn, as a result of which the lid is firmly pressed against the jar, and the gasket under the lid provides a hermetic closure.

The OEE indicator was used to evaluate the efficiency of the analysed machine. The formulas given in the paper [9] were used to calculate OEE and its components. The OEE calculation was made for a period of 1 year of operation of the analysed machine in relation to one of the product. The machine worked for 90 days closing glass containers filled with tomato concentrate (analysed product). The standard cycle time of the analysed machine was assumed at 0.006 pcs/min (standard performance is 10,000 pcs/h). The standard efficiency of the tested machine depends on the next machine in the line, which is the pasteuriser. The pasteuriser is a bottleneck in the confectioning process (the maximum capacity of this device is 10,000 pcs/h). The closing machine work results were summarized for a period of 90 days. It was the starting point for the proper analysis – the analysis of the benefits resulting from the improvement of the OEE indicator on the analysed machine to the World-Class level, i.e. 85%. For this purpose, the example of such analyses which was presented in the work [10] was the base.

Results

The results of the closing machine work were summarized during a 90 - day work period. The result of the summary of the work and the calculation of the OEE index and its components are presented in Table 1.

Table 1. Calculation of the Overall Equipment Effectiveness (OEE) for the closing machine. Summary of the machine's work for a period of 90 days.

Research period [90 days]	Total production time [min]	Planned breaks [min]	Planned production time [min]	Unplanned breaks [min]	Operating time [min]	Availability (A) [%]	All units [pcs]	Standard cycle time [min/pcs]	Real cycle time [min/pcs]	Performance (P) [%]	Number of defects [pcs]	Quality (Q) [%]	OEE [%]
	A	B	C=A-B	D	E=C-D	F=(E/C)*100	G	H	I=(E/G)*100	K=[(G*H)/E] *100	L	M=[(G-L)/G]*100	N=(F*K*M)/ 10,000
Σ	68130	5755	62375	6070	56305	90.27	5028198	0.006	0.011	53.47	384	99.99	48.27

The result of the comparison of OEE indicators of the analysed machine to World-Class OEE indicators levels is shown in Fig. 2.

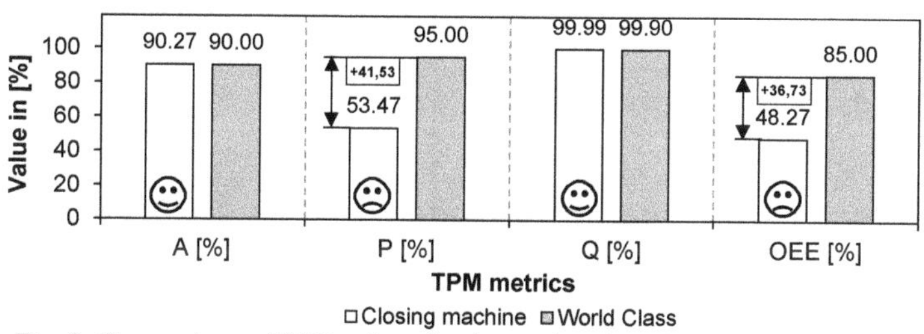

Fig. 2. Comparison of OEE values for the analysed machine with World-Class OEE indicators levels

As it can be seen from the comparison, in the case of two indicators, i.e. "Availability" and "Quality", the examined machine exceeds the world-class level in terms of these indicators. In the case of the "Performance" indicator, the analysed machine had a much lower value (by 41.53%) than the world-class level for this indicator, which resulted in the OEE indicator having a significantly worse result (by 36.73%) than the World-Class OEE level.

What benefits (production, financial) will the company incur by increasing the level of the "Performance" indicator by 41.53% in the case of the analysed machine? It was decided to analyse it using a specially developed sheet for this purpose (Table 2).

In order to improve the "Performance" indicator to the world-class level, i.e. 95%, the capability achieved by the device should be increased by 69.04 pcs/min (i.e. by 77.31%). As a result of increasing the capability by a given value, the capability achieved by the device will be 158.34 pcs/min (≅ 158 pcs/min) and will be lower only by 8.66 pcs/min (≅ 8 pcs/min) from the standard performance for this machine adopted at the level of 166.67 pcs/min (≅ 166 pcs/min). As a result of the capability increase achieved by the device, the World-Class level of OEE can be reached or even exceeded (85.75%). Taking into account that the level of quality will be at the same level (99.99%), the new level of capability will increase the number of defects to the level of 681 pieces from the level of 384 pieces. As a result of the machine's efficiency increase by an average of 69 pcs/min, a significant increase in the number of manufactured products will be achieved by 3,886,838 pieces compared to the base state, which will give a financial benefit of 3,303,812 zloty/90 days, assuming that the profit of 1 piece of the product is 0.85 zloty. The work assumes that the values of the remaining indicators ("Availability" and "Quality") will not change (*ceteris paribus*), during the efficiency of the machine will improve to the assumed level.

Table 2. Impact of performance improvement achieved by the closing machine on the OEE value

Initial state				Condition after improving the performance achieved by the machine		
Time losess						
Total Working Time	A	data	68,130	A'	constant	68,130
Planned breaks	B	data	5755	B'	constant	5755
Planned production time	C	A-B	62,375	C'	constant	62,375
Unplanned breaks	D	data	6070	D'	constant	6070
Operating time	E	C-D	56,305	E'	constant	56,305
Availability [%]	F	E/C	90.27	F'	constant	90.27
Performance losess						
Total units [pcs]	G	data	5,028,198	G'	J' x E'	**8,915,333**
Standard cycle time [min/pcs]	H	data	0.006	H'	constant	0.006
Standard performance of the device [pcs/h]	I	60/H	10,000	I'	constant	10,000
The performance achieved by the device [pcs/min]	J	G/E	89.30	J'	J+69.04	158.34
Perfomance [%]	K	(G*H)/E	**53.47**	K'	(G' x H')/E'	**95.00**
Quality losess						
Number of defects [pcs]	L	data	384	L'	(100-M') x G'	**681**
Quality [%]	M	(G-L)/G	99.99	M'	constant	99.99
OEE [%]	N	**F x K x M**	**48.27**	N'	**F' x K' x M'**	**85.75**

Profit [pcs/90 days]	**O=(G'-L')-(G-L)**	**3,886,838**
Profit of 1 piece of the product [zloty/pcs]	**P**	**0.85**
Profit [zloty/90 days]	**O x P**	**3,303,812**

The benefits resulting from the improvement of the OEE ratio from the current 36.73% level to the level of 85% were also analysed (Table 3).

With the assumed increase in the efficiency of the machine to the level of OEE equal to 85%, i.e. by 36.73% in relation to the current level, a total of 8,854,807.5 pieces will be produced, which will translate into a profit generated from the process of 3,252,618.10 zloty.

Table 3. Production and financial benefits due to the increase in the efficiency of the closing machine measured with OEE to 85%

Analysis of the benefits of improving OEE	Value	Unit of measure
The number of products that can be produced in the calculation period	9,402,935	pcs
The number of products produced in the calculation period	5,028,198	pcs
Loss of production in the calculation period	4,374,737	pcs
The profit of 1 piece of the product	0.85	zloty/pcs
Loss in zloty	3,718,526.45	zloty
The current level of OEE	**48.27**	**%**
Increase in OEE by 36.73%	**85.00**	**%**
The number of products produced after the increase	8,854,807.5	pcs
The profit after an increase in OEE by 36.73% to 85%	**3,252,618.1**	**zloty**

Conclusion

The analysis of the impact of the efficiency improvement achieved by the closing machine on the world-class OEE level on the efficiency of the production process will undoubtedly affect the imagination of managers in the analysed company. It shows what the effects of actions significantly reducing minor-stops and declines in work pace can bring – the main causes for performance losses of the analysed machine. And they can bring a significant - expressed in an additional number of produced products and financially (in zlotys). Additionally, more than 3 million zlotys due to the improvement of performance and, ultimately, improvement of OEE encourages to take action to reduce the level of performance losses.

Assumptions were made when analysing the benefits of improving the OEE indicator on the closing machine. It was assumed that in the analysed period after the change, i.e. increase in the efficiency achieved by the device (thus the "Performance" indicator), the level of other indicators, i.e. "Availability" and "Quality" will not change (decrease) compared to the initial state. Research, in turn, shows that the improvement of only one of the components of the OEE indicator "at the expense" of the others, makes it possible for them to be reduced, which in effect will not improve the value of OEE. A significant increase in the efficiency of the analysed machine may lead to major problems with quality, more frequent failures, which will affect lower values of the "Quality" and "Availability" indicators and, as a result, will generate a lower than expected OEE level (less than 85%).

Is it possible to achieve such a significant improvement in the efficiency of the closing machine? We are talking here about the increase in the machine utilization rate by 41.54%, which is over 77% in relation to the initial state! In the short term, certainly not. In the long-term perspective (realistically over 2 years) it is possible to get closer to this result. Such a low level of performance of the analysed machine must have its serious and complex causes, which should be identified and gradually eliminated. So what are the reasons for the low efficiency of the analysed machine and, more precisely, its low performance rate? These causes belong to each of the 5M categories, i.e. man, machine, material, method, management. Among them, the following should be distinguished: blocking the jars on the plate conveyor, blocking the screw caps in the feeder on the closing machine, overturning the jars on the conveyor, time intervals in feeding the jars, frequent current adjustments and checks due to operator inexperience, poor technical condition of machine sub-assemblies, system regulations (production planning that forces frequent retooling the line for a new assortment), weariness, distraction of the employee and incorrect service. These are just some of the causes that have been identified and which need elimination.

It should be emphasized that improving the level of performance of the analysed machine should be comprehensive. It cannot limit itself in these activities only to the closing machine itself. It should be kept in mind that the analysed machine is only one of the machines in the packaging line.

Machines in a line are connected vessels. Losses in one machine in the line are compensated in the next machine. "Taking care of" only the closing machine will certainly not significantly improve its efficiency. Only a comprehensive approach to improving and reducing losses on all the machines in the line can contribute to a rapid increase in the performance and efficiency of the closing machine, including a big role of machine operators, employees of the machine maintenance department and employees of the management division. Without the involvement of all employees, success in this area is not possible.

References

[1] M. Laszkiewicz, Utrzymanie ruchu to strategiczne aktywa firmy, Inżynieria & Utrzymanie Ruchu Zakładów Przemysłowych, Marzec 2005.

[2] J. Selejdak, R. Ulewicz, M. Ingaldi, The Evaluation of the Use of a Device for Producing Metal Elements Applied in Civil Engineering, METAL 2014. 23rd International Conference on Metallurgy and Materials, TANGER Ltd., Ostrava, 2014, pp. 1882-1888.

[3] R. Stasiak-Betlejewska, R. Ulewicz, The Effectiveness of Selected Machinery and Equipment in the Woodworking Joinery, Path Forward for Wood Products: a Global Perspective. Proceedings of Scientific Papers, WoodEMA, Zagreb, 2006, pp. 149-156.

[4] P. Szataniak, F. Novy, R. Ulewicz, HSLA steels - comparison of cutting techniques, METAL 2014. 23rd International Conference on Metallurgy and Materials, TANGER Ltd., Ostrava, 2014, pp. 778-783.

[5] M. Krynke, K. Knop, K. Mielczarek, Using Overall Equipment Effectiveness Indicator to Measure the Level of Planned Production Time Usage of Sewing Machine, Production Engineering Archives, Vol. 5, No. 4, 2014, pp. 6-9. https://doi.org/10.30657/pea.2014.05.02

[6] E. Kardas, S. Brozova, P. Pustejovska, S. Jursova, The evaluation of efficiency of the use of machine working time in the industrial company - case study, Management Systems in Production Engineering, Vol. 25, Iss. 4, 2017, pp. 241-245. https://doi.org/10.1515/mspe-2017-0034

[7] J. D. M. Mendez, R. S. Rodriguez, Total productive maintenance (TPM) as a tool for improving productivity: a case study of application in the bottleneck of an auto-parts machining line, International Journal of Advanced Manufacturing Technology, Vol. 92, Iss. 1-4, 2017, pp. 1013-1026. https://doi.org/10.1007/s00170-017-0052-4

[8] R. Niwas, M. S. Kadyan, J. Kumar, MTSF (mean time to system failure) and profit analysis of a single-unit system with inspection for feasibility of repair beyond warranty, International Journal of System Assurance Engineering and Management, Volume 7(1), 2016, pp. 198-204. https://doi.org/10.1007/s13198-015-0362-6

[9] J. Czerska, Jak wyznaczyć wskaźnik efektywności maszyn OEE?, Zarządzanie produkcją w praktyce, Wydawnictwo Wiedza i Praktyka, Styczeń/Luty 2007.,

[10] J. Selejdak, K Knop, Analysis of Advantages Resulted from Improvement of an OEE Factor, in: S. Borkowski, T. Lipiński (Eds.), Quality. Technological Improvement, Tripsoft, Trnava, 2011.

Influence of the Local Plasticise Zone in the Elastic Deformation Environment on the Fatigue Strength of the Construction Elements

BIŃKOWSKI Robert

Railway Research Institute (Instytut Kolejnictwa), ul. J. Chłopickiego 50, 04-275 Warszawa, Polska

rbinkowski@ikolej.pl

Keywords: Draw Hook, Overstress, Fatigue Strength, Yield Point, FEM, Strain Gauge Measure

Abstract. This article presents the problem of the overstress of the construction elements and its influence on the increase of the fatigue strength of the coupling device's draw hook for railway vehicles. The analysis has been done by the FEM computer simulation and during bench tests of the draw hook measures using strain gauges.

Introduction

The aim of this article is to present the influence of overstress, based on the yield point, of the construction elements on their fatigue strength in further exploitation.

The term „overstress" means applying local plasticise zones of the material in the environment of elastic deflection with a small acceptable change of the geometry of the overstressed element.

Certainly, overstress effect is possible in elements with a large variety of stresses under load, and the best results can be obtained in the case of local concentrations.

The draw hook, which is a part of the railway vehicles coupling device, will be analysed further.

This element, working directly with screw coupling, is responsible for longitudinal transmission forces generated by the train (Fig. 1) that consist of a wide range of loadings as occasional large forces and permanent fatigue loads.

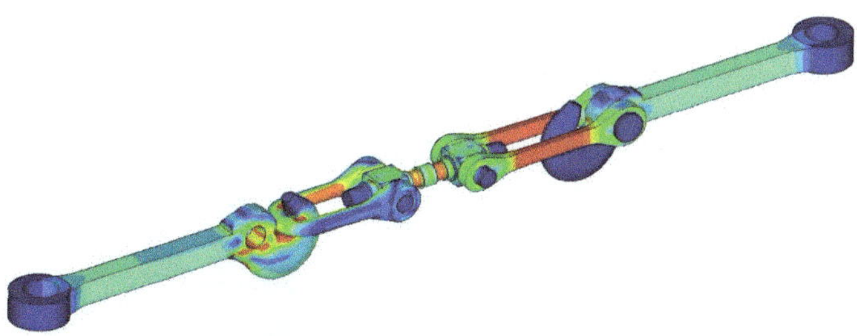

Fig. 1. General view of reduced stresses of the coupling device consisting of two draw hooks and screw coupling under maximum load.

The draw hook, as many other construction parts in the field of railways, is an interesting element to analyse due to the fact that its final form was created by evolution on the basis of trial and error, by increasing the cross section, using stronger steel and other, possibly unforeseen effect, such as the overstress or residual stress arising from technological processed thermal treatment and forging.

The analysis of the draw hook overstress was done during standard fatigue tests, required to obtain certificate for particular manufacturer, conducted by the Material and Structure Laboratory of the Railway Research Institute in Warsaw. The procedure of the standardised fatigue test of a draw hook [1] specifies the range of forces and number of cycles that are the following: 40 – 341 kN, total 1.5 million cycles, 41 – 1057 kN total of 2.15 thousand cycles in 700 cycles blocks, of the first scope and 1 cycle of the second. After the completion of the fatigue tests, the draw hook is the subject of a static test under the load of 1500 kN in this case. The main task in the analysis of the draw hook overstress is to determine the stress in the indicated zone after unloading.

The strain gauge method supported by the FEM [4] was used for the measurements. Numerical calculations were aimed, among others, at locating the areas of greatest measurable stress accumulation and determination of maximum overstress force in such a way that the strain gauges are not damaged by excessive deformation.

Analysis of the issues using the FEM

On the basis of the simulations, the safe overstress force was determined as 750 kN for a particular strain measuring location.

Fig. 2 and 3 show the location of the most stressed sections of the hook in the reduced stress form and plastic deformation area for the load of 750 kN and 1500kN.

Fig. 4 shows the main directions and senses for the loading of1500 kN and after unloading in the area of the strain gauge measurement.

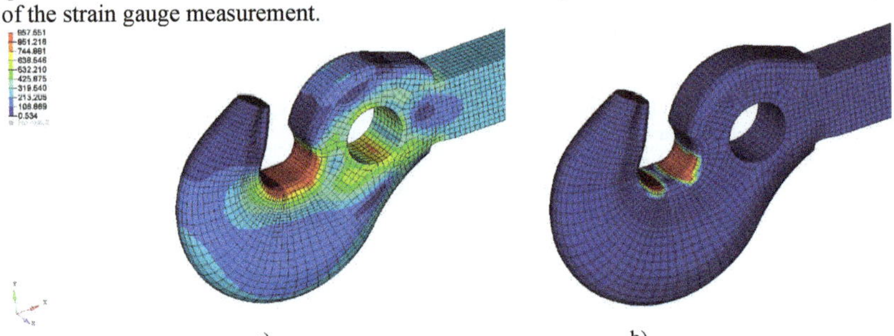

a) b)

Fig. 2. Reduced stresses (a) and plastic deformation area (b) for the loading force of 750 kN.

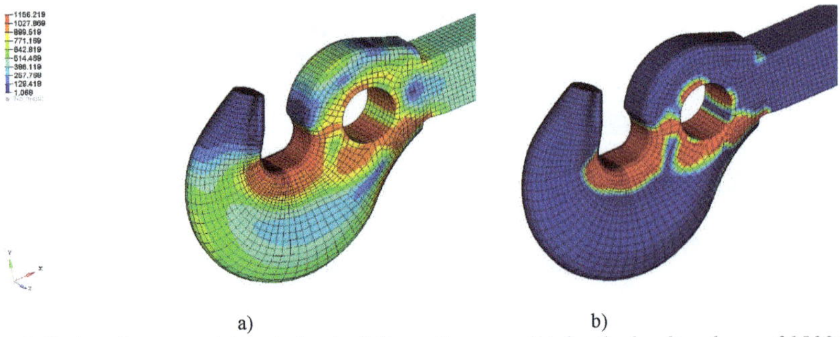

a) b)

Fig. 3. Reduced stresses (a) and plastic deformation area (b) for the loading force of 1500 kN.

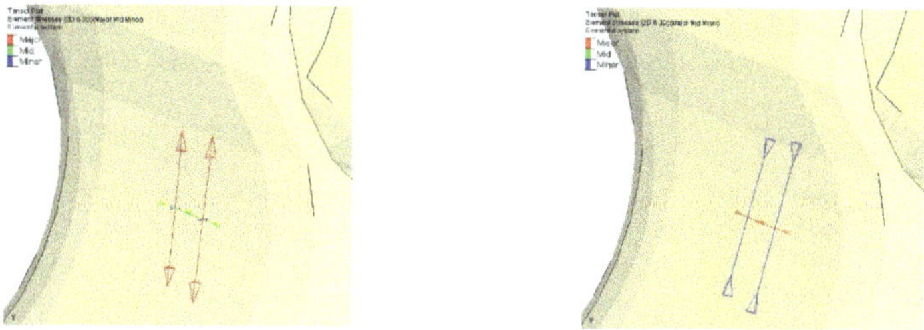

a) b)

Fig. 4. The directions and senses of the main stresses at the place of the strain gauge location for the loading of 1500 kN (a) and after unloading (b).

The place of location of the strain gauge is shown in Fig. 5. It should be noted that the measuring area is not the most stressed place of the entire hook.

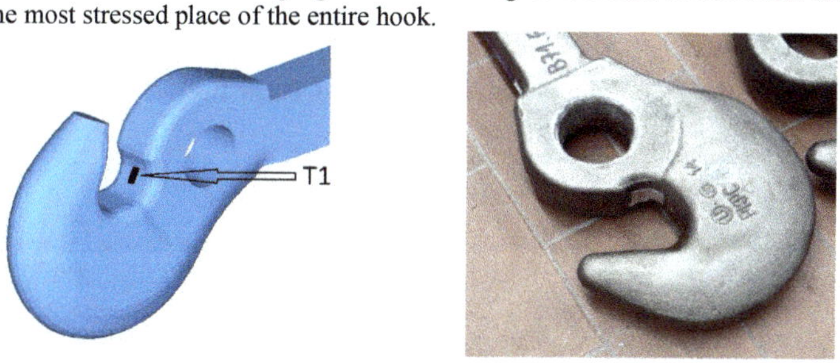

Fig. 4 Location of the T1 strain gauge in the draw hook.

Charts 1 and 2 show calculated tension forces and stresses as a function of deformation at the strain gauge location for maximum loads as above.

The stress level in the measurement direction after unloading at the considered area can be seen in the charts.

These values are -70 MPa for the loading of 750 kN and -720 MPa for the loading of 1500 kN.

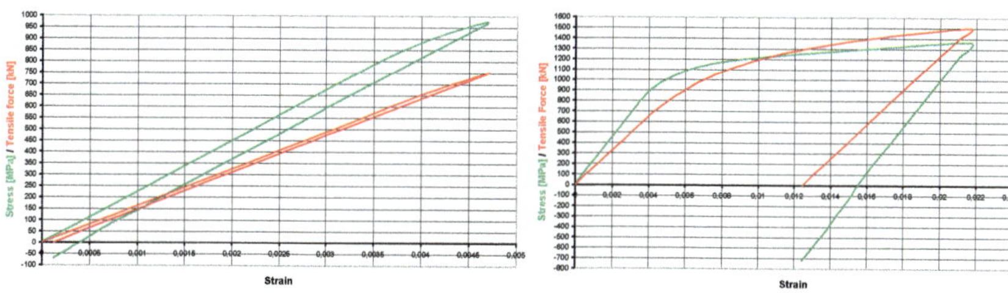

a) b)

Chart 1. The tension forces and stresses as a function of deformation for the loading of 750 kN (a), 1500 kN (b), calculated values.

This is the effect which could be expected due to the fact that after unloading the entire volume of not plastic deformed material tries to return to its original shape, and a small plasticised area opposes generating significant stresses, the contrary to the stress of the overstressed area. The new state of balance arises.

The bench tests

The photo on the left (Fig. 5) shows the draw hook in the test bench during the test with the loadings of 0–750–0 kN. The determination of stresses in the overstressed area using strain gauge measurement causes some problems arising from material plasticise and strengthening effect.

By analysing Chart 1, made as a FEM simulation, a simple geometric dependency can be illustrated in Chart 2a, where as before, when placing the course of loading force and stresses in function of deformation on the chart, and finding the intersection point of stress function in unloading phase, it is possible to find the stresses that are looked for.

The stress at the measuring place after overstressing with the force of 750 kN is -49 MPa, Chart 2b.

Fig. 5. Draw hook in the test bench.

The test with the load of 1500 kN, as expected, led to a strain gauge damage. Therefore, in order to estimate the stress after overstressing the load of 1500 kN, the strain gauge measurement method was used to determine residual stress [3].

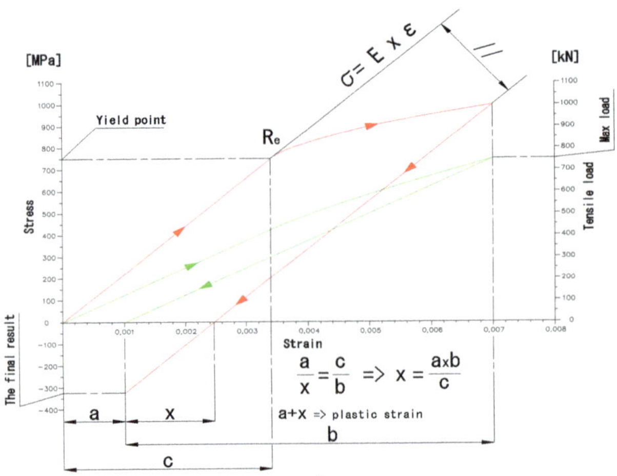

a)

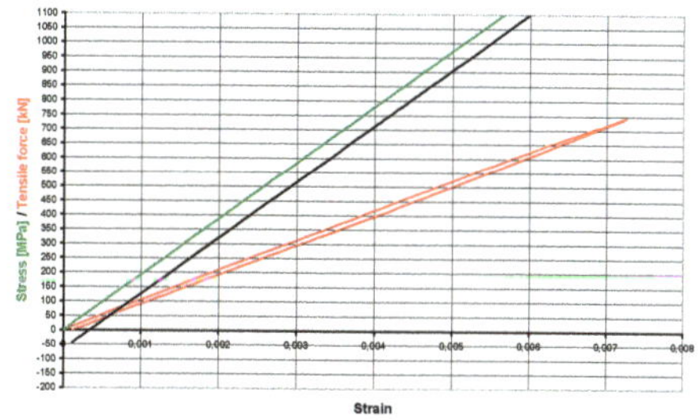

b)

Chart 2. Method of determining stress after overstress using strain gauges (a) Tensile force and stress as a function of deformation for the load of 0-700-0 kN. Experimental values.

At the same place of the hook, after removing the old strain gauge, the new one was glued, and then the analysed part of the hook was cut releasing the stresses from overstress (Fig. 6.).

Fig. 6. Residual stress measuring method using strain gauges of the draw hook.

The stress value calculated from deformations, after temperature stabilisation, was 844 MPa, which shows that they were at the same level before cutting but with the opposite sign.

The influence of the overstress effect on the fatigue strength

Generating compression stresses in the most intense areas, which are in the opposite direction to the stresses from the main loads, makes it possible to increase the strength of cyclical loads.

Most of the fatigue theories assume that the maximum stress, which after the overstress treatment is much lower, has the dominant influence on the crack formation. Using the fatigue Palmgren-Miner material model (1), (the accumulation of stress), which assumes that if the stress amplitude σ_{ai} repeats during N_i cycles, and the number of cycles to destroy specified from S-N curve in this amplitude is $N_{f,I}$, the distruction will occur when:

$$\sum \frac{N_i}{N_{f,i}} = 1 \qquad (1)$$

It can be shown that the hook subjected to the fatigue loading of only 40 – 341 kN without occasional loads overstressed in the rage of 41 – 1057 kN is not able to survive the test specified in the standard [1], Fig. 7. Due to the fact that the S-N curve is binding for the symmetric curve correction cycle and taking into account the cycle asymmetry, the Goodman method was used.

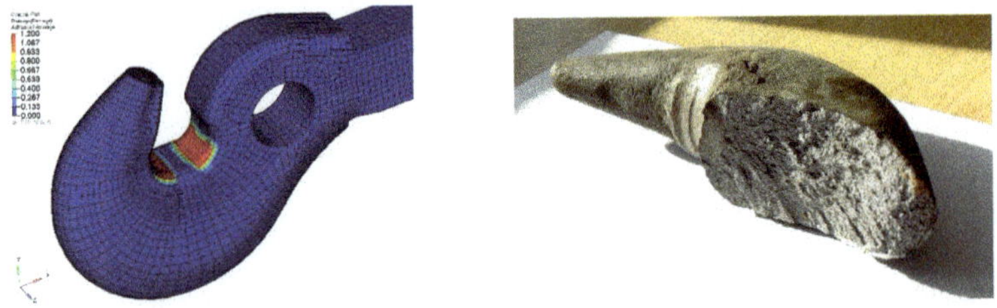

a) b)

Fig 7. The areas of the fatigue damages under the loading of the 40 – 341 kN forces without overstresses, determined with the stresses cumulation method for 1,5 million cycle (a), breakthrough (b).

In order to better illustrate the influence of the overstress on stress from fatigue loads in the range of 40 – 341 kN of the analysed part of draw hook, the Goodman chart is shown [2], which admittedly relates to permanent fatigue, but it gives a clear picture of the quality changes for the hooks overstressed with the forces of 750, 1057 and 1500 kN and in the absence of overstress (Chart 3).

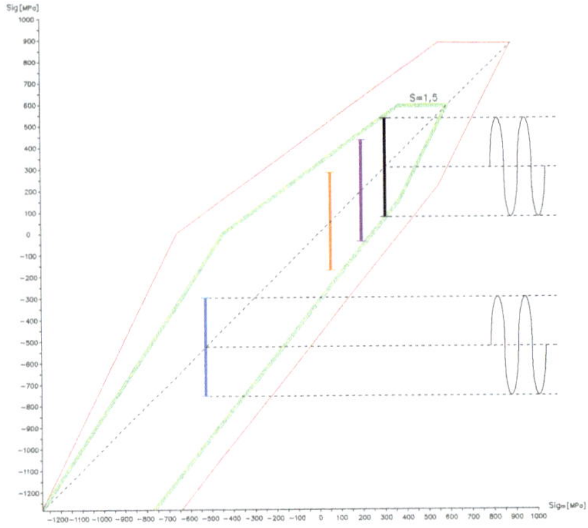

Chart 3. Fatigue stress range in the T1 strain gauge location for hook: black colour – without overstress, purple colour – overstress with the force of 750 kN, orange colour – overstress with the force of 1057 kN, blue colour – overstress with the force of 1500 kN.

Materials Research Proceedings **5** (2018) 123-127 doi: http://dx.doi.org/10.21741/9781945291814-22

Summary

On the basis of the results, it can be concluded that large convergence of numerical calculation and tests on the bench can be noticed, which makes it possible to validate calculations in areas impossible to verify on a real object. The draw hook is a good example of achieving a certain optimum by evolution, at least when it comes to the value of the stresses which are at such a level that without overstressed effect there is high risk of fatigue cracks during the test and exploitation.

Advantages of implementing additional stresses in the areas of the local plasticised material are demonstrated by the comparison of 1.34 safety coefficients for overstress with the force of 1057 kN and 1.03 without overstress, after applying the appropriate stress ranges from fatigue loads on the Goodman graph.

In the era of ubiquitous design optimisation of the constructions supported with significant engineering software development in this respect, somewhat forgotten overstress technique can bring quite measurable results.

References

[1] PN-EN 15566+A1:2011 Railway application – Railway rolling stock – Draw gear and screw coupling.

[2] ERRI B 12/RP 17 8th Edition.

[3] I. Mikłaszewicz, J. Siwiec, Research of residual stress in the railway rail joints, Railway Reports, Volume 61, Issue 177 (2017) 45-50.

[4] Radioss Theory Manual 13.0 version, Altair Engineering Inc. (2014).

Terotechnology 2017 Materials Research Forum LLC
Materials Research Proceedings **5** (2018) 130-135 doi: http://dx.doi.org/10.21741/9781945291814-23

The Influence of Variable Cross-Section of System Modeling Mining Support Structure on the Value of Critical Load under Active and Passive Specific Load

SZMIDLA Janusz[a] and JURCZYŃSKA Anna[b*]

Institute of Mechanics and Machinery Design Fundamentals, Częstochowa University of Technology, Częstochowa, Poland

[a]j.szmidla@imipkm.pcz.pl, [b]a.jurczynska@imipkm.pcz.pl

Keywords: Stability, Specific Load, Passive and Active Load, Non-Prismatic

Abstract. Theoretical considerations and results of numerical computations regarding a slender column modeling mining supporting structure were presented in this paper. The system was subjected to a selected case of a specific load (follower force directed towards the positive pole), active and passive. The numerical simulations presented in this article refer to the influence of a cross-section variable along the axis and the geometry of a structure implementing an external load on the value of critical force.

Introduction

The widespread use of slender elements (beams, rods and columns) in mechanical engineering, mining industry or civil engineering leads to the necessity of paying special attention to the issue of their vibrations and stability. A significant problem of the exploitation of such structures is a search for and implementation of new structural solutions, relating both to models of systems generating a conservative external load and to the appropriate shaping of rods in order to increase their critical load. The works [1, 2] propose design solutions of loading heads that can generate active and passive (as a supporting structure) external loads. The article [2] where the schemes of a loading structure implementing a specific load (L. Tomski's load, [3]) are presented is of particular significance. The described structures under such an external loading are characterized by much greater critical forces compared to the classic Euler's load.

A separate group of publications is devoted to the issue of stability and free vibrations of systems of a variable cross-section. The paper [4] presents a scientific problem of shape optimization of a slender column subjected to an axial load with a concentrated force and continuous load using the Rayleigh – Ritz method. The study of the influence of size and location of a crack in a beam with a variable cross-section on its eigenvalue and stability constitutes the subject of the publication [5]. The issue of the stability of columns with a variable cross-section was also discussed in the works [6, 7]. This paper presents a model of a non-prismatic column modeling a mining support structure subjected to a selected case of a specific load, active and passive.

Physical Model

A slender column of a cross-section variable along an axis modeling mining support structure presented in Fig. 1 is taken into consideration in this work. In order to model the variable cross-section, the column is considered as a set of prismatic segments of constant length l and thickness h, whose width is described by approximation functions, while maintaining the conditions of a constant total volume and total length of the system. The performed analysis includes a shape approximation using:

- a linear function;
- a polynomial of degree 2.

Materials Research Proceedings **5** (2018) 130-135 doi: http://dx.doi.org/10.21741/9781945291814-23

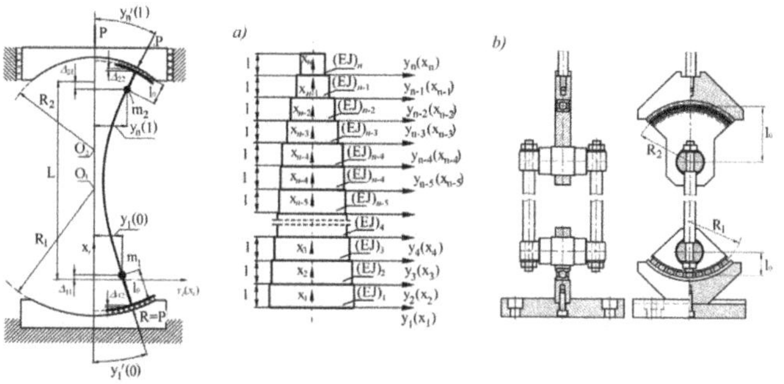

Fig. 1. Physical model of a non-prismatic column subjected to a passive and active specific load: a) division into segments, b) real construction solutions.

The analyzed column is subjected to the follower force directed towards the positive pole (the case of the specific load defined by L. Tomski), active and passive. The active load is implemented through the upper structure realizing the load, built of two elements of a circular outline (constant curvature, a slice of a rolling bearing) – loading and receiving heads. The supporting structure constructed in the same way generates the passive load. Both constructions are real objects whose validity has been confirmed experimentally. For structural reasons, theoretical considerations and numerical calculations include also a rigid element of a length of l_0, to ensure a proper mounting of the column.

Problem Formulation

On the basis of the presented description of a physical model of a slender column of a variable cross-section modeling a mining support structure, taking into account the Bernoulli – Euler's theory, the potential energy of the analyzed system was defined. The potential energy V is a sum of the energy of the bending elasticity of the particular steps of the column V_1 and the potential energy derived from the vertical (V_2) and horizontal (V_3) components of the active and passive external load:

$$V_1 = \frac{1}{2}\sum_{i=1}^{n}(EJ)_i \int_0^1 \left[y_i^{II}(x_i)\right]^2 dx_i; \tag{1}$$

$$V_2 = -\frac{P}{2}\sum_{i=1}^{n}\int_0^1 \left[y_i^{II}(x_i)\right] dx_i + P\frac{l_0}{2}\left[\left(y_1^{I}(0)\right)^2 + \left(y_n^{I}(1)\right)^2\right], \tag{2}$$

$$V_3 = \frac{P}{2}\left[R_1\left(y_1^{I}(0)\right)^2 + R_2\left(y_n^{I}(1)\right)^2\right]; \tag{3}$$

$$V = V_1 + V_2 + V_3. \tag{4}$$

where: EJ – flexural stiffness, $y_i(x_i)$ – function of the deflection of the i-th segment, R_1, R_2 – radii of the heads implementing the passive and active load, l_0 – length of the rigid element.
The boundary problem was formulated on the basis of the minimum of the potential energy principle (static criterion of the stability loss):

$$\delta V = 0. \tag{5}$$

The geometric boundary conditions and the continuity conditions referring to the deflection and the angles of the deflection in the mounting of the system and in the points of joining individual segments are expresses by the following formulas:

$$y_1(0) = -(R_1 - l_0)\, y_1^{I}(0), \qquad y_n(1) = (R_2 - l_0)\, y_n^{I}(1), \tag{6-7}$$

$$y_j(1) = y_{j+1}(0), \qquad y_j^{I}(1) = y_{j+1}^{I}(0). \tag{8-9}$$

Formulas (Eq. 6-7) result directly from the design assumptions of the loading structure. Performing the variational calculus of the potential energy (Eq. 4), substituting to the equation which defines the principle of the minimum of the potential energy (Eq. 5) and then taking into account the geometric boundary conditions (Eq. 6-7) and the continuity conditions (Eq. 8-9) makes it possible to determine:

- the differential equations of displacements:

$$(EJ)_i\, y_i^{IV}(x_i) + P y_i^{II}(x_i) = 0; \tag{10}$$

- the natural boundary conditions:

$$(R_1 - l_0)\, y_1^{III}(0) + y_1^{II}(0) = 0, \qquad (R_2 - l_0)\, y_n^{III}(1) + y_n^{II}(1) = 0; \tag{11-12}$$

- the natural continuity conditions:

$$(EJ)_j\, y_j^{II}(1) = (EJ)_{j+1}\, y_{j+1}^{II}(0), \qquad (EJ)_j\, y_j^{III}(1) = (EJ)_{j+1}\, y_{j+1}^{III}(0). \tag{13-14}$$

The solution of the differential equations of the displacements is written in the following form:

$$y_i(x_i) = A_i \sin(k_i x_i) + B_i \cos(k_i x_i) + C_i x_i + D_i. \tag{15}$$

After substituting the solution (Eq. 15) with the geometrical and natural boundary conditions as well as continuity conditions, a system of homogeneous equations is obtained. Equating zero to the characteristic determinant of the system of equations defines a transcendental equation for the value of the critical force of the considered column.

Results of Numerical Simulations

The results of the numerical calculations are presented in the dimensionless form, introducing the following parameters:

- the parameters describing the approximating functions:

$$p^* = \frac{p}{L}, \quad q^* = \frac{q}{L}, \quad Z^* = \frac{b_l - b_n}{L} \cdot 100\%; \qquad (17\text{ a-c})$$

- the parameter of the critical load:

$$\lambda_{kr} = \frac{P_{kr} L^2}{(EJ)_{pr}}; \qquad (18)$$

- the parameter of the increase of the critical load:

$$\delta_{\lambda_{kr}} = \frac{\lambda_{kr} - (\lambda_{kr})_{pr}}{(\lambda_{kr})_{pr}} \cdot 100\%. \qquad (19)$$

The values with the subscript "pr" in the formulas (Eq. 18, 19) refer to the prismatic (of constant cross-section) comparative system of the same volume.

Figure 2 shows the influence of the shape approximation of the non-prismatic rod by means of the approximating functions on the value of the critical load increment in relation to the equal values of the radii of the heads generating the active and passive load. On the basis of the presented graphs, it is possible to determine the ranges of the values of the parameter of the convergence of the system (in the case of the shape approximation by the linear function, Fig. 2a) or the parameters of the position of the vertex of the parabola (in the case of the shape approximation by the polynomial of degree 2, Fig. 2b), for which there is an increase in the value of the critical force of the structure compared to the reference system of the constant cross-section. The maximum increase of the critical load achieved was about 10%.

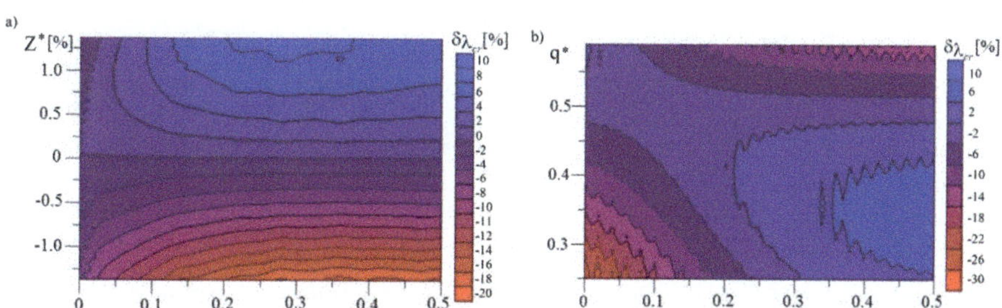

Fig. 2. Value of the critical load increase of the non-prismatic column as a function of the variable geometry of the structures performing the active and passive load: a) when the outline is approximated by the linear function; b) when the outline is approximated by the parabolic function.

Materials Research Forum LLC
doi: http://dx.doi.org/10.21741/9781945291814-23

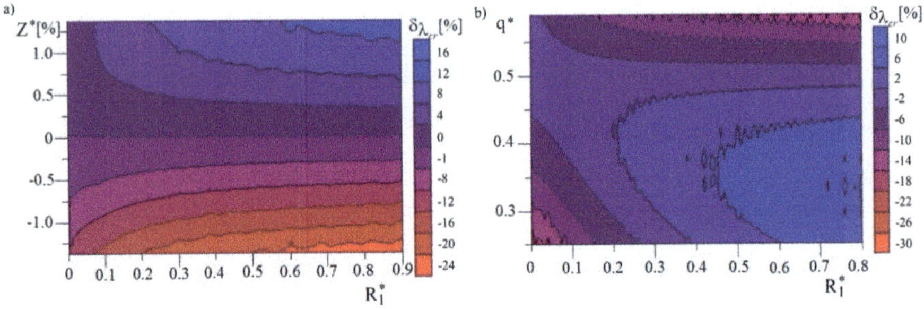

Fig. 3. Value of the increment of the critical load of the non-prismatic column depending on the variable geometry of the support structure realizing the passive load (when $R_2^* = 0.4$): a) when the outline is approximated by the linear function; b) when the outline is approximated by the polynomial of degree 2.

Similar data is presented in Figure 3 that illustrates a detailed analysis of the influence of the cross-section variable along the axis on the increment if the critical load value as a function of the variable geometry of the support structure implementing the passive load, at a specified radius of the heads performing the active load. The distribution of critical load increments is similar to that shown in Figure 2. In the case of the linear approximation function, it has been shown that in comparison with the comparative system of the constant cross-section, the critical load can be increased to 16%, while the approximation with parabolic function to 10%.

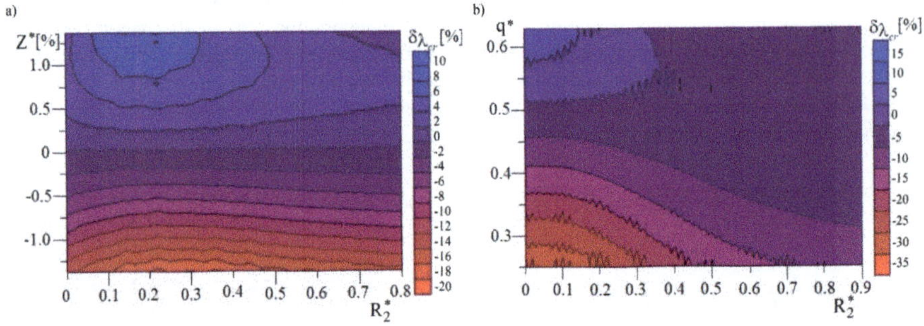

Fig. 4. The value of the increase of the critical load of the non-prismatic column depending on the variable geometry of the structure implementing active load (while $R_1^*=0.3$): a) when the outline is approximated by the linear function; b) when the outline is approximated by the parabolic function.

Figure 4 shows the increase of the critical load depending on the coefficients of the approximating functions and the variable radius of the upper heads realizing the active load, with the given geometry of the support structure. In the discussed case there was a slightly different distribution of critical load increments with maximum values of 10% in the approximation of the outline by a linear function and 15% when considering the polynomial of degree 2.

Summary

The work aimed at performing a numerical analysis of the issue of the stability of a slender column modeling a mining support structure subjected to the follower force directed towards the positive pole, active and passive. Based on the performed computations, the influence of the variable geometry of the structures implementing the selected case of the specific load and the parameters defining approximate functions on the value of the increase of the critical load in comparison with the comparative system of the constant cross-section and the same volume. The ranges of the convergence parameter and the coordinates of the vertex of the parabola parameters, for which the column with variable cross-section is characterized by a higher critical load than the reference system have been defined and the maximum recorded increase of the critical load value was about 16%. The experimental verification of the assumptions made in the mathematical model should be the next stage of the work.

Acknowledgements

The study has been carried out within statutory research BS/MN 1-101-302/17/P of the Institute of Mechanics and Machinery Design Fundamentals of Częstochowa University of Technology.

References

[1] A. Kasprzycki, L. Tomski, Chosen constructional solutions of the structures loading the Euler columns in vibrational and stability studies, Archives of Civil and Mechanical Engineering, vol. XI(2) (2011) 321-332. https://doi.org/10.1016/S1644-9665(12)60146-0

[2] L. Tomski, A. Kasprzycki, J. Szmidla, [in Polish] Chosen constructional solutions of the structures implementing potential active and passive load, XXII Symposium of Machine Design Fundamentals, Referaty sesyjne i plakatowe, vol. 4 (2005) 353-358

[3] R. Bogacz, B. Imiełowski, L. Tomski, Stability and vibration of column structures subjected to generalized concentrated load, Proceedings of International Symposium Dynamics of Continua, Physikzentrum Bad Honnef (1996), 45-54

[4] R. Drazumeric, F. Kosel, Shape optimisation of beam due to lateral buckling problem, International Journal of Non-Linear Mechanics, vol. 47 (2012) 65-74. https://doi.org/10.1016/j.ijnonlinmec.2011.12.004

[5] K. Sokół, A. Kulawik, Optimization of slender systems by means of genetic algorithms, Journal of Applied Mathematics and Computational Mechanics, vol. 13(1) (2014) 115-124. https://doi.org/10.17512/jamcm.2014.1.12

[6] J. Szmidla, A. Jurczyńska, [in Polish] the stability of the slender non-prismatic column subjected to the follower force directed towards the positive pole, Modelowanie inżynierskie, vol. 60 (2016) 67-73

[7] J. Szmidla, A. Jurczyńska, Free vibrations of non-prismatic slender system subjected to the follower force directed towards the positive pole, Vibrations in Physical Systems, vol. XXVII (2016) 363-368.

Materials Research Proceedings **5** (2018) 136-141 doi: http://dx.doi.org/10.21741/9781945291814-24

A Change of Mechanical Properties of the Self-hardening UNIFONT 90 Due to Temperature

MAZUR Magdalena[1, a *], KUCHARIKOVÁ Lenka[2,b], TILLOVÁ Eva[2,c] and CHALUPOVÁ Mária[2,d]

[1]Department of Management, Czestochowa University of Technology, Al.Armii Krajowej 19B, 42-200 Czestochowa, Poland

[2]Department of Materials Engineering, Faculty of Mechanical Engineering, University of Žilina, Univerzitná 8215/1, 010 26 Žilina, Slovakia

[a]magdalena.mazur@wz.pcz.pl, [b]lenka.kucharikova@fstroj.uniza.sk, [c]eva.tillova@fstroj.uniza.sk, [d]maria.chalupova@fstroj.uniza.sk

Keywords: Self-Hardened Aluminum Alloy, UNIFONT 90, Mechanical Properties, Temperature Effect, Morphology of Microstructural Components

Abstract. UNIFONT 90 is aluminum cast alloy used for engine and vehicle constructions, hydraulic units and mold making without heat treatment (self-hardening alloy). When using this alloy for high-temperature castings, it is necessary to study its structure and mechanical properties depending on the temperature. The contribution describes the effect of temperature in the range from – 196 °C up to + 400 °C on mechanical properties (especially Brinell hardness and bending impact toughness) of the UNIFONT 90. The mechanical properties were measured on samples after cooling and heating for holding 20 minutes at temperature. Fracture surfaces and changes in microstructure of material were evaluated by means of light microscopy and scanning electron microscopy (SEM). The results showed that with increasing temperature, the material was more plastic and making corresponding fracture surface.

Introduction

Aluminum alloys for industry application are divided into two groups. The first are non heat-treatable alloys, achieving their strength by the amount of cold work applied to the alloy. Mechanical properties of non heat-treatable alloys are obtained by hot/cold working mechanisms during their production. The initial strength of these alloys depends on the hardening effect of elements such as Mn, Si, Fe and Mg, individually or in various combinations. The next improvement of the strength can be reached by cold working process, as the aluminum has high ductility and therefore can be easily deformed. The second, heat-treatable or precipitation hardening alloys, reach their strength and other properties following heat treatments. The properties are dependent on heat treatments and age hardening conditions. Significant strengthening can be achieved by heating and cooling, especially with age-hardening process (solution heat treatment, quenching, age hardening by artificial or natural aging). The initial strength of these alloys is also influenced by the addition of alloying elements to pure aluminum [1-6].

Generally, mechanical and microstructural properties of aluminum cast alloys are dependent on the composition; melt treatment conditions, solidification rate, casting process and the applied thermal treatment. By implementing adaptable alloying and process technology, mechanical properties will therefore be radically enhanced, leading to larger application fields of complex cast aluminum components such as safety details [6-9].

The present study is a part of a larger research project, which was conducted to investigate and provide a better understanding temperature effect on the properties of experimental material.

Experimental material and procedure
The experimental material UNIFONT 90 (AlZn10Si8Mg) was prepared from scrap in foundry UNEKO, s.r.o. Zátor, Ltd. Czech Republic. The experimental bars with dimensions of ø 20 mm and length of 300 mm (blue- Fig. 1) were casted into a sand mold (Fig. 1). The experimental specimens for bending impact test were produced with turning and milling operation without notch (55x10x10 mm) from the experimental bars (Fig. 1). The experimental material belongs to self-hardening alloys that are particularly used when good strength values are required without the need for heat treatment. Self-hardening starts when the castings are removed from the mold and the final mechanical properties are achieved after storage of approximately 7 to 10 days at room temperature. In the process of thin-walled castings, self-hardening completed after 7 days and the thick-walled castings after 10 days. During this time, it is not possible castings processed. Castings of the UNIFONT 90 alloy reach about 50 % after the first day of the casting, and about 80% of the final mechanical properties after the second and third day, which must be taken into account in production conditions. Curing time is reduced by annealing the castings from 100 to 120 °C with the time of holding from 10 to 16 hours depending on the thickness of the cast. Castings made from this alloy are perfectly weldable by standard procedures and can withstand prolonged cyclic loading [3, 10].

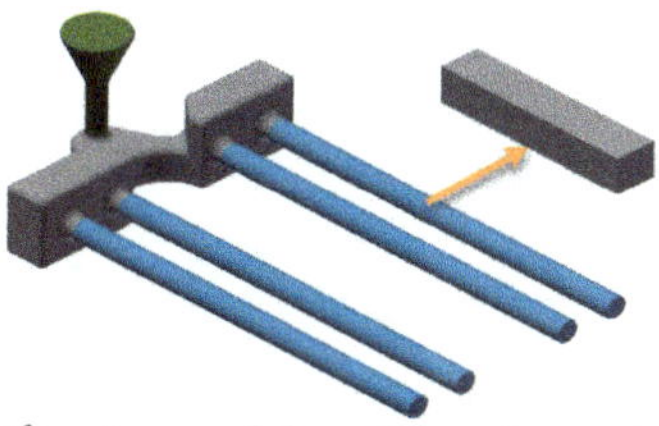

Fig. 1 The scheme of casting module and experimental specimen for impact bending test

The chemical composition of the experimental material was checked out using arc spark spectroscopy, and it is given in Table 1. The experimental material is assigned to aluminum alloys series 7xxxx because of higher amount of Zn, but some international authors include it between Al-Si alloys on behalf of the structure [3, 10].

Table 1 Chemical composition of experimental bars, wt. %

Zn	Si	Cu	Fe	Mn	Mg	Ti	Ni
9.6	8.64	0.005	0.1143	0.181	0.452	0.0622	0.0022
Cr	Hg	Ca	Cd	Bi	P	Sb	Al
0.0014	0.0006	0.0002	0.0001	0.0003	0.0001	0.0007	remainder

The resistance of materials against the formation of fracture depends mainly on temperature, kinds of mechanical stress and geometry of products. The bending impact toughness is a major property which characterizes resistance of materials against the formation of cleavage fracture. Temperature can affect the structure of materials and can lead to changes in properties too. Therefore, Brinell hardness and impact bending toughness were measured. These mechanical properties were measured according to the following standards: STN EN ISO 6506-1 and STN EN 10045-1. Hardness measurement for experimental secondary aluminum alloy was performed using a Brinell hardness (HBW) tester with a load of 250 kp (1kp = 9.81 N), 5 mm diameter ball and a dwell time of 10s. Impact bending toughness (K) was measured on the Charpy hammer with the nominal value of 300J.

Materials Research Forum LLC
doi: http://dx.doi.org/10.21741/9781945291814-24

Each property was measured on samples whose temperatures were in the following range: - 196 °C, - 20 °C, + 21 °C - room temperature in time of the test and then elevated temperatures of 50 °C, 100 °C, 150 °C, 200 °C, 250 °C, 300 °C, 350 °C and 400 °C. The higher temperature was reached by the leaving the samples in the electric resistance furnace for holding time of 20 minutes at temperature. The evaluated K and HBW reflect average values of at least six separate specimens.

The samples for metallography observation and assessment of fracture surfaces were prepared from selected specimens (after testing) and were studied using a NEOPHOT 32 light microscope and a VEGA LMU II scanning electron microscope. The metallography samples were prepared by standard metallographic procedures.

Experimental results

Mechanical properties

Multiple types of tests are used for the description of material behavior at elevated and low temperatures. These test are used to monitor the influence of parameters (temperature, time, load, and so on) on mechanical properties. However, the experimental material was also used for the application of the impact bending test for different elevated temperatures. The impact bending test was realized on samples immediately after being removed from the electric furnace for ensuring the endurance of temperature. Fig. 1a shows the measurements results.

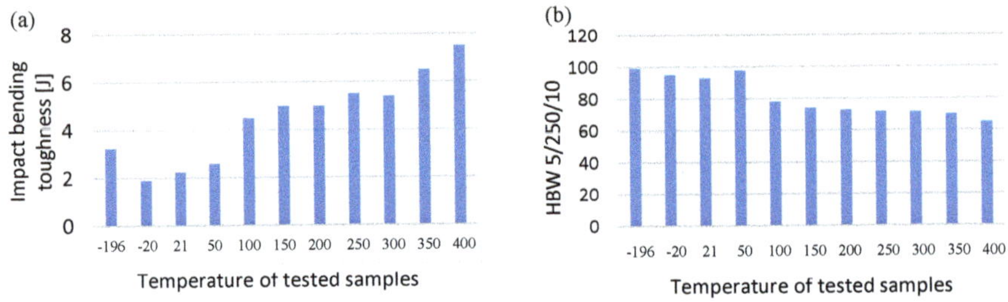

Fig. 1 Results of mechanical properties.
(a) impact bending test; (b) Brinell hardness

The energy required to crack the test samples at room temperature equalled 2.26 J. Then the results show an increase in impact bending toughness with increasing temperature. Two time higher value of energy required for cracking is at 100 °C (4.5 J). The impact bending toughness is similar in the interval of temperature 100 ÷ 300 °C. Three times higher value of energy required for cracking is at 350 °C (6.5 J). Temperature -20 °C leads to the decrease in impact bending toughness in comparison to room temperature, but at the temperature of liquid nitrogen (-196 °C) the value of this toughness increases. The results of the impact bending test show that this material will not have rapid transition from ductile to cleavage fracture.

Hardness was measured on cooled samples after the bending impact test. Fig. 1b shows the results of the Brinell hardness measurements after holding at the temperatures. It can be seen that the measured Brinell hardness decreases with increasing temperature. The interval of – 196 ÷ 50 °C has the highest and similar values. The lowest hardness was measured at 400 °C (64.9 HBW). The results also show similar hardness value in the interval of 100 ÷ 400 °C. The difference between these two intervals is about 25.3 %. Based on these results, it is anticipated that strength tensile will be decreasing too [11].

Changes in microstructure

The microstructure of the recycled UNIFONT 90 consists of a primary phase, α-solid solution, a eutectic mixture of α-matrix and spherical Si-phases and various types of intermetallic phases. The α-matrix precipitates from the liquid as the primary phase in the form of dendrites and is nominally comprised of Al and Zn. Al_2CuMg, Mg_2Si, Al (FeMn) Si and ternary eutectic Al-$MgZn_2$-Cu were observed from the intermetallic phases. Small amounts of AlFeMnMgSi and AlFeMnMgNi intermetallic phases were observed in a few areas [10].

(a) (b) (c) (d)

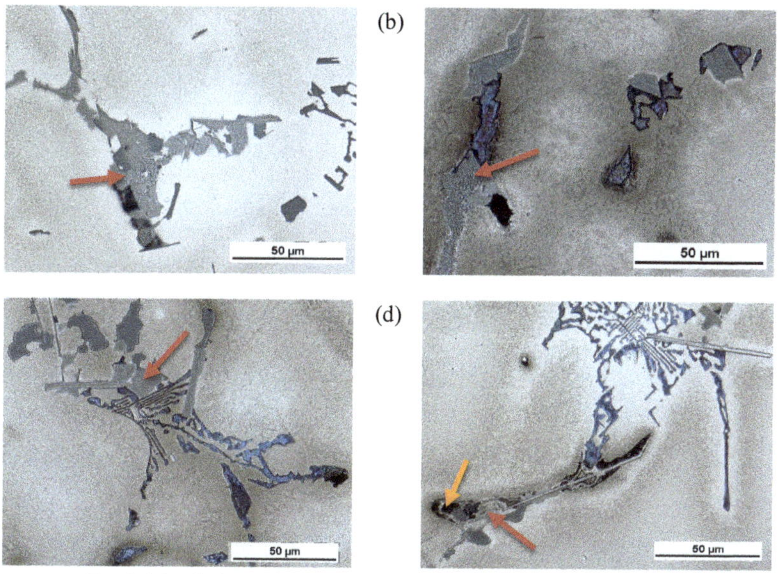

Fig. 2 Morphology changes of Al-$MgZn_2$-Cu, etch. 0.5 % HF.
(a) 21 °° C; (b) 100 °C; (c) 300 °C; (d) 400 °C

The testing temperature has influence on microstructural components. Eutectic silicon particles were coarsened around dendrites of matrix, but no other changes were observed. The metallography observation and quantitative analysis have shown that morphology, size, and distribution of phases Al (FeMn) Si, Mg_2Si and Al_2CuMg were not affected by the increasing temperature. The major changes occurred in the Al-$MgZn_2$-Cu ternary eutectic phases (red arrow in Fig. 2). In the as-cast state (21 °C) has this phase compact shape (Fig. 2a), but with increasing temperature up to 300 °C decreasing size of this phase (Fig. 2b, c). The temperature of 400 °C led to greater changes in morphology of this phase which is fragmented in smaller isolated particles and in a few areas cavities were visible (orange arrow in Fig. 2d). This morphology changes probably relate with the melting point of zinc (419.5 °C) [12, 13].

Materials Research Proceedings **5** (2018) 136-141 doi: http://dx.doi.org/10.21741/9781945291814-24

Fracture behavior of experimental material

The fracture surfaces have a mixed character (ductile and cleavage). The character of fracture surfaces at 21 °C and -20 °C is similar (Fig. 3b). The ductile fracture has the same character and ratio. The temperature of -196 °C led to increasing ratio of cleavage fracture thanks to intermetallic phases (Fig. 3a). In a few regions ductile fractures with smaller dimples and plastically transformed matrix (α-phase) ridges were observed in comparison with the state at room temperature. Higher amounts of plastically transformed matrix ridges and grosser dimples were observed in ductile fracture surfaces with the increase of testing temperature (Fig. 3).

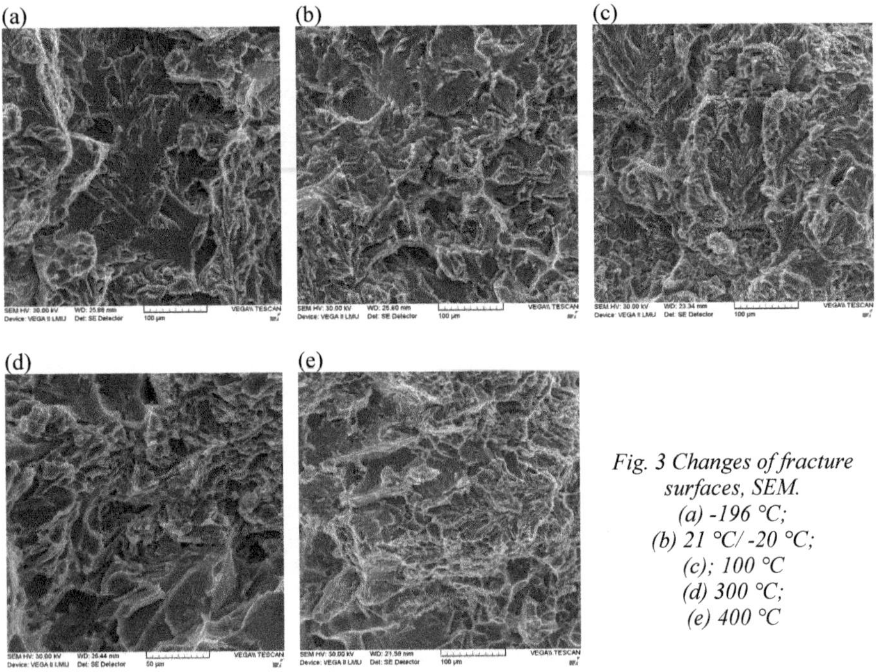

Fig. 3 Changes of fracture surfaces, SEM.
(a) -196 °C;
(b) 21 °C/ -20 °C;
(c); 100 °C
(d) 300 °C;
(e) 400 °C

In general, mixed character cleavage and ductile fracture characters were observed in all surfaces where predominantly there was ductile fracture. Increasing the temperature increases the plasticity of the material on the fracture surfaces, which is reflected by increasing the depth of the ductile dimples and the occurrences of larger amounts of plastically transformed matrix ridges.

Conclusion

Mechanical tests and evaluation of microstructural component changes indicate that different temperature has influence on the increasing impact bending toughness and decreasing hardness. The results show resistance of the material against higher temperature. Mixed fracture (ductile and cleavage) of specimens was observed. Also, the results confirm increasing plasticity with increased temperature. The hardness decreases especially as a result of the changes of ternary eutectic phase's morphology during heating at higher temperatures, but the value was about 25.3% lower at higher temperatures.

Acknowledgements
This work has been supported by Scientific Grant Agency of Ministry of Education of Slovak republic N° 1/0533/15 and EU project ITMS 26220220154.

References

[1] J. L. Cavazos, R. Colás, Quench sensitivity of a heat treatable aluminum alloy, Materials science and engineering A363 (2003) 171-178.

[2] J.S. Robinson, W. Redington, The influence of alloy composition on residual stresses in heat treated aluminum alloys, Materials Characterization 105 (2015) 47-55.

[3] M. Rosso, I. Peter, C. Castella, R. Molina, Optimization of composition for self-hardening AlZn10Si8Mg alloys, Materials today: Proceedings 2 (2015) 4949-4956.

[4] T. Lipiński, The structure and mechanical properties of Al-7%SiMg alloy treated with a homogeneous modifier, Applied Crystallography XXI Book Series: Solid State Phenomena 163 (2010) 183-186.

[5] R. Ulewicz, P. Szataniak, F. Novy, Fatigue properties of wear resistant martensitic steel, METAL 2014: 23rd International conference on metallurgy and materials (2014) 784-789.

[6] T. Dokšanocić, I. Džeba, D. Markulak, Variability of structural aluminum alloys mechanical properties, Structural safety 67 (2017) 11-26.E.

[7] S. Liu, Q. Li, H. Lin, L. Sun, T. Long, L. Ye, Y. Deng, Effect of quench-induced precipitation on microstructure and mechanical properties of 7085 aluminum alloy, Materials and Design 132 (2017) 119-128.

[8] N. Berndt, P. Frint, M. Böhme, M. F.-X. Wagner, Microstructure and mechanical properties of an AA6060 aluminum alloy after cold and warm extrusion, Materials Science and Engineering A 707 (2017) 717-724.

[9] S. Li, H. Dong, L. Shi, P. Li, F. Ye, Corrosion behavior and mechanical properties of Al-Zn-Mg aluminum alloy, Corrosion Science 123 (2017) 243-255.

[10] E. Ďuriníková, E. Tillová, M. Chalupová, Phase and structure characteristics of recycled AlZn10Si8Mg cast alloy, Manufacturing technology 11 (2011) 11-17.

[11] E. Rincón, H. F. López, M. M. Cisneros, H. Mancha, M.A. Cisneros, Effect of temperature on the properties of an as-cast aluminum alloy A319, Materials science and engineering A 453-453 (2007) 682-687.

[12] F.C. Porter, Zinc Handbook: Properties, processing, and use in design. Madison Avenue, New York, 1991.

[13] A. Dudek, B. Lisiecka, B, R. Ulewicz, The effect of alloying method on the structure and properties of sintered stainless steel, Archives of Metallurgy and Materials 62(1) 2017, pp.281-287.

Selection of Components of Upholstery Systems

RADZISZEWSKA-WOLIŃSKA Jolanta Maria

Instytut Kolejnictwa, 50, Chlopicki Street, 04-275 Warsaw, Poland

jradziszewska-wolinska@ikolej.pl

Keywords: Polyurethane Foam, Rextile Fireblocker, Fire Safety of Rolling Stock, Passenger Seat, Cone Calorimeter, Smoke Properties

Abstract. The article discusses methods and results of fire smoke tests of modified flexible polyurethane foams and upholstery systems carried out as part of the KOLEJMAT project at the Instytut Kolejnictwa (Railway Research Institute). The search for optimal components for ensuring fire safety of passenger rolling stock has been described. A different influence of fireproof nonwovens on the combustion process of graphite and non-graphite foam systems has been demonstrated. The results of tests of heat emission, smoke properties and toxicity are presented. The best two upholstery systems were selected.

Introduction

The development of rail vehicles designed to reduce their weight, increase operational speed and increase passenger comfort included also the construction of passenger seats. Wagon seats represent a significant mass of non-metallic materials and in the event of a possible fire - they determine the intensity and speed of fire propagation in the vehicle. These elements can therefore pose a direct threat to passengers [1,2]. Therefore, the development of passenger seats, in addition to taking into account the improvement of their ergonomics, also includes the search for non-combustible components. The above was also the aim of the KOLEJMAT project entitled: " Manufacturing Technology Basics of Passenger Seat with Use of Advanced Materials" [3], implemented in the years 2013-2016 as part of the NCBiR (National Center for Research and Development) Applied Research Program. The leader of the project consortium was the Faculty of Materials Science and Engineering at the Warsaw University of Technology, and the Partners: Transport Vehicles' Upholstery Manufacturer - TAPS - Maciej Kowalski in Łódź and Instytut Kolejnictwa (Railway Research Institute) in Warsaw. The project included the development of the following:

- modern constructions of forms, constituting the load-bearing base of the upholstery system (modification of plywood),
- armrest and table knot (selection of modified selected composite),
- upholstery system, including modifications of the PUR foam constituting the elastic element of the system.

European requirements in the field of fire properties of upholstery systems

Currently designed for European rolling stock, wagon seats and their component materials are subject to fire tests according to the EN 45545-2: 2013 requirements [4]. Table 1 presents the scope of the required tests for upholstery and requirements depending on the level of danger of HL vehicles, resulting from the combination of their design categories with the operating category according to EN 45545-2: 2013 [4].

As can be seen in Table 1, upholstery systems should meet the requirements for the following three parameters:

MARHE (Maximum Average Rate of Heat Emission) is the maximum average rate of heat release during 20 minutes of the test determined according to ISO 5660-1: 2016 [5] using a cone calorimeter. Its operation consists in using the principle of oxygen calorimetry, according to which, for most materials, each kilogram of oxygen consumed by the burning material causes the release of 13.1 [MJ] of heat (energy) with an accuracy of ± 5% [1, 6]. The use of a paramagnetic oxygen analyzer allows the exact determination of the amount of heat released and the rate of its release.

Table 1. R21 requirements for upholstery according to EN 45545-2:2013[4]

No.	Test method	Parameter	Requirement for:		
			HL1	HL2	HL3
1	ISO 5660-1: 25 [kWm^{-2}]	MARHE [kWm^{-2}]	≤ 75	≤ 50	≤ 50
2	EN ISO 5659-2: 25 [kWm^{-2}]	D_s max [dimensionless]	≤ 300	≤ 300	≤ 200
3	EN ISO 5659-2: [kWm^{-2}] + EN 45545-2 Annex C	CIT_G [dimensionless]	≤ 1,2	≤ 0,9	≤ 0,75

D_s. max (Maximum Specific Optical Density) - determined by the method according to PN-EN ISO 5659-2: [7] by the single-chamber test. The test consists in determining the optical density of the generated smoke (Ds) from the exposed surface of the test sample subjected to thermal radiation with the power of 25 [kW / m2] in the smoke chamber.

CIT_G (Conventional Index of Toxicity) determined in the fourth and eighth minute as *CIT(4)* and *CIT(8)* of the test, by FTIR spectroscopy according to PN EN ISO 5659-2 [7] taking into account EN 45545-2 [4] Annex C and ISO 19702 [8]. The applied method of infrared spectrophotometry allows for the qualitative and quantitative determination of the composition of fumes generated during the thermal decomposition of the test sample.

Laboratory tests of fire properties

2 types of PUR foams currently used at the Industrial Partner (PUR DB1 and PUR DB2) poured with use of an injection molding machine were the starting material for the selection of the optimal upholstery system. In addition, after the analysis of the bibliography regarding fire retardant polyurethane foams [9-12], three types of hand-poured foams containing expandable graphite as flame retardant [13] were prepared: PUR FRM S1, PUR FRM S2 and PUR FRM S3.

The study began with the determination of the MARHE parameter. Fig. 1 presents an example diagram of the heat release rate for one of the tested samples, showing the course of combustion in a cone calorimeter. However, the obtained results of MARHE in the form of a diagram were placed in Fig. 2. Tests made using a cone calorimeter have shown that the non-graphite foams (DB1, DB2, DB3) flamelessly melt during the measurement and emitted little heat (MARHE below 5 kW/m^2). On the contrary, graphite foams (FRM S1, FRM S2 and FRM S3) revealed MARHE at the level of (42-45) kW/m^2. The above means that too large additions of expandable graphite were used, which, as demonstrated by the parallel physical and mechanical tests, also deteriorated the static and fatigue mechanical properties of these foams.

However, in order to indicate the direction of further investigation, the effect of various fireblocker materials on all the above-mentioned polyurethane foams was evaluated. The following fireproof nonwovens were selected to the evaluation:

- BWF Protec type SFO from preoxided fibers with the weight of 150 g/m^2,
- DARA type ZWFR made of para-aramid and viscose fibers with the weight of 106 g/m^2,
- Duflot type WSM 132D from a mixture of polyacrylic, polyacrylonitrile and paraamide fibers with the weight of 250 g/m^2.

Materials Research Proceedings **5** (2018) 142-147 doi: http://dx.doi.org/10.21741/9781945291814-25

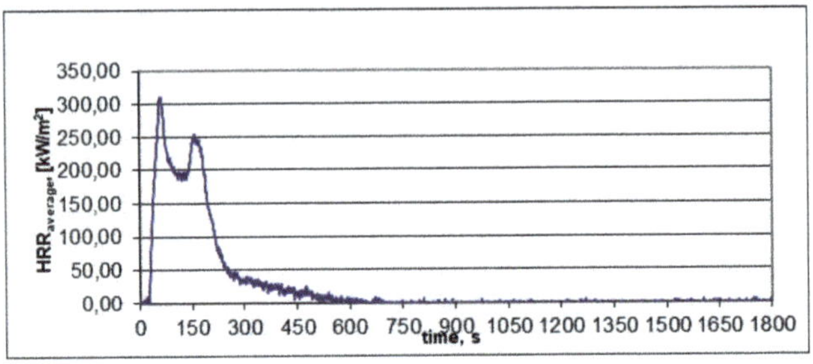

Fig.1 An example of the HRR (Heat Release Rate) chart for the FRM S1 foam.

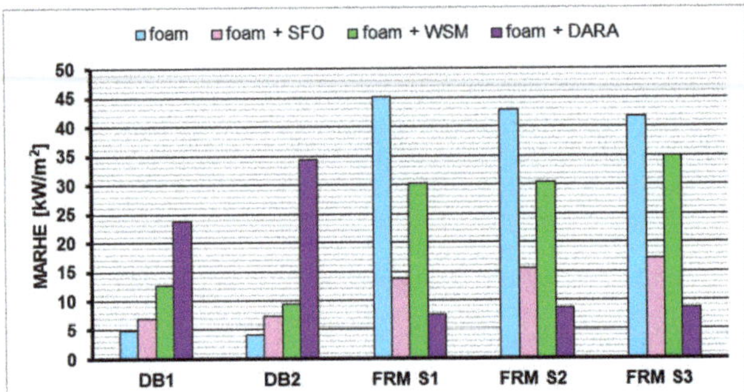

Fig.2 MARHE values for the PUR foams and foam sets with various fireproof nonwovens.

The use of nonwovens caused different behavior of the systems, depending on the type of foam applied in them. In the case of non-graphite foams, non-woven fibers from preoxidated fibers (SFO) and from a mixture of polyacrylic, polyacrylonitrile and paraamide fibers (WSM 132D) caused a slight increase in heat emission. However, the system protected with a nonwoven fabric made of paraaramid and viscose fibers (DARA-ZWFR) generated much more heat than the other systems. In connection with the above, for the examination of the full upholstery system, including the upholstery fabric, the DB1 foam was selected in a system with a SFO fire-resistant non-woven fabric made of preoxidized fibers.

However, systems with graphite foams behaved differently. In that case, MARHE values for all the systems were lower than for the foams alone. At the same time, the best results occurred for the paraaramid and viscose fiber spacers (DARA-ZWFR), due to which the MARHE value decreased about five times to the level of (7-9) kW/m^2. Non-woven fibers from preoxidated fibers (SFO) reduced the amount of heat released to the level (13-17) kW/m^2, and those made of a mixture of polyacrylic, polyacrylonitrile and paraamide fibers (WSM 132D) - to the level of (30-34) kW/m^2. On the basis of the above results, for testing of the full upholstery system of graphite foam together with upholstery fabric, a DARA-ZWFR fireproof nonwoven fabric made of para-aramid and viscose fibers was selected. Wherein, the Industrial Partner provided for testing a modified version of the foam, marked as the FRC.

Each of the two foam-nonwoven sets selected above was subjected to the tests of full upholstery systems in connection with the following upholstery fabrics:

- 100% trewira, Quadrat art. 14460, by Schoeppf, (Germany)

Materials Research Proceedings **5** (2018) 142-147
doi: http://dx.doi.org/10.21741/9781945291814-25

- 80% wool and 20% polyamide, manufactured by Runotex (Poland),
- 30% wool and 70% polyester, produced by Kneitz (Austria).

The results obtained are shown in Figures 3-5, where the following determinations of the systems studied were used:

1 - DB – SFO – Kneitz,
2 - DB - SFO – Schoeppf,
3 - DB – SFO – Runotex,
4 - FRC - DARA – Kneitz,
5 - FRC - DARA – Schoeppf,
6 - FRC - DARA – Runotex.

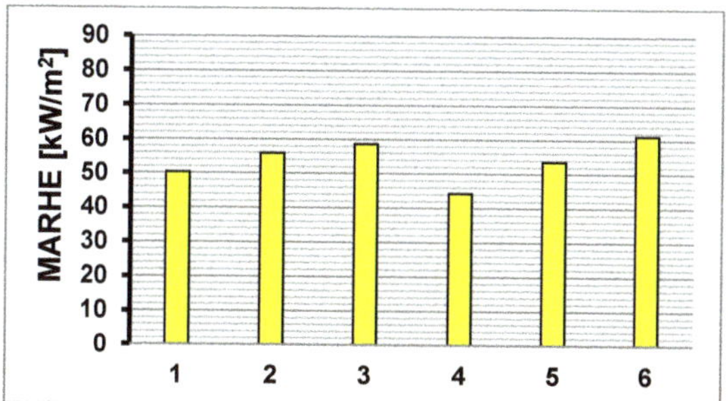

Fig.3 MARHE values for upholstery systems (foam - fireproof nonwoven - upholstery fabric).

As can be seen from the chart (Fig. 3), fulfillment of MARHE parameter requirements for hazard category HL3 (MARHE < 50 kW/m^2) was obtained only for upholstery consisting of the FRC foam, DARA / ZWFR 107 nonwoven (mixture of paraaramid and viscose fibers) and Kneitz upholstery (30 % wool and 70% polyester). In that case, the value was 44.2 kW/m^2. However, the value at the limit of the requirements for HL3 for this parameter reached the following layout: DB-free foam, SFO non-woven fiber from preoxidized fibers and above - Kneitz upholstery fabric (30% wool and 70% polyester).

Further, for all the six variants of the upholstery system, tests of smoke properties and toxicity were carried out. The results are shown in diagrams (Fig. 4 and 5).

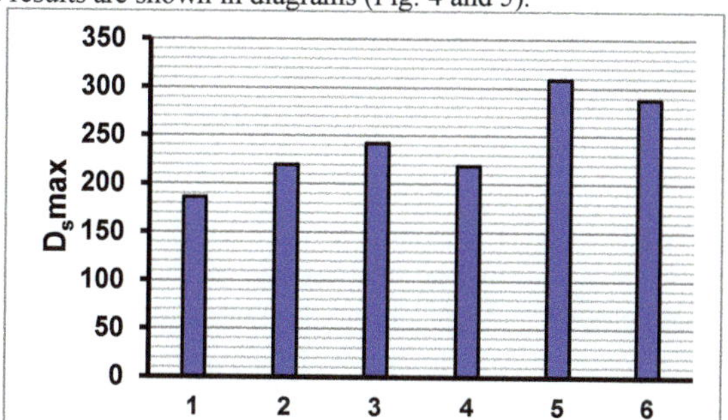

Fig.4 $D_s$max values for upholstery systems (foam - fireproof nonwoven - upholstery fabric).

Materials Research Proceedings 5 (2018) 142-147 doi: http://dx.doi.org/10.21741/9781945291814-25

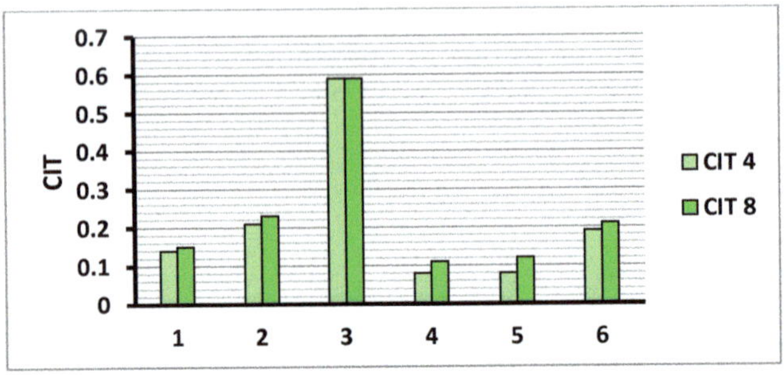

Fig.5 CIT_G (4) and CIT_G (8) for upholstery systems (foam - fireproof nonwoven - upholstery fabric).

As results from the analysis of all the obtained results for upholstery systems, the following two systems were considered the best to be subject to further modification:

- FRC graphite foam, DARA-ZWFR 107 nonwoven, upholstery 30% wool and 70% polyester, Kneitz production (values: MARHE - 44.2 kW/m^2, $D_s$max - 218.7, CIT (4) - 0.08 and CIT (8) - 0.11,
- DB foam, SFO nonwoven, upholstery fabric 30% wool and 70% polyester, Kneitz production (values: MARHE - 50.25 kW/m^2, $D_s$max - 185.9, CIT (4) - 0.14 and CIT (8) - 0.15.

However, because during a fire the greatest threat is created by smoke limiting visibility and hindering the conduct of rescue and extinguishing operations as well as often causing panic, the properties of the upholstery system with non-graphite foam should be considered as better.

Summary

1. The combustion tests using a conical calorimeter showed the following:
   - for standard polyurethane foams, the best MARHE results when protected with a pre-oxidised fiber, the values being higher than for the foam itself,
   - for polyurethane foams with expandable graphite, the best MARHE results when protected with a nonwoven made from a mixture of para-aramid and viscose fibers - the MARHE value was reduced five times in relation to the level of this parameter for the foam itself.
2. Studies on various variants of upholstery systems have shown that the best results have been achieved for the following two systems:
   - graphite foam FRC, fire-retardant material (mixture of paraaramid and viscose fibers), upholstery fabric (30% wool and 70%, polyester),
   - for DB foam, non-woven (preoxidized fiber, upholstery fabric (30% wool and 70% polyester).

References

[1] J.M. Radziszewska-Wolińska, D. Milczarek – KOLEJMAT - Podstawy technologii wytwarzania elementów fotela kolejowego z zastosowaniem zaawansowanych materiałów - cel projektu i wyniki badań wstępnych, Prace Naukowe, Transport, 101 (2013) 113-124 (in Polish),

[2] J.M. Radziszewska-Wolińska – The Influence of the Ignition Source on Passenger Seat Burning, Railway Reports, 60 (2016) 67-78

[3] J. M. Radziszewska-Wolińska, M. Jurczyk-Kowalska, S.Krauze - Influence of materials modification to meetthe requirements of EN 45545-2 for railway passenger seat, 11th annual

FIRE PROTECTION OF ROLLING STOCK 2015, 18-19 of March, 2015, Berlin, Germany (PowerPoint presentation),

[4] EN 45545-2:2013 Fire protection of railway vehicles – Part 2: Requirement for fire behaviors of materials and components.

[5] ISO 5660-1:2015 Reaction-to-fire tests — Heat release, smoke production and mass loss rate — Part 1: Heat release rate (cone calorimeter method) and smoke production rate (dynamic measurement)

[6] R. D. Peacock, R. W. Bukowski, S.H. Markos.: Evaluation of Passenger Train Car Materials in the Cone Calorimeter, Fire and Materials, 23 (1999) 53-62. https://doi.org/10.1002/(SICI)1099-1018(199903/04)23:2%3C53::AID-FAM669%3E3.0.CO;2-I

[7] PN-EN ISO 5659-2: 2013 Plastics - Smoke generation - Part 2: Determination of optical density by a single-chamber test.

[8] ISO 19702: 2015 Guidance for sampling and analysis of toxic geses and vapours in fire effluents using Fourier Transform Infrared (FTIR) spectroscopy.

[9] G. Matuschek : Thermal degradation of different fire retardant polyurethane foams, Thermochimica Acta, 263 (1995) 59 –71. https://doi.org/10.1016/0040-6031(94)02386-3

[10] A. Konig, U. Fehrenbacher, T. Hirth: Flexible Polyurethane Foam with the Flame-retardand Melamine, Journal of Cellular Plastics, 44 (2008) 469-480. https://doi.org/10.1177/0021955X08095502

[11] R. Bashirzadeh, A. Gharehbaghi, , An Investigation on Reactivity, Mechanical and Fire Properties of Pu Flexible Foam , Journal of Cellular Plastics, 46 (2009) 128-159.

[12] A. Konig, E. Kroke: Methyl-DOPO—a new flame retardant for flexible polyurethane foam, Polymers and Advanced Technologies, 22 (2011) 5–13. https://doi.org/10.1002/pat.1728

[13] P. Wei, S. Bocchini, G. Camino, Flame retardant and thermal behavior of polylactide /expandable graphite composites, Polimery, 5 (2013) 361. https://doi.org/10.14314/polimery.2013.361

The Surface Treatment of Sintered Stainless Steel

LISIECKA Barbara[1,a *], DUDEK Agata[2,b]

[1] Institute for Material Engineering, Faculty of Processing Engineering and Materials Technology, Czestochowa University of Technology, Armii Krajowej Street 19, 42–200 Czestochowa, Poland

[2] Institute for Material Engineering, Faculty of Processing Engineering and Materials Technology, Czestochowa University of Technology, Armii Krajowej Street 19, 42–200 Czestochowa, Poland

[a]lisiecka.barbara@wip.pcz.pl, [b]dudek.agata@wip.pcz.pl

Keywords: Surface Alloying, Gas Tungsten Arc Welding (GTAW), Sintered Duplex Stainless Steel (SDSS), Powder Metallurgy (PM)

Abstract. Sintered duplex stainless steels are characterized by a two–phase structure consisting of approximately equal content of ferrite and austenite, which makes them very interesting materials for numerous applications. This study proposes a method to improve functional properties by means of alloying the surface of the SDSSs by the gas tungsten arc welding (GTAW) process, which it is an efficient method of surface modification in engineering materials. The results of a SEM/EDX optical metallographic microscope, XRD analysis and hardness are presented. The main assumption of this study was to analyze the microstructure and hardness of a surface layer of SDSS with different proportions of individual powders subjected to the alloying process by means of the GTAW method.

Introduction

Sintered duplex stainless steels (SDSSs) manufactured using the powder metallurgy technology (PM) belong to the most dynamically growing class of engineering materials and they are relatively widely used in many sectors, mainly in highly industrialized countries. The automotive industry is the biggest consumer of sintered products (ca. 73% of all sinters). In the case of SDSS, it is possible to obtain a sintered structure with different proportions of two basic structural components, austenite and ferrite. Consequently, SDSSs are distinguished by very good mechanical strength as well as corrosion resistance, which results mainly from their chemical composition, contribution of individual phases of powders and distribution of the main alloying elements (Cr/Ni) between these phases [1–5].

Surface treatment technology is used to obtain a surface layer with appropriate properties [6–7]. Many times it happens that properties of the surface layer are independent of properties required from the native material (base material). Increasing development of materials, which should be adequately dependent on more and more new technologies of surface treatment is an extremely important challenge. Therefore, it is possible to obtain elements that are characterized by both good volume and surface properties [8].

The purpose of a surface alloying treatment and rapid solidification process is to induce favourable structural changes in the surface layer [9]. Appropriate selection of a treatment method and process parameters affects the achievement of structural changes of the surface, which will lead to the improvement of functional properties of SDSSs [10]. The interest in various welding techniques using concentrated sources of thermal energy has increased, which allows for modifying the surface layer not only of conventional corrosion–resistant steels, but also of sintered duplex stainless steels [11–13].

Laser technologies are the most widespread method used in surface alloying treatments. They are characterized by high precision, ease in automation, but for economic reasons (e.g. high cost of equipment), they belong to expensive processes. Furthermore, it is also imperative to maintain additional EU–OSHA (European Agency for Safety and Health at Work) requirements due to the presence of a characteristic laser radiation. The use of a gas tungsten arc welding (GTAW) is a promising proposal for hardening the surface layer of SDSSs, which they are showing porosity. The ability to modify the surface in terms of hardness and coefficient of friction is a valuable advantage which encourages to use this method. Low costs of equipment and ease of use are significant advantages of using the gas tungsten arc welding (GTAW) method [14–15].

The gas tungsten arc welding (GTAW) method consists in melting and joining the surface of a welded metal by heating it with an electric arc. The electric arc is established between two poles of the current: a non–consumable tungsten electrode (cathode) and a melted metal (anode). The shielding gas (argon) protects a non–consumable electrode and a weld pool and it affects arc voltage and the shape of a weld. Moreover, the current type and intensity, arc voltage and surface scanning rate significantly affect the final properties of the modified surface layer [16–17].

Materials and Methods

The specimens of sintered steels were made of water–atomized powders of 316 L steel and ferritic 409 L steel manufactured by Höganäs (Sweden). The chemical composition of steel powders was presented in Table 1. Both powders selected nominal particle size of 150 μm.

Table 1. Chemical composition of steel powders (% wt.)

Powder grade	Cr	Ni	Mo	Si	Mn	C	S	Fe
316 L	16.80	12.00	2.00	0.90	0.10	0.022	0.005	Balance
409 L	11.86	0.14	0.02	0.82	0.14	0.020	0.010	Balance

Powders mixed at different proportions of ferritic and austenitic steel powders were used in the investigations to give three different series of samples (see Table 2). The powders were compacted uniaxially with the addition of 1% Acrawax C lubricant at 720 MPa. The molded pieces were sintered at the temperature of 1250°C for 30 minutes and then cooled down with a cooling rate of 0.5°C/s. The whole process was carried out in a reducing atmosphere using hydrogen in order to significantly limit oxidation of the batch and protect the chromium content from reduction.

Table 2. Percentage contribution of individual powders

Powder grade	Series number		
	1	2	3
316 L	100%	80%	20%
409 L	0%	20%	80%

The surface layers alloying treatment of the SDSS coating was performed by means of the GTAW technology (gas tungsten arc welding). The electric arc established between a non–consumable electrode of tungsten and work–pieces to be joined was the source of the heat. The alloying treatment of the sintered steel was carried out with a constant surface scanning rate of 340 mm/min and changing welding current intensity, with its values ranging from 30 to 60 A. Argon was the shielding gas, with the flow set at ~ 14 l/min.

The analysis of the microstructure was performed using an Olympus GX41 optical microscope and a Jeol JSM–6610LV scanning microscope. The macroscopic evaluation of sintered duplex stainless steels after alloying process was carried out using an Olympus SZ61 stereo microscope. The Vickers methodology (with the load of 980.7 mN) was employed to measure the microhardness of SDSS obtained by Shimadzu HMV–G Series.

The identification of phase composition of the alloyed surface of SDSS was carried out using an X–ray diffractometer (Seifert 3003 T–T) with a cobalt lamp with a characteristic radiation wavelength of $\lambda_{coK\alpha} = 0.17902$ nm. Other parameters were the following: supply voltage: $U_r = 30 \div 40$ kV, current intensity $I_r = 30 \div 40$ mA, angle range: $2\theta = 10 \div 120°$, measurement step: 0.1°, pulse integration time $t_r = 10$ s.

Results and Discussion

Macroscopic examinations were used to evaluate how the alloying process affects the surface quality of sintered duplex stainless steel. The morphology of the surface after the alloying process for each SDSS with changing welding current intensity obtained by the GTAW method are presented in Figure 2.

Fig. 2. Morphology of the surface of SDSS after alloying 50 A:
a) 100% 316 L; b) 80% 316 L + 20% 409 L; c) 20% 409 L + 80% 316 L

The microstructures obtained for SDSS after the surface alloying treatment were observed using metallographic sections etched with aqua regia. Figure 3 illustrates the microstructure of the surface after alloying at 30 A obtained for 100% 316 L steel by the Olympus GX41 optical microscope. Microstructural examinations revealed a homogeneous cellular–dendritic structure in the surface layers after the alloying treatment. The occurrence of columnar crystals oriented according to the direction of heat transfer was caused by the fast heat transfer and high gradient of temperature. The transient zone obtained by the contact of the alloyed layer with base material was the location of nucleation and growth of primary structure crystals.

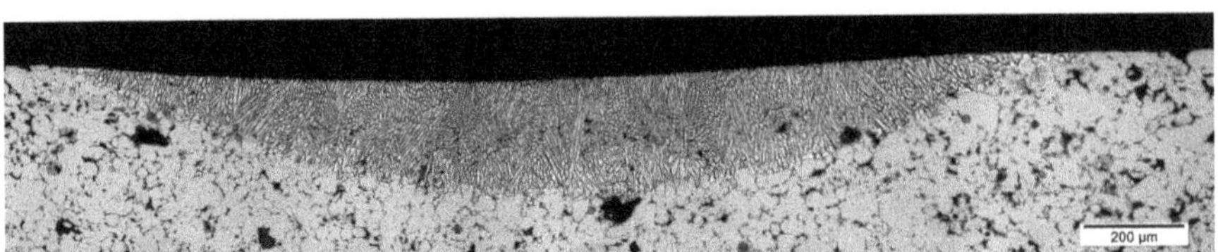

Fig. 3. Microstructure of the entire alloyed zone of 100% 316 L after alloying 30 A

Figure 4 are presents the microstructure of the surface after the alloying process for SDSS obtained by the Jeol JSM–6610LV scanning microscope.

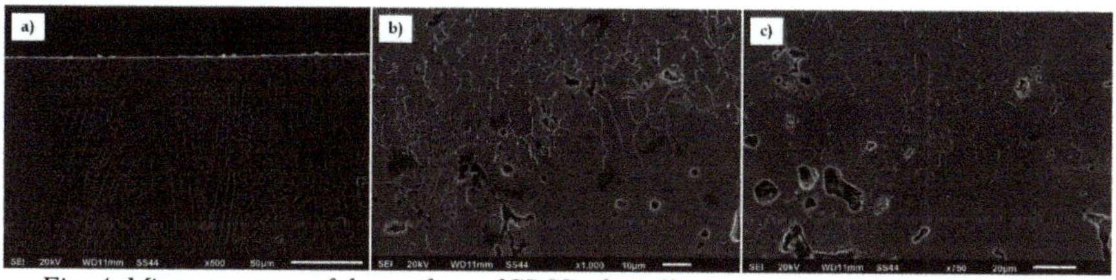

Fig. 4. Microstructure of the surface of SDSS after alloying at current intensity of 40 A: a) 100% 316 L; b) 80% 316 L + 20% 409 L; c) 20% 409 L + 80% 316 L

The analysis of chemical composition for each SDSS after the surface layer alloying treatment is presented in Table 3. The aim of the analysis was to compare areas obtained for SDSS after the surface alloying treatment, determination of the migration process of the alloying elements (Cr, Ni) during the crystallization process and determination of the degree of homogeneity of the chemical composition in the surface layers.

The examination of the surface layer alloying demonstrated migration of Cr and Ni during sintering caused by the diffusion process. The effect of diffusion was a creation of mutual diffusion areas between the ferritic and austenitic phases. The formation of the intermediate zone is a result of Ni diffusing to austenite, which leads to different phase transformations.

The remelting pathway for the alloying parameters was non–homogeneous in terms of chemical and phase compositions. Furthermore, the surface revealed oxide phases, likely to have been created at the stage of solidification of the alloyed mass. The oxide phases were confirmed by the investigation of the analysis of phase composition.

Table 3. EDX–analysis of chemical composition of the surface in sintered duplex stainless steels after alloying at current intensity of 40 A

SDSSs	Element	Weight %		
		Alloying zone	Heat affected zone	Native material
100% 316 L	Cr	17.23	16.26	14.57
	Fe	65.07	63.98	57.35
	Ni	12.09	13.13	10.63
80% 316 L + 20% 409 L	Cr	15.26	15.72	16.88
	Fe	68.73	70.07	66.96
	Ni	9.60	7.67	11.28
20% 409 L + 80% 316 L	Cr	15.07	14.88	14.43
	Fe	75.49	73.37	69.96
	Ni	6.11	7.35	7.10

Hardness measurements by the Vickers method were performed in order to evaluate mechanical properties. The results represent the mean of three measurements to the alloying zone, heat affected zone and native material obtained after alloying at current intensity of 40 A (see Table 4). The contribution of individual phases has a direct effect on mechanical properties of the whole sinter. The hardness results obtained in the study directly show improvement in strength properties after the surface alloying treatment.

X–ray phase analysis was conducted to identify the phase composition of the alloying surface layer. The results of the analysis are presented in Figure 6. The phase analysis for individual specimens showed the austenitic and ferritic phases with the content proportional to the powders used.

Table 4. Hardness of sintered stainless steel

SDSSs	Hardness [HV 0.1]		
	Alloying zone	Heat affected zone	Native material
100% 316 L	185.0	172.7	90.4
80% 316 L + 20% 409 L	196.3	173.0	107.0
20% 316 L + 20% 409 L	353.8	334.0	284.0

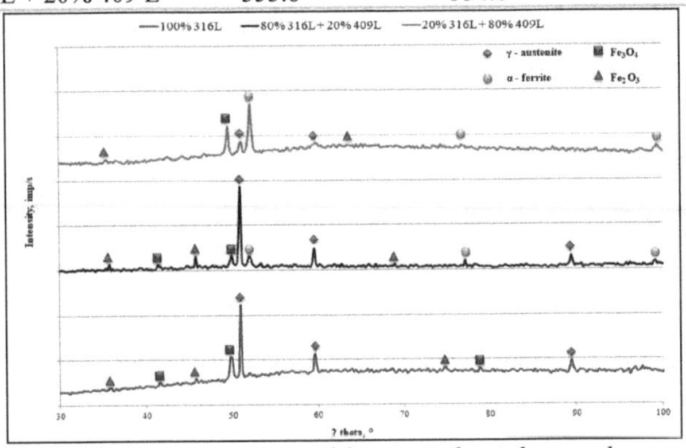

Fig. 7. Diffractograms of the sintered stainless steels

The phase composition analysis for individual specimens revealed the presence of the ferritic phase (cell parameters: a = b = c = 0.286 nm, α = β = γ = 90°) and the austenitic phase (cell parameters: a = b = c = 0.359 nm, α = β = γ = 90°), which crystallized in a cubic cell. Additionally, the analysis revealed the presence of iron oxides: Fe_3O_4 phase crystallizes in a cubic cell (cell parameters: a = b = c = 0.840 nm, α = β = γ = 90°) and Fe_2O_3 phase crystallizes in a cubic cell (cell parameters: a = b = c = 0.835 nm, α = β = γ = 90°). The peak intensities obtained on the diffractogram depend on the amount of individual phases in sintered steels.

Conclusions

The gas tungsten arc welding (GTAW) method proposed in the study represents an alternative solution compared to the laser technique. A modification of surface layers of SDSSs leads to an increase in mechanical properties of sintered steels and their wear resistance. The study shows that the increase in current intensity causes an increase in hardness, which leads to the increased wear resistance. Additionally, studies have shown that the wear resistance of SDSSs depends on the microstructure and hardness of the alloying surface layer. Consequently, this allows for a wider use of these materials, mainly in terms of their use in industry.

References

[1] J. Capus, Powder metallurgy, Progress and the eco-friendly car, Met. Powder Rep. 66 (2011) 16–18. https://doi.org/10.1016/S0026-0657(11)70058-3

[2] F. Martin, C. Garcia, Y. Blanco, M.L. Rodriguez-Mendez, Influence of sinter-cooling rate on the mechanical properties of powder metallurgy austenitic, ferritic, and duplex stainless steels sintered in vacuum, Mater. Sci. Eng., A 642 (2015) 360–365. https://doi.org/10.1016/j.msea.2015.06.097

[3] A. Dudek, A. Wrońska, L. Adamczyk, Surface remelting of 316L + 434L sintered steel: microstructure and corrosion resistance, J. Solid State Electrochem. 18 (2014) 2973–2981. https://doi.org/10.1007/s10008-014-2483-2

[4] T. Tański, Z. Brytan, K. Labisz, Fatigue behaviour of sintered duplex stainless steel, Procedia Eng. 74 (2014) 421–428. https://doi.org/10.1016/j.proeng.2014.06.293

[5] M.M. Asif, K.A. Shrikrishna, P. Sathiya, S. Goel, The impact of heat input on the strength, toughness, microhardness, microstructure and corrosion aspects of friction welded duplex stainless steel joints, J. Manuf. Process. 18 (2015) 92–106. https://doi.org/10.1016/j.jmapro.2015.01.004

[6] M. Gwoździk, Analysis of crystallite size changes in an oxide layer formed on steel used in the power industry, Acta Phys. Pol., A 130 (2016) 935–938. https://doi.org/10.12693/APhysPolA.130.935

[7] M. Gwoździk, Z. Nitkiewicz, Studies on the adhesion of oxide layer formed on X10CrMoVNb9-1 steel, ACME 14 (2014) 335–341. https://doi.org/10.1016/j.acme.2013.10.005

[8] N. Kurgan, Effects of sintering atmosphere on microstructure and mechanical property of sintered powder metallurgy 316L stainless steel, Mater. Des. 52 (2013) 995–998. https://doi.org/10.1016/j.matdes.2013.06.035

[9] N. Radek, J. Pietraszek, B. Antoszewski, The average friction coefficient of laser textured surfaces of silicon carbide identified by RSM methodology, Adv. Mater. Res. 874 (2014) 29–34. https://doi.org/10.4028/www.scientific.net/AMR.874.29

[10] Z. Zhang, H. Jing, L. Xu, Y. Han, L. Zhao, Investigation on microstructure evolution and properties of duplex stainless steel joint multi-pass welded by using different methods, Mater. Des. 109 (2016) 670–685. https://doi.org/10.1016/j.matdes.2016.07.110

[11] A. Wrońska, A. Dudek, Characteristics of surface layer of sintered stainless steels after remelting using GTAW method, ACME 14 (2014) 425–432. https://doi.org/10.1016/j.acme.2013.10.011

[12] E. De Las Heras, D.A. Egidi, P. Corengia, D. Gonzalez-Santamaria, A. Garcia-Luis, M. Brizuela, G.A. Lopez, M. Flores Martinez, Duplex surface treatment of an AISI 316L stainless steel; microstructure and tribological behaviour, Surf. Coat. Technol. 202 (2008) 2945–2954. https://doi.org/10.1016/j.surfcoat.2007.10.037

[13] S. Buytoz, M. Ulutan, In situ synthesis of SiC reinforced MMC surface on AISI 304 stainless steel by TIG surface alloying, Surf. Coat. Technol. 200 (2006) 3698–3704. https://doi.org/10.1016/j.surfcoat.2005.02.178

[14] A.N. Md Idriss, S. Mridha, T.N. Baker, Laser and GTAW torch processing of Fe–Cr–B coatings on steel Part I – melt features, Mater. Sci. Technol. 30 (2014) 1209–1213. https://doi.org/10.1179/1743284713Y.0000000435

[15] S. Kou, Welding Metallurgy, Wiley, New Jersey, 2002. https://doi.org/10.1002/0471434027

[16] P. Paulraj, R. Garg, Effect of welding parameters on pitting behaviour of GTAW of DSS and super DSS weldments, JESTECH 19 (2016) 1076–1083.

[17] E. Klar, P.K. Samal, Powder Metallurgy Stainless steels: Processing, Microstructures and Properties, Ohio ASM International, Materials Park, 2007.

Materials Research Proceedings 5 (2018) 154-159 doi: http://dx.doi.org/10.21741/9781945291814-27

Enhanced Surfaces Used in Research on Flow Boiling Heat Transfer in Minichannels

STRĄK Kinga[1, a *], PIASECKA Magdalena[2,b] and STRĄK Dariusz[3,c]

[1,2,3] Kielce University of Technology, Al. Tysiaclecia Panstwa Polskiego 7, 25-314 Kielce, Poland

[a]kzietala@tu.kielce.pl, [b]tmpmj@tu.kielce.pl, [c]dstrak@tu.kielce.pl

Keywords: Laser Texturing, Vibration-Assisted Laser Texturing, Electromachining Texturing, Porous Surface, Flow Boiling, Heat Transfer, Minichannel

Abstract. This paper reviews enhanced surfaces used to study boiling heat transfer in minichan-nels with rectangular cross-sections. Experiments were conducted for refrigerants flowing along a vertical minichannel, asymmetrically heated by one of the enhanced surfaces. The enhanced surfaces used in the experiments were the following: laser textured and vibration-assisted laser textured surfaces with regular cavities, electro-discharge machined surfaces with irregular cavities, fibrous surfaces with capillary-porous structures, and powder-coated surfaces. Heat transfer efficiency was evaluated based on local heat transfer coefficients calculated at the heated surface-fluid interface in the minichannel. Vibration-assisted laser surface texturing provided the most efficient heat transfer compared to the other enhanced surfaces and the smooth surface.

Introduction

The current trend towards miniaturization has accelerated the development of heat transfer solutions for energy supply and cooling of heat-generating machine components. Effective cooling methods employ heat transfer processes accompanied by a phase change. The heat transfer coefficients obtained during these processes are higher than those provided by forced convection heat transfer without a phase change. One of the methods of heat transfer enhancement involves heat transfer surface modification. The modification causes the increase of heat transfer coefficient values through increasing the number of active nucleation sites during flow boiling [1-10] or pool boiling [10-12]. Modified surfaces exhibit reduced surface superheating and improve heat dissipation by increasing the area of heat removal. Heat exchangers with enhanced surfaces find application over a wide range of industries from microelectronics (cooling of large scale integrated circuits, microprocessors, memory modules) to power engineering (gas micro turbines). The literature describes numerous processes that are used to modify metallic surfaces [13-16]. In their previous articles, the authors of this paper reported their findings concerning various enhanced heat transfer surfaces with the flow of various working fluids through rectangular heat exchangers with minichannels. The surfaces studied included laser-textured surfaces with regular recesses [2,7], vibration-assisted laser-textured surfaces [3,4,6], surfaces with irregular recesses obtained in the process of spark erosion [1,2,5,6,10,11], surface with capillary-fibrous structure [1] and powder surfaces produced by sintering or soldering [8,9]. Based on the research results for flow boiling in minichannels, this article discusses the enhanced surfaces used, along with their production methods and effects on heat transfer processes.

Characteristics of smooth surface (base)

All the enhanced heated plates were made on smooth plate (base) with a Haynes-230 alloy plate, about 0.10 mm and 0.45 mm thick or a Hastelloy X alloy plate, about 0.65 mm thick. All the tested alloy plates were manufactured by Haynes Int. Inc. (USA) and chosen for their electrical resistivity, assuming minor resistivity changes with temperature. The Hastelloy X and Haynes-230 alloy plates are made mainly of Ni-Cr-Fe-Mo and Ni-Cr-W-Mo alloys, respectively. In addition to excellent high-temperature strength and oxidation resistance [6]. Images of the smooth and enhanced tested surfaces were presented in Fig. 1.

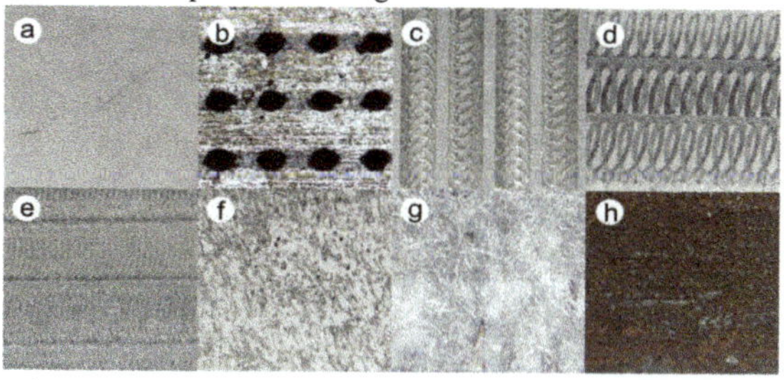

Fig. 1. Images of the tested surfaces: a) smooth, b) laser surface texturing, c-e) vibration-assisted laser texturing no: 1 (c), 2 (d), 3 (e), f) electromachining texturing, g) capillary-fibrous structure, h) porous structure produced by Fe powder soldering

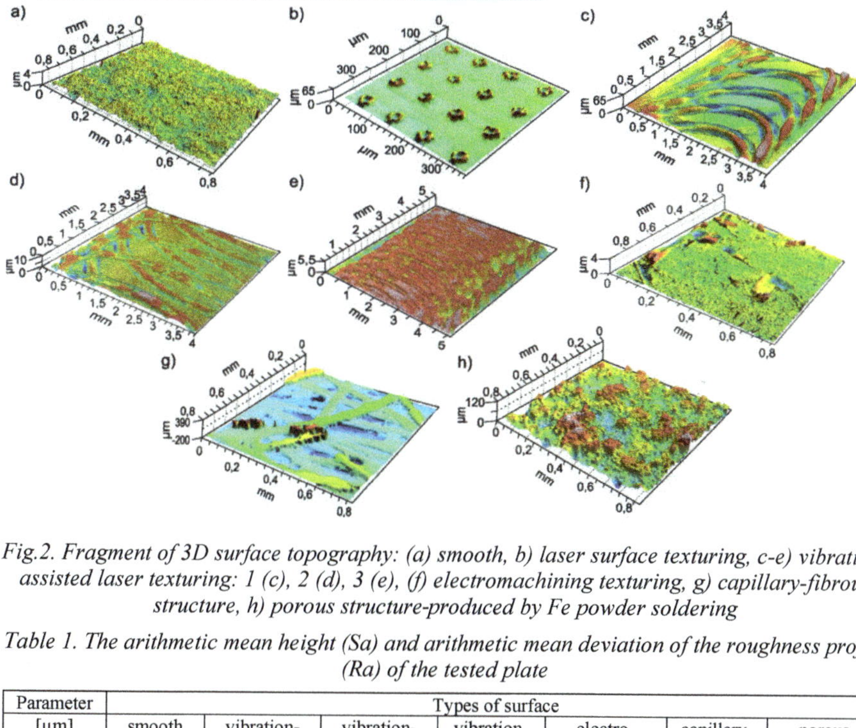

Fig.2. Fragment of 3D surface topography: (a) smooth, b) laser surface texturing, c-e) vibration-assisted laser texturing: 1 (c), 2 (d), 3 (e), (f) electromachining texturing, g) capillary-fibrous structure, h) porous structure-produced by Fe powder soldering

Table 1. The arithmetic mean height (Sa) and arithmetic mean deviation of the roughness profile (Ra) of the tested plate

Parameter [μm]	Types of surface						
	smooth	vibration-assisted laser 1	vibration-assisted laser 2	vibration-assisted laser 3	electro-machining	capillary-fibrous structure	porous, produced by soldering
Sa	0.346	7.934	1.981	3.589	0.983	41.044	13.919
Ra	0.175	3.812	1.364	1.025	0.615	21.407	11.283

The study of roughness parameters was conducted at the Geometric Computer Measurement Laboratory at the Kielce University of Technology. The profilometer measured using the Taylor

Hobson CCI to ISO 25178-2:2012 enables measurement and analysis of the geometric structure of the tested surface (Tab. 1).

Characteristics of surfaces with laser and vibration-assisted laser surface texturing

During the laser microprocessing, the material is removed by the laser beam's energy. An impulse beam time depends on the radiation intensity and exposure time. The cavities on the surface were performed with a laser drilling with a diode pumped Nd:YAG laser producing UV radiation of the wavelength of 355 nm and were evenly distributed every 100 μm in both axes. One of four tested enhanced surfaces [2] produced on the 0.10 mm thick Haynes plate had the best properties in terms of heat transfer, Fig. 2b. The diameter of a single cavity had 10 μm, its depth was 3 μm, 5-7 μm high layers of melted metal deposited annularly around the cavities, forming structures as "craters" [2, 7].

One Hastelloy X alloy (0.65 mm thick) surface and two Haynes-230 alloy (0.45 mm thick) surfaces processed using the vibration-assisted laser surface texturing were studied. The surfaces were prepared with the CO_2 Trumpf Lasercell 1005 owned by the Centre for Laser Technologies of Metals at the Kielce University of Technology. The Hastelloy X alloy surface, surface no 1, (Fig. 2c) had the laser paths aligned perpendicularly to the flow in the minichannel [6]. The texturing process followed the programmed trajectory of the laser head movement with the following processing parameters: laser power of 2500 W, scanning velocity of 4 m/min, and argon blow intensity of 10 l/min.

The two other vibration-assisted laser surfaces were made of the Haynes-230 alloy (0.45 mm thick) and had the laser paths arranged along the flow, with a lower laser path density in the case of surface no 2 (Fig. 2d) and a higher laser path density in the case of surface no 3 (Fig. 2e). The processing parameters for these surfaces were as follows: laser power of 1500 W or 1250 W, scanning velocity of 4 or 2 m/min, and argon blow intensity of 10 l/min [3, 4].

Characteristics of the surface with electromachining texturing

During the spark erosion process in electromachining, the eroded material is transferred from the anode to the cathode. As a result, the cathode surface is covered with pure anode material or a layer created by the mutual impacts between electrodes and the medium between electrodes. The intensity of the transfer of the material depends mainly on impulse energy [2, 9]. Sample cavities in the 0.10 mm thick Haynes-230 alloy surface section were made by electromachining, using an electric-etcher and branding-pen manually controlled and oriented in selected directions. Finally, one method was selected to produce cavities (Fig. 2f). The depth of the cavity craters was usually below 2 μm. The layer of melted metal of the plate and the electrode material, a few μm high, reaching locally 5 μm, accumulated around the cavities.

Characteristics of the surface with the capillary-fibrous structure

The surface with the capillary-fibrous structure was soldered to the 0.1 mm thick Haynes-230 plate surface, manufactured using stainless steel fibers 0.1 mm in thickness and about 2 cm in height. The length of the capillary structure, see figure 2.b (III), was usually up to 2 cm. They were spot welded with a microwelder and rolled to a thickness of 0.15 – 0.2 mm (Fig. 2g) [1].

Characteristics of the porous surface produced by soldering and sintering iron powder

The porous surface was produced by soldering to the tested surface (made of the Haynes-230) according to the method described in the Polish patent [17]. The main parameters of the microstructured surface are the following: the diameter of a solder granule - approx. 15÷80 μm (with an average being 40÷65 μm), the density of the solder paste – 2990 kg/m^3, the thickness of the soldered layer – approx. 100 μm and its maximum height – 200 μm (Fig. 2h). One of the porous surface types was produced with metal foams formed by the reduction of metal oxides during sintering. The mixture was sintered in a dissociated ammonia atmosphere. The iron foam was prepared according to the method described in the Polish patent [18]. The porous layers were formed by the sintering technique, the powder mixture and the reduction of iron powder ASC 100.29, ASC 100.24, DISTALOY SE and iron oxide of different granulation. The powder mixture was sintered in a dissociated ammonia atmosphere at 1180 °C for 45 minutes.

Experimental stand and the heat transfer coefficient calculation

The experimental stand (Fig. 3) is composed of several systems: the test loop in which Fluorinert FC-72 circulates as the working fluid, the calibration loop, the data and image acquisition system, the supply and control system and the lighting system, described in [4]. Local heat transfer coefficients were determined using the one-dimensional method, presented in [4, 5].

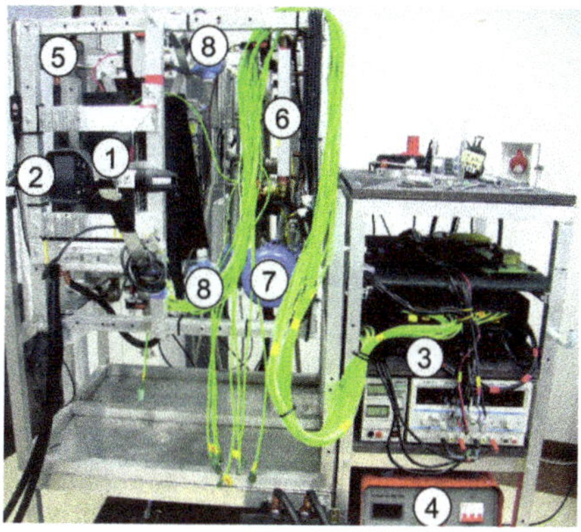

Fig. 3. The image of the main systems of the experimental setup, with visible elements marked: 1 - test section, 2 - infrared camera, 3 - data acquisition station, 4 - inverter welder, 5 - shunt, 6 - mass flowmeter, 7 - gear pump, 8 - pressure sensor

Results

The results confirm that considerable heat transfer enhancement takes place when selected enhanced heated surface is used in the minichannel flow boiling and that it depends on the type of surface enhancement. The analysis of the experimental data revealed that the values and distributions of the heat transfer coefficient (α) versus the distance from the minichannel inlet (x). The experiments were conducted using the test module with Fluorinert FC-72 flow along the minichannel 1.7 mm deep, 16 mm wide, and 180 mm long, vertically oriented [6].

Figure 4 shows examples of the results represented as the heat transfer coefficient vs. the distance from the minichannel inlet obtained for the surface with the vibration assisted laser surface texturing no 2 and the surface with the capillary-fibrous structure.

The results for all the surfaces used in the experiments indicated that:

- the surfaces with the vibration assisted laser surface texturing no 2 provided the highest values of the heat transfer coefficient, which could be due to the laser path alignment along the fluid flow in the minichannel and lower density of the laser path leading to more intense formation of nucleation sites,
- the surface with the capillary-fibrous structure showed the lowest values of the heat transfer coefficient, probably due to the instability of the boiling fluid flow and its turbulization.

Materials Research Forum LLC
doi: http://dx.doi.org/10.21741/9781945291814-27

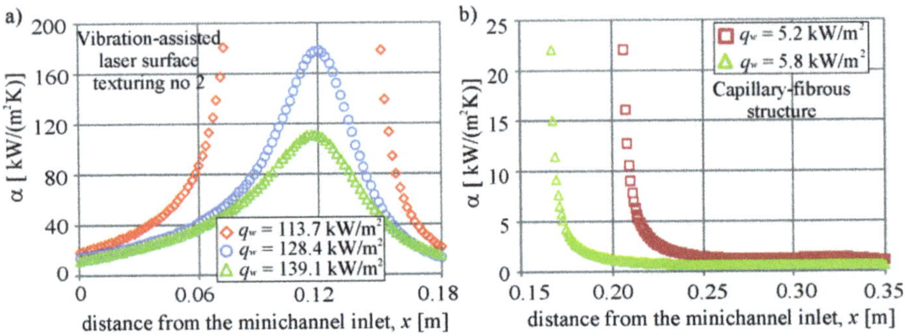

Fig. 4. Heat transfer coefficient vs. the distance from the minichannel inlet: a) surface no 2 with vibration assisted laser surface texturing, data for the saturated boiling region, experimental parameters: average inlet pressure p_{in} = 161 kPa, average mass flux G = 440 kg/(m²s), average inlet liquid subcooling $\Delta Tsub$ = 45 K, b) surface with capillary-fibrous structure; experimental parameters: average inlet pressure p_{in} = 118 kPa, average mass flux G = 161 kg/(m²s), average inlet liquid subcooling $\Delta Tsub$ = 42 K

Conclusions

This paper reports the test results for FC-72 flow boiling heat transfer in a vertical, asymmetrically heated minichannel with different surface modifications.

The heat transfer efficiency was evaluated on the basis of heat transfer coefficient values at the interface between the enhanced heating surface and the cooling liquid flowing through the minichannel. The laser-vibration-assisted texturing surface no 2 provided the most effective heat transfer. Compared to the other analyzed surfaces, the lowest coefficients were provided by the surface with the capillary-fibrous structure.

It can be concluded that the application of the enhanced surfaces allows achieving effective heat transfer. The enhanced surfaces can be successfully applied to compact heat exchangers with minichannels.

Acknowledgments

The research reported herein was supported by a grant from the National Science Centre, Poland (No DEC-2016/23/N/ST8/01247).

References

[1] M. Piasecka, K. Strąk, Effect of the heating surface enhancement on the heat transfer coefficient for a vertical minichannel, EPJ Web of Conferences 114, 02095 (2016). https://doi.org/10.1051/epjconf/201611402095

[2] M. Piasecka, Laser texturing, spark erosion and sanding of the surfaces and their practical applications in heat exchange devices, Adv. Mater. Res. 874 (2014) 95–100. https://doi.org/10.4028/www.scientific.net/AMR.874.95

[3] K. Strąk, M. Piasecka, B. Maciejewska, B. Grabas, A study of the flow boiling heat transfer in a minichannel for a heated wall with surface texture produced by vibration-assisted laser

machining, J. Phys. Conf. Ser. 745, 32123 (2016). https://doi.org/10.1088/1742-6596/745/3/032123

[4] K. Strąk, M. Piasecka, B. Maciejewska, Spatial orientation as a factor in flow boiling heat transfer of cooling liquids in enhanced surface minichannels, Int. J. Heat Mass Transf. 117 (2018) 375–387. https://doi.org/10.1016/j.ijheatmasstransfer.2017.10.019

[5] M. Piasecka, K. Strąk, B. Maciejewska, Calculations of flow boiling heat transfer in a minichannel based on liquid crystal and infrared thermography data, Heat Transf. Eng. 38(3) (2017) 332–346. https://doi.org/10.1080/01457632.2016.1189272

[6] M. Piasecka, K. Strąk, B. Grabas, Vibration-assisted laser surface texturing and electromachining for the intensification of boiling heat transfer in a minichannel, Arch. Metall. Mater. 62(4) (2017) 1983–1990. https://doi.org/10.1515/amm-2017-0296

[7] M. Piasecka, Flow boiling heat transfer in a minichannel with enhanced heating surface, Heat Transf. Eng. 35 (10) (2014) 903–912. https://doi.org/10.1080/01457632.2014.859505

[8] B. Maciejewska, K. Strąk, M. Piasecka, The solution of a two-dimensional inverse heat transfer problem using the Trefftz method, Procedia Eng. 157 (2016) 82–88. https://doi.org/10.1016/j.proeng.2016.08.341

[9] W. Depczyński, A. Piasecki, M. Piasecka, K. Strąk, Impact of Fe powder sintering and soldering in production of porous heating surface on flow boiling heat transfer in minichannels, E3S Web of Conferences 19, 03012 (2017), 1-6.

[10] M. Piasecka, Heat transfer research on enhanced heating surfaces in flow boiling in a minichannel and pool boiling, Ann. Nucl. Energy. 73(2014) 282-293. https://doi.org/10.1016/j.anucene.2014.06.041

[11] R. Pastuszko, M. Piasecka, Pool boiling on surfaces with mini-fins and micro-cavities, J. Phys. Conf. Ser. 395, 012137 (2012). https://doi.org/10.1088/1742-6596/395/1/012137

[12] R. Kaniowski, R. Pastuszko, Ł. Nowakowski, Effect of geometrical parameters of open microchannel surfaces on pool boiling heat transfer, EPJ Web of Conferences 143, 02049 (2017). https://doi.org/10.1051/epjconf/201714302049

[13] W. Depczyński, Kazala R., Ludwinek K., Jedynak K., Modelling and microstructural characterization of sintered metallic porous materials, Materials 9 (7), 567 (2016). https://doi.org/10.3390/ma9070567

[14] N. Radek, J. Konstanty, Cermet ESD coatings modified by laser treatment, Arch. Metall. Mater. 57 (2012) 665-670. https://doi.org/10.2478/v10172-012-0071-y

[15] N. Radek, J. Pietraszek, B. Antoszewski, The average friction coefficient of laser textured surfaces of silicon carbide identified by RSM methodology, Adv. Mat. Res. 874 (2014) 29-34. https://doi.org/10.4028/www.scientific.net/AMR.874.29

[16] J. Borowiecka-Jamrozek, J. Lachowski, Modelling of retention of a diamond particle in matrices based on Fe and Cu, Procedia Eng. 177 (2017) 289-296. https://doi.org/10.1016/j.proeng.2017.02.227

[17] M. Piasecka, PL. Patent 225512 B1, dec. (2016).

[18] W. Żórawski. R. Chatys, W. Depczyński, PL. Patent 199720 B1, dec. 2008.

Terotechnology 2017 Materials Research Forum LLC
Materials Research Proceedings 5 (2018) 160-165 doi: http://dx.doi.org/10.21741/9781945291814-28

FEM Thermal Analysis of Thermal Weldings of Tramway Rails, 60R2 and 60E1 Type

KOWALCZYK Dariusz

[1]Instytut Kolejnictwa, Kolejnictwa Materials and Structure Laboratory, Chlopickiego Street 50, 04-275 Warsaw, Poland

dkowalczyk@ikolej.pl

Keywords: FEM Thermal Analysis, Cracking Tram Rail of Thermite Welding, 60E1 Rail, 60R2 Rail, Tram Rail

Abstract. In many newly built tram tracks by means of the thermite method, after putting them into operation, cracks appear in tram rails in the areas of splice rails [1]. In the case of solutions used in the tram infrastructure, there are requirements for railways in accordance with PN-EN 14811 + A1: 2010 [2]. For rail connections intended for the construction of tramway tracks, there are no precise records of such requirements, and their scope may be determined by the Contractor. The article presents and discusses examples of broken rail joints in tramway tracks. Thermal analyses were carried out using the FEM method of cooling welds / thermite connections of the 60R2 and 60E1 tram rails [3-5]. The microstructures of rails and welded joints in the area of the heat affected zone were observed. In the light of the quite common problems related to the cracking of rails and rail joints in tramway tracks, it seems reasonable to introduce durability and quality tests of rail joint connections in order to improve safety in rail transport, reduce cracks in newly built and renovated tracks, and reduce costs related to repairs during operation. In this article, thermal analysis of cooling of thermal welds based on the FEM method was performed.

Introduction

There are various techniques for the construction of railways. The classic track, i.e. in which the rails are connected by splinters, requires careful maintenance due to the dynamic impact of the rolling stock wheels on the rail contacts. The first tram tracks, e.g. in Warsaw, were built and put into service as early as 1866, and electrified in 1908. Until 1899, as many as 20 cities in the area of the present Polish territory had tramlines [6]. The development of new solutions used in the construction of railways has led to the possibility of using the so-called contactless track [7], by introducing rails connected in a permanent manner as a result of using, for example, thermite welds. This solution has made it possible to introduce higher travel speeds, increase safety and increase travel comfort. Modern solutions used in the construction of high-speed rails and modern tram tracks include properly designed tracks, arcs with appropriate radii, cant selection, selection of appropriate rails, application of other infrastructure elements such as modern fastening systems, etc.

The most commonly used method of joining rails in tramway tracks is the method of thermite welding. Thermite or ferromite is a mixture of fine iron and aluminum oxide fractions with alloy additions. After igniting this mixture, with the so-called immediate enthusiasm, giving a focused source of heat, a highly exothermic chemical reaction that generates temperature of over 2500[°C] is initiated. The SoWoS method is the basic method of thermite welding of rails in tramway tracks, i.e. welding without a riser with upper preheating of the ends of the rails. The technology of welding rails using the thermite SoWoS method consists of the following processes: preparation and setting of the contact of rails to be welded, inserting and sealing the mold, filling and setting the crucible, heating the ends of the rails, welding, cooling the joint, removing the mold and finishing [7].

Rail connections - rail connectors and areas around them - are usually places with an increased risk of cracks, defects, damage etc. This is due to the thermal impact on the rails in which a heat affected zone is created that introduces changes to the material microstructure. In the construction of new tram tracks or modernized tramways, the 60R2 grooved rails are being increasingly used. The advantage of this type of rails is the ability to make a track surface in the system of integrated track and road surface, track in a continuous watering, green track with elastic fastening, green track on a concrete slab with flexible point fixing of rails. Fig. 1 shows examples of solutions using the 60R2 rail.

Fig. 1 Green track and pavement integrated using the 60R2 type rail.

One of the advantages of the 60R2 type rail, made in a grooved and shaped system, is the possibility of building a modern surface of tramway tracks shared with the roadway, i.e. track-road surfaces, as well as the so-called green track, which is more and more frequently used in the construction. The 60R2 rail profile, different from the classical one (60E1, 49E1), has been subjected to the FEM calculations in the field of thermal analysis during the cooling process of rail thermite weld.

Research

Observations were made on the naked eye of broken rail joints made by the 60R2 thermite welding. Samples were cut from the parts of the joint and metallographic specimens were prepared from them. Observations were made on a VHX-5000 Series Digital Microscope - a digital optical microscope equipped with the Z20 and Z2000 lenses for magnification from 50 to 1200x. Sample photos of microstructures are shown in Fig. 3.

Fig. 2 Cracked 60R2 rail in the area of weld occurrence.

Materials Research Proceedings **5** (2018) 160-165 doi: http://dx.doi.org/10.21741/9781945291814-28

Fig. 3 HAZ microstructure in the area of a) overheating and b) normalization, c) ferrite separation at the grain boundaries.

As evidenced by observations of rail joint microstructures, they are characterized by a large difference in grain size. This indicates a large difference in the temperature gradient during the cooling process. The technological process of making thermite welds should be controlled in such a way as to limit the formation of unfavorable areas of microstructure occurring in the heat-affected zone, such as grain growth, austenite separation at grain boundaries and thermal stresses. The microstructure affects the mechanical properties of the material [7]. This indicates that locally in the areas of microstructure change there are various strength properties of the material (the weld area, 60R2 and 60E1 rails and heat affected zone).

Thermal analysis of cooling thermite welds by the FEM method

Models of the 60R2 and 60E1 rails in accordance with EN 13674-1: 2011 were made in the SOLIDWORKS 2018 program, boundary conditions and FEM calculations were performed in the ALTAIR HYPERWORKS 2017 program [9]. The purpose of the FEM calculations was thermal analysis of the system. Comparative analysis of the distribution of temperature fields during cooling of rail joints in the same variable thermodynamic conditions. The temperature of the thermite joint in the mold and in the area of the rails subjected to the action of the hot medium (liquid weld alloy) can range from over 1500 [°C] to 2500 [°C]. The FEM calculations included analysis of results, heat flow, temperature gradient, and absolute values of temperature.
The FEM calculation results for the rail cooling process (Fig. 4) after 60 seconds.

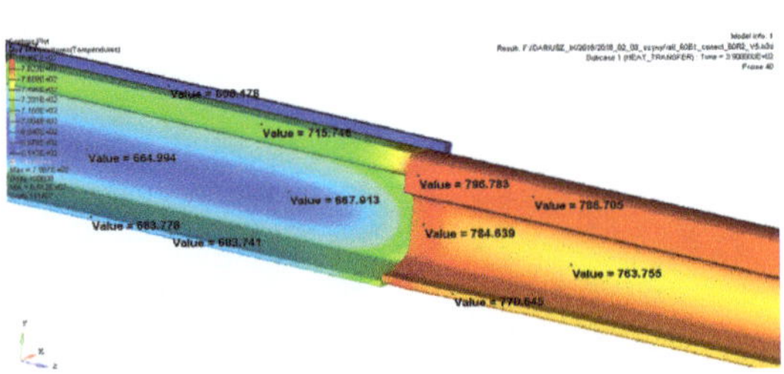

Fig. 4. Temperature distribution on the joint surface after 60 seconds.

Fig. 5. Example of fracture of a rail joint (a thermite weld) of the 60R2 rail.

The temperature distribution on the rail joint surface after 380 seconds of the cooling process using a temperature filter (in this case set at 723 [°C]).

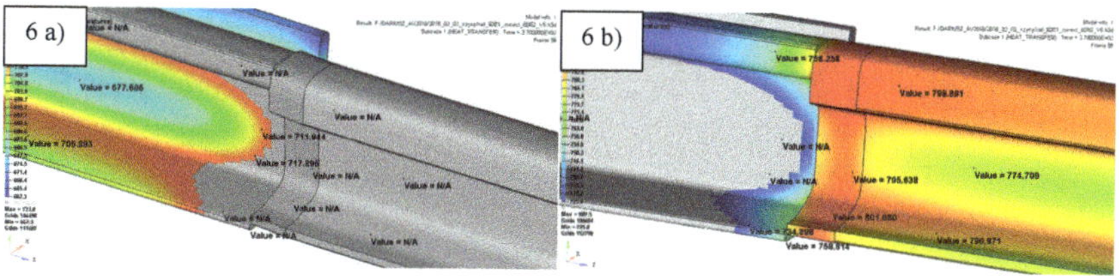

Fig. 6. Temperature distribution on the surface rail joint (a thermite weld)) after 380 seconds. a) a filter was applied showing areas with a temperature below 723 [°C], b) a filter showing areas with temperature above 723 [°C] was used.

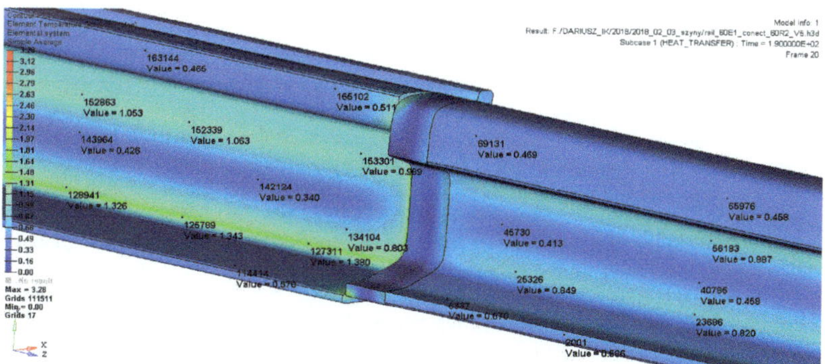

Fig. 7. Temperature gradient in the rail during the cooling process.

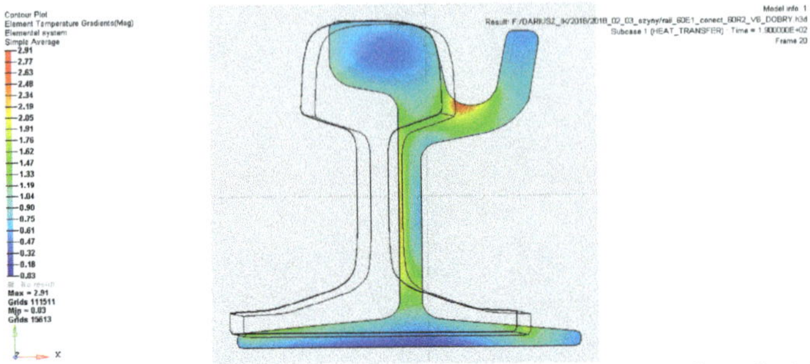

Fig. 8. Temperature gradient in the 60R2 rail, cross section 10 mm away from the joint (thermite weld).

Table 1 presents the temperature [°C] values of the selected nodes of the rail.
Table 2 presents the temperature [°F] values of the selected nodes of the rail.

Table 1. Temperature values for individual nodes in the rail [°C]

Time [Frame, second]	60R2 head rail	60R2 web rail	60E1 head rail	60E1 web rail	Temperature difference in the rail 60R2	Temperature difference in the rail 60E1	Temperature difference in the rail 60R2/60E1
1 - 10 s	1000 [°C]	1000 [°C]	1000 [°C]	1000 [°C]	0 [°C]	0 [°C]	0 [°C]
6 - 60 s	958 [°C]	929 [°C]	967 [°C]	951 [°C]	29 [°C]	16 [°C]	38 [°C]
40 – 400 s	715[°C]	664 [°C]	783 [°C]	763 [°C]	51 [°C]	20 [°C]	119 [°C]
100 - 1000 s	417 [°C]	381 [°C]	540 [°C]	522 [°C]	36 [°C]	18 [°C]	123 [°C]

Table 2. Temperature values for individual nodes in the rail [°F]

Time [Frame, second]	60R2 head rail	60R2 web rail	60E1 head rail	60E1 web rail	Temperature difference in the rail 60R2	Temperature difference in the rail 60E1	Temperature difference in the rail 60R2/60E1
1 - 10 s	1832 [°F]	1832 [°F]	1832 [°F]	1832 [°F]	0 [°F]	0 [°F]	0 [°F]
6 - 60 s	1756 [°F]	1704 [°F]	1772 [°F]	1743 [°F]	84 [°F]	60 [°F]	100 [°F]
40 – 400 s	1319 [°F]	1227 [°F]	1441 [°F]	1405 [°F]	123 [°F]	68 [°F]	246 [°F]
100 -1000 s	782 [°F]	717 [°F]	1004 [°F]	971 [°F]	96 [°F]	64 [°F]	253 [°F]

Conclusions of the FEM analysis

The thermal analysis of FEM showed that the 60R2 rail is cooled faster than the 60E1 rail under the same thermodynamic conditions. In the 60R2 rail itself, there are larger temperature differences between the rail head and the web than in the 60E1 rail. From the sample measurements during the cooling process, the temperature differences for the 60R2 rail are over 50[°C], for the 60E1 rail the difference is 20[°C]. In the 60R2 / 60E1 rail joint, temperature differences can be as high as 125[°C]. In Fig. 6a, a temperature filter for the entire joint was used to visualize temperature differences in the changes occurring in the material microstructure during the cooling process - eutectoidal transformation [7]. In the 370 - second cooling period (Fig. 6b), the entire area of the web and the 60R2 rail foot (gray) is a part of the joint in which the course of eutectoid transformation has already been completed (the conversion of austenite into perlite + cementite Fe3C (III)). In the remaining area of the joint - a microstructure (austenite + ferrite) occurs in accordance with the phase-iron balance diagram of iron-cementite. A higher temperature gradient in the 60R2 rail during the cooling process indicates that these joints are more demanding in the field of their manufacturing technology, and therefore more it is difficult to obtain a homogeneous material microstructure. Fig.5. shows a broken 60R2 rail joint, the shape of the crack run is in correlation with the ISO runlines (designated in Fig.6) defining the eutectoid transformation occurring at 723[°C]. Fig. 7 and Fig. 8 show the temperature gradients in the joint during the cooling process (heat release rate). The highest temperature drops occur in the area between the foot and the web of the 60R2 rail. The inhomogeneity of the temperature gradient over the cross-section of the 60R2 rail (in the area of the joint / thermite weld) results from the offset of the 60E1 rail profile and the maintenance of the proper rail running plane, thus the 60E2 rail is not symmetrically aligned with the 60R2 rail. This also causes the asymmetry of the cooling of the joint in this area.

Summary

Observations of the microstructure of the 60E1 and 60R2 rail areas in the area of the heat affected zone of the thermite weld indicate the existence of a structure with a large grain size differentiation. The FEM thermal analyses have shown that in the 60R2 rail there are greater temperature differences than in the 60E1 rail under the same thermodynamic cooling conditions. There are large differences in temperatures between the 60R2 rail head and its foot.

This indicates that the 60R2 rail, in addition to its advantages related to the possibility of building modern integrated tramway tracks, is a more difficult rail in the manufacture of thermite welds. The technology of making the right thermite weld joint is a process that determines its durability, reliability and safety in rail transport.

References

[1] I.Mikłaszewicz, Wykonywanie i badanie kolejowych złączy szynowych, Instytut Kolejnictwa, Warszawa, Problemy Kolejnictwa, 158 (2013) 35 (in Polish).

[2] PN-EN 14811+A1:2010; Railway applications. Track. Special purpose rail. Grooved and associated construction

[3] Pramod Bhatt, Thermal Analysis of Hollow pipe using Altair Hyperworks,International Jurnal of Engineering Research & Technology, 6 (2012).

[4] Practical Aspects of Finite Element Simulation A Study Guide - Academic Program -3rd Edition Released 05 (2015)

[5] D. Kowalczyk, R. Bińkowski, Zastosowanie obliczeń FEM do określania rozkładu pól naprężeń w szynie typu 49E1 w warunkach eksploatacyjnych wyjątkowych. Infraszyn, Projektowanie budowa i utrzymanie infrastruktury w transporcie szynowym, (2015) 157-168 (in Polish).

[6] Information on tram in Poland; https://pl.wikipedia.org/wiki/Tramwaje_w_Polsce,

[7] Kądziołka A, „Termitowe spawanie - rys historyczny, zmiany, postęp", IV Ogólnopolska Konferencja Naukowo-Techniczna, Spawalnictwo dróg szynowych – jakość, niezawodność, bezpieczeństwo, Bochnia (2010) (in Polish).

[8] Dobrzański L., A., Podstawy nauki o materiałach i metaloznawstwo. Materiały inżynierskie z podstawami projektowania materiałowego, WN-T, (2002) (in Polish).

[9] Altair Engineering, Inc. OptiStruct 13.0 User's Guide.

Materials Research Proceedings 5 (2018) 166-171 doi: http://dx.doi.org/10.21741/9781945291814-29

Characteristics of Multimaterial Joints of Nickel-Based Superalloys

STRZELCZAK Katarzyna[1, a *], DUDEK Agata[2, b]

[1]Czestochowa University of Technology, Institute for Material Engineering, Faculty of Production Engineering and Materials Technology, Czestochowa, Poland, EU

[2] Czestochowa University of Technology, Institute for Material Engineering, Faculty of Production Engineering and Materials Technology, Czestochowa, Poland, EU

[a]strzelczak.katarzyna@wip.pcz.pl, [b]dudek@wip.pcz.pl

Keywords: Brazing, Welding, Inconel 625, Inconel 718, Aerospace Industry

Abstract. In the process of designing advanced constructions, it is extremely important to choose not only materials but also methods of joining them. The brazing technique allows for obtaining high-quality combinations of materials radically different in chemical composition and properties. In turn, the welding method guarantees obtaining a weld whose composition is a resultant combination of materials and is characterized by a low degree of deformation. In this study, a microstructure SEM+EDX analysis and hardness test for welding and brazing multimaterial joints of Inconel® 625 and Inconel® 718 were performed.

Introduction

Due to the application, nickel-based alloys are divided into structural alloys and alloys with special physical properties such as magnetically soft alloys, high-temperature alloys, corrosion-resisting alloys. Furthermore, nickel superalloys are used as shape memory alloys (NiTi) and alloys with special purpose in plastic forming [1-3]. However, the most important feature of all nickel superalloys is their heat resistance, corrosion resistance and the ability to work in extreme environmental conditions. For this reason, these materials are most often used in the aerospace industry for the construction of combustion chambers, gas turbines and flue gas exhaust, i.e. elements that require high-temperature and heat-proof materials [4-5]. Due to the occurrence of vibrations within these elements, they must also be characterized by high fatigue strength and corrosion resistance due to their interaction with combustion products. In the aerospace industry, the most commonly used nickel alloys are the following: Inconel 718, Inconel 625, Hastelloy X, Udimet R41 and Nimonic 90 [6]. In this study, the authors focused on the characteristics of Inconel® 625 (Ni-Cr-Mo alloy) and Inconel® 718 (Ni-Cr-Fe alloy) alloys.

In addition to the selection of appropriate construction materials, the choice of the method of joining is important. Both factors affect the efficiency of the structure. Glue, rivet, welded, threaded and brazed are the most commonly used joints in the aerospace industry. In the recent years, there has been a significant development of beam techniques (lasers and electrons) in the processes of joining materials [7-8]. The choice of joining method depends on such factors as type of materials to be joined, geometrical dimensions of elements, operational requirements and conditions of connections [9]. This paper presents issues in the field of welding and brazing. Brazing allows to obtain high-quality joints of materials that differ in chemical composition and properties or work in difficult operating conditions [10]. The process intended for the joining of nickel alloys is high temperature brazing and vacuum brazing. However, in the aerospace industry, mainly vacuum brazing is used for these types of alloys [11]. Joints soldered with this method are characterized by high precision, high strength, no oxide inclusions, dissolved gases and blisters. The necessity to

Materials Research Proceedings **5** (2018) 166-171 doi: http://dx.doi.org/10.21741/9781945291814-29

maintain a gap between the brazed elements and conducting the process in a vacuum furnace constitute limitations of this method. [12].

In the electron beam welding technique, the diffusion processes in the weld pool and mixing of molten nickel- based superalloys (Inconel 625 and Inconel 718) lead to the homogenization of the chemical composition of the liquid metal. During the solidification of the weld metal, the segregation of niobium and molybdenum occurs, which leads to the formation of eutectics at 1150°C and 1250°C. As a result of the eutectic formation taking place at 1150°C, austenite, brittle and intemetallic Laves phase are formed [13], while at the temperature of 1250°C niobium carbide and austenite are formed. The presence of carbides and the Laves phase leads to deterioration of crack resistance and plasticity of the material. Susceptibility to cracking during solidification as well as fracture cracks increases [14-16]. In order to reduce the impact of these phenomena, an oscillated electron beam is used instead of an unoscillated one during the welding process. Such a solution results in a faster cooling rate of the weld, and thus less niobium segregation and a smaller volume of the Laves phase. The finer, unrelated particles of the Laves phase, the better the properties of the welds [17].

Materials and methods

The research was carried out for specimens of welded and brazed Ni-based superalloys – Inconel® 625 and Inconel® 718. Compositional specification of both Inconel superalloys is included in Figure 1.

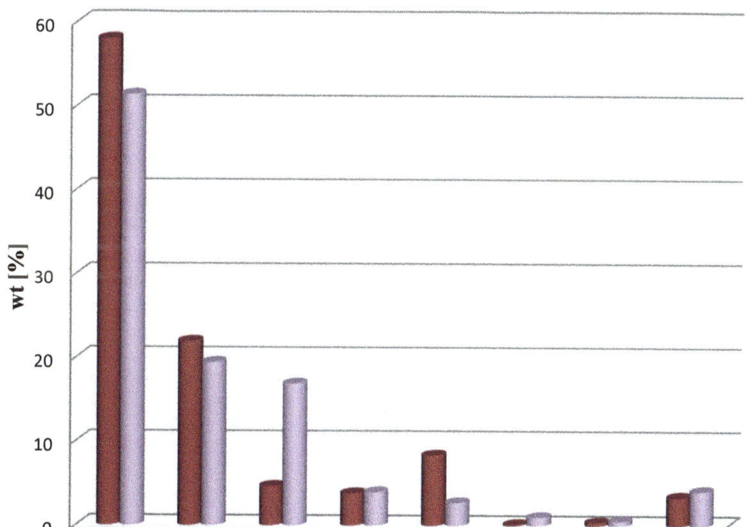

	Ni	Cr	Fe	Nb	Mo	Ti	Al	C
Inconel 625	57.91	21.88	4.64	3.82	8.33	0.16	0.41	3.42
Inconel 718	51.34	19.41	16.88	3.96	2.72	1.02	0.56	4.13

Fig. 1. Chemical composition of Inconel® 718 and Inconel® 625 (in wt. %).

Inconel® 625 in the form of a tube was brazed into a hole made in an Inconel® 718 plate. The surface of the materials was cleaned and chemically degreased. The solder consisted of 20% nickel and 80% gold. The brazing process was carried out in a vacuum oven. The multimaterial weld was obtained by oscillated electron beam welding, as a result of welding of two plates. The structure and shape of the joints were determined using an Olympus SZ61 stereoscopic microscope. Microstructural investigations were performed using an Axiovert 45 optical microscope and a Joel JSM 5400 scanning electron microscope with an EDS analyzer. A Vickers microhardness tester Shimadzu HMV-G -21DT with the load of 980.7 mN was used for the specimens.

Results and discussion

The stereoscopic images of welded and soldered multimaterial joints are shown in Figures 2a and 2b respectively.

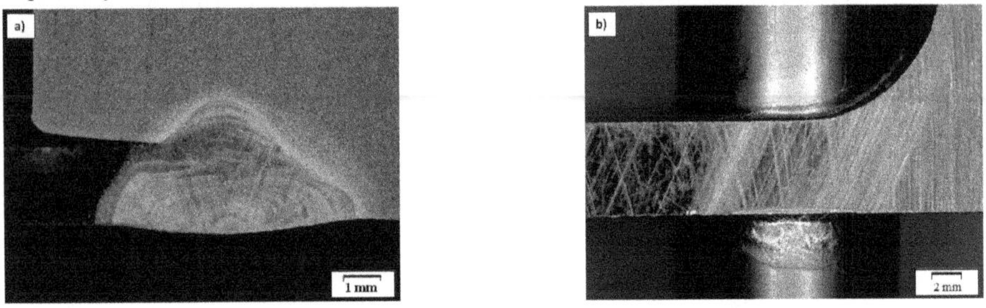

Fig. 2. a) stereoscopic image of the welded joint b) stereoscopic image of the brazed joint

In the image of the weld joint, the lack of the flash weld, crystallization cracks and microcracks in the Heat-Affected-Zone (HAZ) are evident, while in the case of the brazed joint, there is no dissolution of solder, cracks and porosity.

The microstructure of the parent metal Inconel® 625 is presented in Figure 3. In some grains, annealing twins are present, which is a characteristic defect for metals and alloys with a face centered cubic structure. The microstructure of the second native material Inconel® 718 is characterized by fine graining and the presence of the δ phase. The precipitation of the δ phase is always preceded by the precipitation of the γ'' phase. Due to the fact that niobium is the main component of both phases, the increase of the δ phase takes place with the γ'' phase depletion. Therefore, the presence of the δ phase causes loss of hardenability of the alloy.

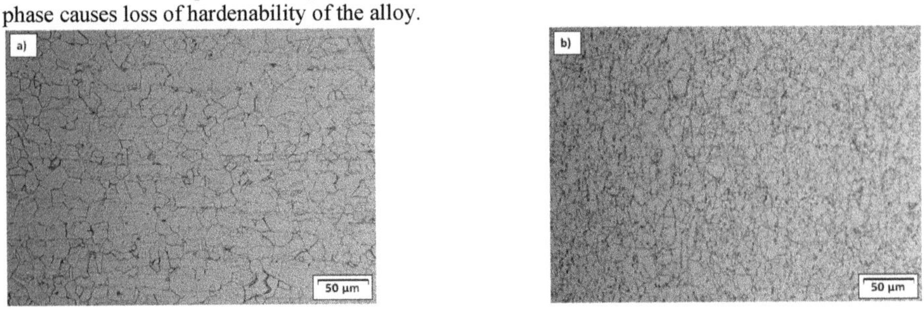

Fig. 3. The microstructure of a) Inconel® 625 b) Inconel® 718

Materials Research Proceedings **5** (2018) 166-171 doi: http://dx.doi.org/10.21741/9781945291814-29

Using a scanning electron microscope, a linear EDS analysis of both connections was made. The points in which the measurement was made are marked with points in the microstructural images. In each case, 11 points were taken into account for the analysis (Figure 4).

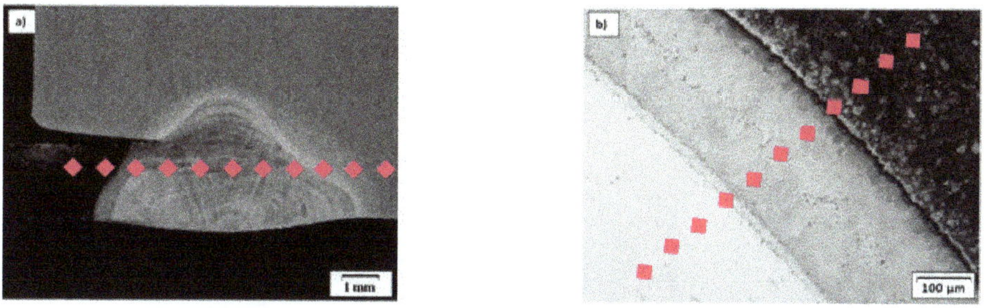

Fig. 4. The microscopic image of a) weld joint b) brazed joint. The place of performance of SEM / EDS analysis are indicated in the figure points.

a)

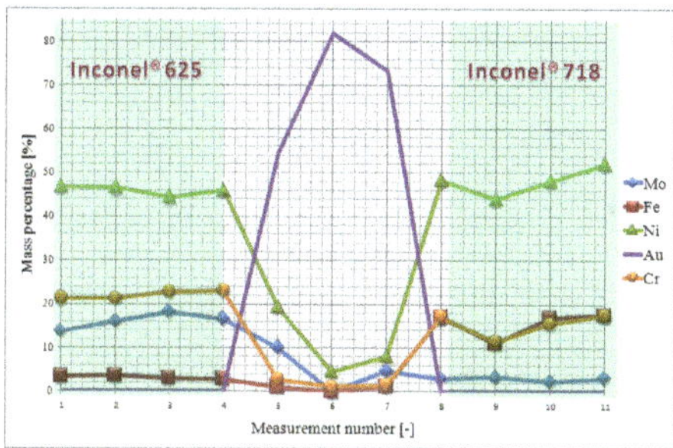

b)

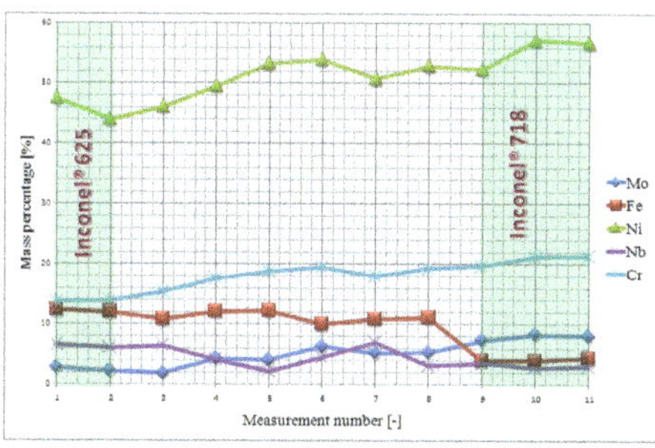

Fig.5. Diagram of linear SEM / EDS analysis of a) brazed joint b) weld joint

In the brazing process, iron, chromium and molibdenum diffused into the solder consisting of 80% Au and 20% Ni (Figure 5). The average molybdenum content in the solder is 4.91%, chromium - 1.83% and iron - 0.78%. The presence of gold in parent materials after the soldering process is not revealed. In the case of electron beam welding, the composition of the welded joint is a resultant combination of nickel superalloys.

The hardness of parent materials and multimaterial joints was measured using the Vickers method (Figure 6).

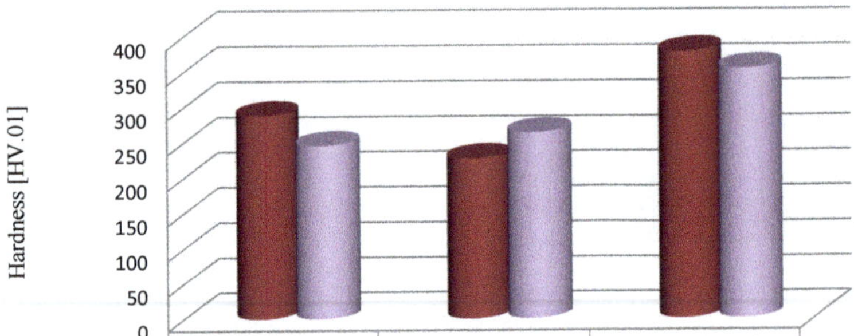

	Inconel® 625	multimaterial joint	Inconel® 718
weld joint	289	227	378
brazed joint	246	264	354

Fig. 6. Average hardness of parent materials and multimaterial joints

For each joint, Inconel® 718 is the material with the highest hardness value. The brazed joint has the lowest hardness value compared to the parent materials. This value is the result of only the joining of nickel superalloys with the use of a solder of a different composition. However, the weld has an intermediate value between the parent materials. The test hardness for the weld revealed that its hardness is the resultant value for individual hardness of the joined materials.

Conclusion

The brazed joint is characterized by the lack of solder dissolution, cracks and porosity, whereas in the welded joint, no defects such as flash weld, crystallization cracks and microfissuring in the Heat-Affected-Zone are revealed.

The linear SEM/EDS analysis conducted for both joints indicated the presence of diffusion phenomena of molybdenum, chromium, and iron into the structure of the brazed joint. The structure of the electron beam welding joint is a result of a combination of parent materials and diffusion phenomena is not observed.

The Vickers hardness test proved that Inconel® 718 has the highest hardness value. The brazed joint has a lower hardness value than parent materials, because only the phenomenon of joining with the solder occurs. The weld connection is characterized by an indirect value of strength properties with reference to the parent materials.

References

[1] F. C. Campbell, Element of Metallurgy and Engineering Alloys, ASM International, 2008.

[2] X.X. Xing, X.J. Di, B.S. Wang, The effect of post-weld heat treatment temperature on the microstructure of Inconel 625 deposited metal, J. Alloys Compd. 593 (2014) 110-116. https://doi.org/10.1016/j.jallcom.2013.12.224

[3] P. Ferro, A. Zambon A, F. Bonollo, Investigation of electron-beam welding in wrought Inconel 706-experimental and numerical analysis, Mater. Sci. Eng., A. 392 (2005) 94-105. https://doi.org/10.1016/j.msea.2004.10.039

[4] L. Dobrzański, Metalowe materiały inżynierskie, WNT. Gliwice-Warszawa, 2004.

[5] M. Wojtewicz, K. Nadolny, W. Sienicki, Analiza stanu wiedzy i techniki w zakresie szlifowalności stopów na bazie niklu. Mechanik. 8-9 (2015) 330 – 339.

[6] S. I. Kwon, S. H. Bae, J. H. Do, C. Y. Jo, H. U. Hong, Characterization of the microstructures and the cryogenic mechanical properties of elektron beam welded Inconel 718, Metall. Mater. Trans. A 47A (2016) 777-787. https://doi.org/10.1007/s11661-015-3269-6

[7] N. Radek, J. Konstanty, Cermet ESD coatings modified by laser treatment. Archives of Metallurgy and Materials, 57 (2012) 665-670. https://doi.org/10.2478/v10172-012-0071-y

[8] J. Pietraszek, N. Radek, K. Bartkowiak, Advanced statistical refinement of surface layer's discretization in the case of electro-spark deposited carbide-ceramic coatings modified by a laser beam, Solid State Phenom. 197 (2013) 198-202. https://doi.org/10.4028/www.scientific.net/SSP.197.198

[9] A. Rudawska, J. Cisek, L. Semotiuk, Wybrane aspekty wytrzymałości połączeń lejowych oraz lutowanych stosowanych w konstrukcjach lotniczych, Technologia i Automatyzacja Montażu. 2 (2012) 61-66.

[10] T. Piwowarczyk, E. Harapińska, T. Wojdat, Trendy rozwojowe technologii lutowania i metod kontrolnych, Przegląd Spawalnictwa. 88/9 (2016) 18-24. https://doi.org/10.26628/ps.v88i9.651

[11] J. Bielanik, Lutowanie próżniowe elementów kół wirnikowych. turbosprężarki z martenzytycznej stali typu 13-4. Report of the Warsaw University of Technology for the TURBOTECH Company 2002.

[12] T. Radomski, A. Ciszewski, Obróbka oraz łączenie tytanu i jego stopów. WNT. Warszawa, 1968.

[13] M. Rozmus –Górniakowska, M. Blicharski Microsegregation and precipitates In inconel 625 arc weld overlay coatings on boiler pipes. Arch Metall Mater. 60 (2015) 2599-2605.

[14] J.M. Cieślak, T.J. Headley, G.A. Konorovsky, A.D. Romig, T. A Kollie, Comparision of the solidification behavior of INCOLOY 909 and INCONEL 718. Metall Mater Trans A. 21 (1990) 479-488. https://doi.org/10.1007/BF02782428

[15] S.W. Banovic, J.N. Dupont, A.R. Marder, Dilution and microsegregation in dissimilar metal welds between super austenitic stainless steel and nickel base alloys. Sci. Technol. Weld. Joining. 7 (2002) 374-383. https://doi.org/10.1179/136217102225006804

[16] M. G. Reddy, C. V. Srinivasa Murthy, S. Rao, K. Prasad Rao, Improvement of mechanical properties of Inconel 718 electron beam welds – influence of welding techniques and postweld heat treatment. Int J Adv Manuf Technol. 43 (2009) 671-680. https://doi.org/10.1007/s00170-008-1751-7

[17] K. Sivaprasad, S. G. S. Ramana, C. V. S. Murthy, G. M. Reddy, Coupled effect of heat input and beam oscillation on mechanical properties of alloy 718 electron beam weldments. Sci. Technol. Weld. Joining. 11 (2006) 127–134. https://doi.org/10.1179/174329306X84337

Terotechnology 2017 Materials Research Forum LLC
Materials Research Proceedings **5** (2018) 172-176 doi: http://dx.doi.org/10.21741/9781945291814-30

A Robotic Method of Laser Cleaning of Antique Materials

Grzegorz WITKOWSKI [1, a *], Szymon TOFIL[2,b] Krystian MULCZYK[2,c]

[1]Kielce University of Technology, Faculty of Mechatronics and Mechanical Engineering, Department of Automation and Robotics, Aleja Tysiąclecia Państwa Polskiego 7, PL- 25314 Kielce, Poland

[2] Kielce University of Technology, Faculty of Mechatronics and Mechanical Engineering Department of Terotechnology and Industrial Laser Systems, Aleja Tysiąclecia Państwa Polskiego 7, PL- 25314 Kielce, Poland

[a]gwitkowski@tu.kielce.pl, [b]tofil@tu.kielce.pl, [c]kmulczyk@tu.kielce.pl

Keywords: Laser Cleaning, Architectural Details, Ablation, Robotics

Abstract. The presented article deals with issues related to laser cleaning of materials, especially the surface of historic elements. Typical types of dirt occurring on the surfaces of historic elements are characterized. A theoretical basis for the interaction of a laser beam on the surface of elements is presented. The proposed method of robotic laser cleaning eliminates the disadvantages of the manual process.

Introduction

Defects associated with external surface contamination are usually a result of the settling of volatile dusts of various origins and the formation of oxides. Surface contamination may also be a result of chemical reactions on the surface of a detail caused by external factors such as acid rainfall. Dirt may also be a result of an act of vandalism consisting in the application of various types of paint. The presence of plant-based soils, such as mosses and lichens, is also a particular type of a defect. It is also quite common for salts and hydrated minerals to penetrate the surface of stone or ceramic elements, which results in the appearance of stains and efflorescence. Often, mineral materials are not homogeneous and consist of many different compounds. Some compounds have a tendency to leak to the surface after hydration. Selected examples of contamination of historical details are shown in Figure 1.

Fig. 1. Examples of defects of architectural details (volatile dusts, limestone blooms, soils of biological origin)

Materials found in the constructions of historical elements require appropriate and delicate renovation treatments. In special cases, such treatments may involve cleaning the surface of a detail or removing top coats to apply new ones [1, 2]. Among the available and currently used methods, many are invasive and may cause too much interference in the structure of a detail. At the moment, scanning a detail with a laser beam is one of the youngest cleaning technologies under development [3, 4]. Through the use of a coherent, high energy coherent beam of a specific wavelength, such a solution guarantees obtaining results unattainable by using other, previously used cleaning technologies.

Laser cleaning process

Laser surface cleaning is a process in which a laser beam with energy capable of removing impurities from it is used [5, 6]. In this process, pulse lasers with a wavelength of 1064 nm and pulsed operation are most commonly used. Typically, pulse times of the order of 10 ns and energy at the level of hundreds of mJ are used, which corresponds to the power in the pulse of about several dozen of MW [7].

A laser beam with a power density of the order of 10^6 W/mm^2 interacting on the contaminated surfaces can affect them in two ways, depending on the thickness and nature of this layer. In the case of a thin layer, the impulse causes its almost immediate evaporation and transition into a plasma state manifested by a strong emission of optical radiation [8, 9]. This mechanism usually occurs on metal surfaces covered with a thin layer of oxides, patina or a thin layer of paint. In turn, thick layers of impurities under the influence of optical radiation pulse undergo very rapid heating, resulting in the phenomenon of thermal expansion, strong shear stress at the boundary layer of contaminants - the substrate that causes detachment from the substrate and falling off in the form of larger fragments [10]. This phenomenon is particularly intense when it comes to brittle and porous coatings such as rust, thick paint coatings or mineral particles glued to the surface, for example fragments of plastering mortar or putty. In both mechanisms, the best results, i.e. the removal of impurities, are obtained with high reflexivity of the substrate, protecting it from damage and low reflexivity of the impurities covering it, which thanks to the high absorption of radiation ensures fast progress of the cleaning process. Selected examples of laser cleaning effects are shown in Figure 2.

Fig. 2. The effects of using the laser cleaning method

Like any technology, also the laser cleaning method requires the fulfillment of specific parameters. Among other things, it is necessary to specify the necessity to meet the perpendicularity of the laser beam to the surface to be machined and to keep a specified distance of the scanning head from the surface of a workpiece. Parameters such as radiation power or pulse frequency are selected depending on the type of dirt of the material being cleaned. In the case of manual manipulation of the scanning head, when cleaning details with complex geometry, it is difficult to meet the two key previously mentioned requirements. Inappropriate beam routing can have far-reaching negative effects, which should be kept in mind when cleaning valuable or unique details. The laser cleaning process also has an adverse effect on personnel operating laser devices.

Developed method of robotic laser cleaning

One of the ways to eliminate disadvantages of the classic, manual laser cleaning process is its partial or full automation or robotization, by placing the scanning head on the device performing the sequence of working movements in the automatic cycle [11]. Among the known solutions, six-axis industrial robots with the RRR configuration provide the greatest manipulation possibilities [12, 13].

The presented method of robotic laser cleaning involves the use of an industrial robot to guide the scanning head of the laser device. Despite numerous benefits, the use of an industrial robot for the laser cleaning process has also some problems. The most important of these is the need to reprogram the robot each time for machining various details that have a different spatial geometry. For industrial robots, a number of standard trajectory planning methods are envisaged, but none of them is beneficial for robotic laser cleaning.

In the presented method, planning the manipulator trajectory is directly based on the knowledge of the geometry of the workpiece. Such a method guarantees each time correct parameters of processing of the cleaned part. It is an offline programming method in which a generation of paths for the robot is done automatically by specialized CAM systems with implemented postprocessors for the robot class.

In order to accomplish this task, it is necessary to pre-acquire the geometry of the workpiece using spatial scanning methods. Laser optical scanners are used for this purpose. These devices operate on the principle of triangulation, generating a cloud of node points of the scanned part. The next stage is solid modeling in order to fill any possible deficiencies in the model that could have arisen in the scanning process. In this stage, the cloud of points is usually transformed into spatial geometry with the possibility of being exported to typical 3D data exchange files via software usually provided by the scanner manufacturer.

The next stage involves automatic generation of paths for the robot manipulating the laser cleaning head. Based on the geometry of the scanned detail, the CAM software prepares the path of the scanning head, while optimizing the sequence of motion of the robot's drives. The final effect of this stage is to generate a program code appropriate for a given robot. A full simulation and visualization of the robot's work movements is also carried out. The sequence of stages of the presented solution is illustrated in Figure 3.

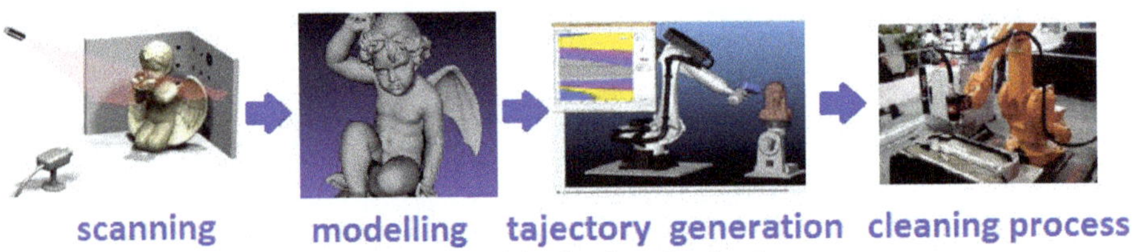

Fig. 3. Stages of the developed technology

The last stage of the presented solution is to implement the created code in the robot control system. After determining the correct speed of movement for the effector, the actual cleaning process of the part under consideration begins. It is very important that the cleaned part is oriented in space according to the orientation adopted during the trajectory planning step. Depending on the thickness of the contamination layer, the process can be carried out repeatedly. If the material contamination is uneven, the operator can remotely control the switching of the laser beam.

Summary

The project concept of a robotic laser cleaning station presented in the article is a result of joint work undertaken with an industrial partner. Based on the experience of manual handling of the cleaning head, the process can be unambiguously characterized as tiring and tedious. Constant exposition of operating personnel to the effects of dangerous laser radiation is also important. Following the example of other well-known solutions, the concept of position automation was developed. However, the method of planning the trajectory for the robot turned out to be a critical problem. It was proposed to use the method of trajectory planning based on the knowledge of the geometry of the cleaned part. The presented laser cleaning process in a robotic manner has also many advantages. The most important ones include improving the quality of the cleaned surface, speeding up the process and eliminating the risk to the operating staff.

Acknowledgement
The project is realized by "Furmanek Renewal" under the European Union Intelligent Development Operational Program 2014-2020 sub-action 3.2.2 "Loans for technological innovation". The presented method is the basis for patent application No. P.421121 in the Patent Office of the Republic of Poland.

References

[1] N. Radek, E. Wajs, M. Luchka, The WC-Co electrospark alloying coatings modified by laser treatment, Powder Metallurgy and Metal Ceramics 47 (2008) 197-201. https://doi.org/10.1007/s11106-008-9005-7

[2] J. Pietraszek, N. Radek, K. Bartkowiak, Advanced statistical refinement of surface layer's discretization in the case of electro-spark deposited carbide-ceramic coatings modified by a laser beam, Solid State Phenom. 197 (2013) 198-202. https://doi.org/10.4028/www.scientific.net/SSP.197.198

[3] M. Scendo, N. Radek, J. Trela, Influence of laser treatment on the corrosive resistance of WC-Cu electrospark coatings, International Journal of Electrochemical Science, 8 (2013) 9264-9277.

[4] J. Marczak, Set of advanced laser cleaning heads and systems. Proceedings SPIE 2009, 7391-7391. https://doi.org/10.1117/12.827541

[5] R. Jóźwicki, J. Zakrzewski, Laser technology and its applications [in Polish: Technika laserowa i jej zastosowania], Warszawa, Oficyna Wydawnicza Politechniki Warszawskiej, 2009.

[6] A. Kujawski, P. Szczepański, Lasers physical bases [in Polish: Lasery podstawy fizyczne] Warszawa, Oficyna Wydawnicza Politechniki Warszawskiej, 1999.

[7] F. Kaczmarek, The basics of lasers work [in Polish: Podstawy działania laserów] WNT, Warszawa, 1983.

[8] T. Burakowski, J. Marczak, W. Napadłek, The essence of ablative laser cleaning of materials [in Polish: Istota ablacyjnego czyszczenia laserowego materiałów] Prace instytutu elektrotechniki WAT, Warszawa, 2004 125-135.

[9] J. Marczak, The issue of the use of laser ablation in removing secondary layers from the surface of works of art and historic objects in architecture [in Polish: Zagadnienie wykorzystania ablacji laserowej w usuwaniu wtórnych nawarstwień z powierzchni dzieł sztuki i obiektów zabytkowych w architekturze] Ochrona zabytków 54 (3) (2001) 233-251.

[10] A. Tam, W. Leung, W. Zapka, W. Ziemlich, Laser-cleaning techniques for removal of surface particulates. Journal of Applied Physics 71 (1992) 3515-3523. https://doi.org/10.1063/1.350906

Materials Research Proceedings 5 (2018) 172-176 doi: http://dx.doi.org/10.21741/9781945291814-30

[11] G. Witkowski, P. Kurp, Design and realization of robotic stations for optical inspection of weld quality [in Polish: Projekt i wykonanie zrobotyzowanego stanowiska do optycznej inspekcji jakości spoin] Nowoczesne zastosowania technologii spawalniczych 2015, 133-142.

[12] J. Honczarenko, Industrial robots [in Polish: Roboty przemysłowe] WNT, Warszawa, 2009.

[13] M.W. Spong, M. Vidyasagar, Dynamics and control of robots [in Polish: Dynamika i sterowanie robotów] WNT, Warszawa, 1997.

Terotechnology 2017 Materials Research Forum LLC
Materials Research Proceedings **5** (2018) 177-182 doi: http://dx.doi.org/10.21741/9781945291814-31

Calculation of Residual Life for P91 Material Based on Creep Rate and Time to Rupture

Przemysław OSOCHA

Cracow University of Technology, Faculty of Mechanical Engineering, al. Jana Pawła II 37, PL-31-864 Kraków, Poland

osocha@mech.pk.edu.pl

Keywords: Creep Rupture, Creep Test, Monkman-Grant, Residual Life

Abstract. The presented paper shows methods of residual life prediction based on creep test data. Real test data for P91 material will be used. Different methods will be shown. Calculations are based on the steady creep rate connection with the time to rupture.

Introduction

Creep is the non-linearity of a material at which the material continuously deforms under a constant load. Also, if a material is subjected to constant deformation, the phenomenon of relaxation will decrease with time as a result of creeping reaction forces and stresses.

Creep is a term that describes the tendency of a material to permanently move or deform to reduce existing stresses. The deformation of a material appears as a result of prolonged stresses close to the yield point or material strength point. Creep is particularly dangerous in materials subjected to high temperatures, especially those close to the material melting point temperature. The creep rate is a function of material properties, time, temperature and applied loads (forces or stresses) and deformations. Long-term operation in creep conditions may lead to deformations that result in device failure. For example, turbine blades may be lengthened so that they come into contact with the casing, causing damage.

Usually, there is no sudden crack in the creep process, but a material subjected to loads in the long term eventually fails. Creep does not occur during a sudden load, but is the effect of cumulative long-term creep distortion. Creep deformation is a time-dependent strain.

Creep dependence on temperature and stress

Increase of temperature values and/or stress values results in an increase of the creep strain and change of the shape of creep strain curves (Fig. 1). Fig. 1 shows the dependence of the change in the creep strain at a constant temperature of 600°C for different loads: 100, 120 and 140 MPa, for P91 steel.

There is a number of empirical formulas of non-linear dependence of creep strain on time. Eq. 1equation presents one of them:

$$\varepsilon = \varepsilon_0 + D\sigma^a\left(1-\exp\left(bt\right)\right) + B\sigma^n t \quad (1)$$

where *B*, *n*, *D*, *a*, *b* are empirical constants, while ε is the creep strain, ε_0 is the initial strain, σ is the stress and *t* is time.

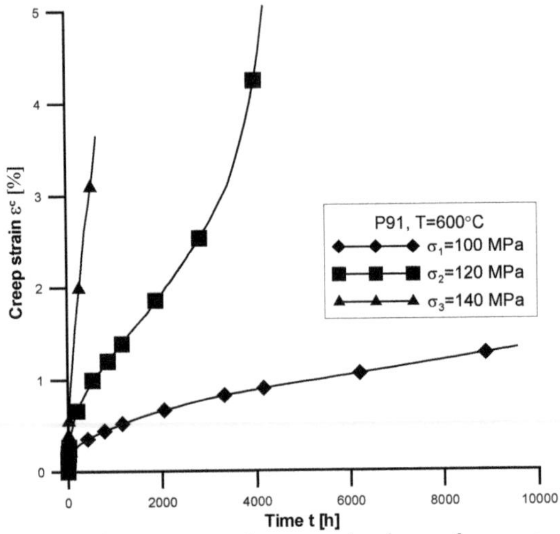

Fig. 1. Influence of temperature change on the shape of creep strain curves as a function of time for P91 steel.

If $t > t_{II}$ then Eq.1 formula takes the form of Eq. 2, where t_{II} is the time of transition from primary to secondary creep stage:

$$\varepsilon = \varepsilon_0 + D\sigma^a + B\sigma^n t \tag{2}$$

And in that case the creep rate is the minimal one, i.e. creep rate in Eq. 3 secondary creep stage:

$$\frac{d\varepsilon}{dt} = B\sigma^n = \dot{\varepsilon}_{II}^c = \dot{\varepsilon}_{\min}^c, \tag{3}$$

where $\dot{\varepsilon}_{II}^c$ is the creep rate in secondary creep stage, and it is minimal creep rate $\dot{\varepsilon}_{\min}^c$.

To take into account the influence of temperature in Eq. 3 equation, the Arrhenius model is used for Eq. 4 creep rate:

$$\dot{\varepsilon}_{\min}^c = A\sigma^n \exp\left(-\frac{Q}{RT}\right). \tag{4}$$

where n is the stress exponent, Q is the creep process activation energy, R is the universal gas constant and T is the absolute thermodynamic temperature.

In order to determine the values of constants present in Eq. 4 equation, it is necessary to perform a series of creep tests at constant temperature as well as a series of tests at constant tension.

For isothermal tests, the exponential factor assumes a constant value and Eq. 4 formula takes the form of the Norton-Bailey formula [1] Eq. 5:

$$\dot{\varepsilon}^{c}_{\min} = B\sigma^{n} \tag{5}$$

Eq. 5 equation can be linearized by logarithmising both sides of Eq. 6:

$$\log \dot{\varepsilon}^{c}_{\min} = \log B + n \log \sigma \tag{6}$$

Log-log charts of minimal creep rate $\dot{\varepsilon}^{c}_{\min}$ in dependence on stress σ are often in the form of two straight lines whose directional coefficient n for small stresses is of smaller value, while for higher stresses n is higher, which indicates the existence of different creep mechanisms for smaller and larger values of applied stress. Data for P91 steel [2] for creep tests at 600°C, 620°C and 640°C is shown in Fig. 2.

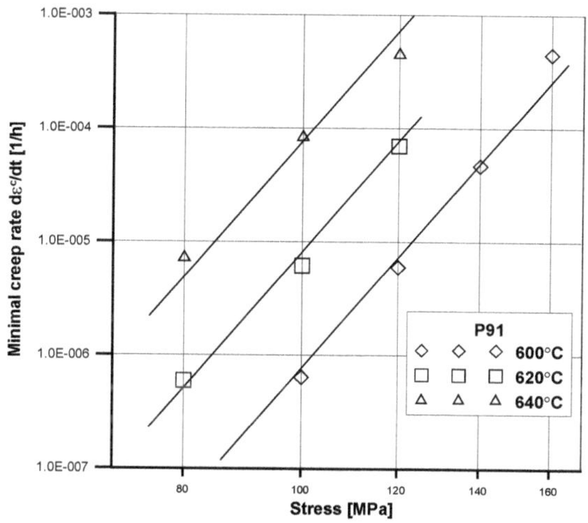

Fig. 2. The minimal creep rate as a function of stress log($\dot{\varepsilon}^{c}_{\min}$)-log(σ) for different temperatures for P91 steel.

Using the linearized Eq. 6 formula, the coefficients of Eq. 5 formula for P91 steel were chosen. The average value of exponent n for the analyzed range of temperatures and stresses equals 12.34. The data from creep tests describing the dependence of minimum creep rate on stress for P91 steel is shown in Fig. 2.

Methods of creep wear assessment

When designing engineering projects with creeping processes involved, the very essential task is to determine the behavior of the material over a long time period. There are three basic methods to solve the problem:

1) destructive tensile tests – charts of strain change in time,
2) dependence of the minimum creep rate on time to rupture – Monkman-Grant formula [3],
3) time compensated by temperature – time-temperature parameters.

Regardless of the method chosen, the following conditions must be met:
a) the duration of the test must equal at least 10% of the planned operating time,
b) creep and/or destruction mechanisms must not change in the analyzed ranges of time, temperatures and stresses.

There is a relationship between the minimum creep rate and time to rupture [4] (i.e. Monkman-Grant formula). This dependence is based on the observation that deformation is a macroscopic symptom of total creep failure. The destruction will occur when the material damage, in the form of creep cavities and cracks formed by the joining of creep cavities, reaches a critical level. This critical level of damage ε_r, leading to rupture, can be predicted on the base of the minimal creep rate and time to rupture Eq. 7.

$$\left(\dot{\varepsilon}_{\min}^{c}\right)^{m} \cdot t_r = K \approx \varepsilon_r \quad (7)$$

After calculating, on the base of experimental data, constants m and K, the Monkman-Grant Eq. 7 formula allow the minimum creep rate $\dot{\varepsilon}_{\min}^{c}$ or time to rupture t_r to be determined.

Predicting creep strain over a long time period using extrapolation is difficult due to a number of factors, such as dispersion among experimental data, especially in terms of sample temperatures and methods of heating; available temperature ranges and trial times; microstructure changes occurring during tests and affecting the strain; and environmental impact such as oxidation, reducing effective force transmitted by the sample.

One of the first attempts to extrapolate data from creep tests was conducted by F. H. Norton (1929), known for the introduction of Eq. 5 creep formula, commonly used in model creep processes.

The relationship between the minimal creep rate (in the secondary creep stage) and the applied stress can be used during the design phase. After establishing the maximum acceptable creep rate in the secondary stage, the corresponding maximum allowable stress can be determined.

The assumption of Larson-Miller that the creep rate in the secondary creep stage is inversely proportional to the time to rupture, has been refined and presented by Monkman-Grant in the form of the following Eq. 8 formula:

$$\left(\dot{\varepsilon}_{\min}^{c}\right)^{m} \cdot t_r = K \quad (8)$$

where K and m are constants.

Eq. 8 equation, known as the Monkman-Grant formula, on the log-log plot of the minimum creep rate as a function of time to rupture for a given material (Fig. 3) should have a directional coefficient of constant value regardless of temperature or applied stress. Eq. 8 equation can be linearized to form the following Eq. 9 equation:

$$\log \dot{\varepsilon}_{\min}^{c} = -\frac{1}{m}\log t_r + \log K \quad (9)$$

Materials Research Proceedings **5** (2018) 177-182 doi: http://dx.doi.org/10.21741/9781945291814-31

Creating a log-log plot of time to rupture, depending on the minimum creep rate, results in a straight line. It is possible to determine the constants K and m this way. The constant m is usually in the range from 0.8 to 1.2.

When the material constants in the Monkman-Grant Eq. 8 formula are determined on the base of experimental data, it is possible to determine the associated minimum creep rate assuming the operational life time of the element, get to know the minimum creep rate (eg. from measurements) and calculate the time to rupture.

Results

To estimate the time to rupture, knowing the creep rate in the secondary creep stage, the minimum creep rate for a given stress and temperature can be measured experimentally or calculated from Eq. 4 formula if constants A and Q are determined. After determining the minimum creep rate, time to rupture for the analyzed material can be determined from the Monkman-Grant Eq. 8 formula.

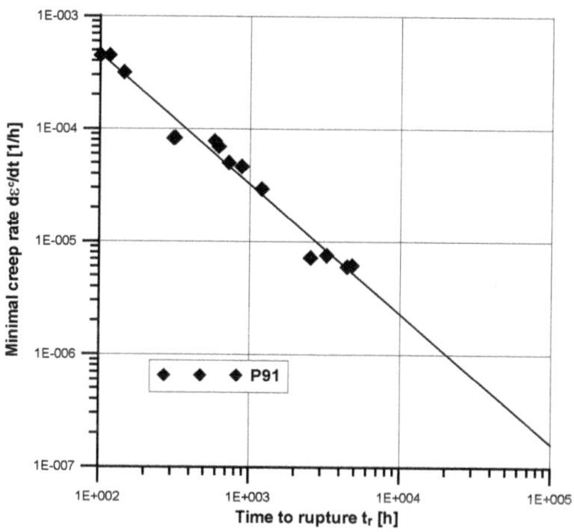

Fig. 3. Monkman-Grant dependence plot for P91 steel.

The Monkman-Grant formula along with the Norton formula can be used for extrapolation. The minimum creep rate associated with the assumed time to rupture can be read from the Monkman-Grant chart (Fig. 3), and then the stress that produces such a minimum creep rate can be determined from the Norton Eq. 5 formula. The Monkman-Grant dependence for P91 steel is shown in Fig. 3.

Using the Monkman-Grant formula and the Norton formula, the maximum allowable stress under the given operating conditions can be determined for the assumed element life time. From the Monkman-Grant Eq. 8 equation, for the P91 steel shown in Fig. 3, it can be stated that for the time to rupture of 100,000 hours, the minimum creep rate is $1.6{\cdot}10^{-7}$ [1 / hour]. In turn, from the Norton Eq. 5 equation shown in Fig. 2, it can be read that this creep speed [5] at 600°C is achieved at a stress level of 88 MPa.

Conclusion

The selected methods presented in this paper allow for the calculation of creep strain or time to rupture for engineering materials like P91 steel. But for the reliability of calculations, a rich set of creep test data is required. The paper presents used theory, all used equations, real material data are presented in the charts. All the calculations are explained step by step. The calculations shown could be used to assure safe operation of various structures operating in creep conditions, like power plants, jet engines, etc. It seems that the promising direction of the method development would be to include methods of the multivariate analysis [6-8], fuzzy description of uncertain knowledge [9-11] as well as specific methods of the materials science analysis [11, 12] and the image analysis [13, 14].

References

[1] A. Jakowluk, Procesy pełzania i zmęczenia w materiałach, WNT, Warszawa, 1993.

[2] ODIN Materials Database (Mat-DB). In: Online Data & Information Network for Energy. European Communities. Available via European. Commission Joint Research Centre (JRC). https://odin.jrc.ec.europa.eu/

[3] D. C. Dunand, B. Q. Han, A. M. Jansen, Monkman-grant analysis of creep fracture in dispersion-strengthened and particulate-reinforced aluminum. Metallurgical and Materials Transactions A, 30(3), (1999) 829-838. https://doi.org/10.1007/s11661-999-0076-y

[4] W.G. Kim, S.H. Kim, W.S. Ryu, Evaluation of Monkman-Grant parameters for type 316LN and modified 9Cr-Mo stainless steels, KSME international journal 16.11 (2002) 1420-1427. https://doi.org/10.1007/BF02985134

[5] P.J. Ennis, A. Czyrska-Filemonowicz, Recent advances in creep resistant steels for power plant applications, OMMI, Vol 1, No.1, April 2002.

[6] E. Skrzypczak-Pietraszek, J. Pietraszek, Phenolic acids in in vitro cultures of Exacum affine Balf. f., Acta Biol Cracov Bot 51 (2009) 62-62.

[7] A.J. Izenman, Modern Multivariate Statistical Techniques. Regression, Classification and Manifold Learning. Springer, New York, 2008.

[8] E. Skrzypczak-Pietraszek, I. Kwiecien, A. Goldyn, J. Pietraszek, HPLC-DAD analysis of arbutin produced from hydroquinone in a biotransformation process in Origanum majorana L. shoot culture, Phytochem. Lett. 20 (2017) 443-448. https://doi.org/10.1016/j.phytol.2017.01.009

[9] J. Pietraszek, Fuzzy Regression Compared to Classical Experimental Design in the Case of Flywheel Assembly, Lect. Notes Artif. Int. 7267 (2012) 310-317.

[10] J. Pietraszek, The Modified Sequential-Binary Approach for Fuzzy Operations on Correlated Assessments, Lect. Notes Artif. Int. 7894 (2013) 353-364. https://doi.org/10.1007/978-3-642-38658-9_32

[11] J. Pietraszek, M. Kolomycki, A. Szczotok, R. Dwornicka, The Fuzzy Approach to Assessment of ANOVA Results, Lect. Notes Artif. Int. 9875 (2016) 260-268.

[10] J. Korzekwa, A. Gadek-Moszczak, M. Bara, The Influence of Sample Preparation on SEM Measurements of Anodic Oxide Layers, Prakt. Metallogr.-Pr. M. 53 (2016) (1) 36-49.

[11] A. Szczotok, B. Chmiela, Effect of Heat Treatment on Chemical Segregation in CMSX-4 Nickel-Base Superalloy, J. Mater. Eng. Perform. 23 (2014) (8) 2739-2747. https://doi.org/10.1007/978-3-319-45243-2_24

[12] A. Szczotok, Metallographic Study of the Casting Made from CMSX-6 Sc Nickel-Based Superalloy, Arch. Metall. Mater. 62 (2017) (2) 581-586.

[13] A. Gadek-Moszczak, History of Stereology, Image Anal. Stereol. 36 (2017) (3) 151-152.

[14] A. Gadek-Moszczak, P. Matusiewicz, Polish Stereology - a Historical Review, Image Anal Stereol 36 (2017) (3) 207-221.

Terotechnology 2017
Materials Research Forum LLC
Materials Research Proceedings **5** (2018) 183-188
doi: http://dx.doi.org/10.21741/9781945291814-32

Laser Padding Welding in Technology of Rebuilding Mechanical Components

Hubert DANIELEWSKI[1,a], Andrzej SKRZYPCZYK[1,b], Włodzimierz ZOWCZAK[1,c], Rastislav NIGROVIC[2,d]

[1]Kielce University of Technology, al. Tysiąclecia P.P. 7, Kielce 25-314, Poland

[2]University of Žilina, Univerzitná 8215/1, Žilina 010 26, Slovakia

[a]hdanielewski@tu.kielce.pl,[b]tmaask@tu.kielce.pl,[c]wzowczak@tu.kielce.pl, [d]rastislav.nigrovic@fstroj.uniza.sk

Keywords: Rebuilding Technology, Laser Padding Welding, Regeneration of Mechanical Components

Abstract. This paper presents results of rebuilding mechanical components using laser technology. Laser technology, including padding welding, is an advanced method of regenerating exploited mechanical parts. Using metallic powder as a filler material and laser beam as a heat source, rebuilding and modification of reliability properties of surface can be done. The article presents results of rebuilding of connecting rod surface using laser technology.

Introduction

Wear of mechanical components working in motion when friction phenomena occur is inevitable [1]. Some components can be regenerated, and further, surface properties can be modified using padding welding technology. Rebuilding technology which uses laser beam as a heat source and metallic powders as an additional material is one of the fastest methods. The amount of heat absorbed in material is relatively small [2, 3]. A metallographic structure changes only into a narrow heat affected zone and stresses occurring in thermal processes are much lower than in other techniques. In some cases, a replacement of exploited parts might be very expensive and difficult, and the operating time of components could be assumed to be similar. During surface regeneration, wearability could be improved. Traditionally, additional materials have similar chemical composition as base materials, however, using metallic powder with different chemical composition properties of padding welding surface could be modified. Appliances metallic powders as additive material facilitates modifications.

Experiment

Laser padding welding experiments of a regenerating surface of connecting rod were carried out with an indigenously developed 6 kW TruFlow 6000 CO_2 laser working on a TruLaserCell 1005 station. The folded laser beam was focused using a 200 mm focal length parabolic mirror. The cladding process involved both scanning of the surface of the substrate with a 3 mm wide defocused laser beam and a simultaneous injection of CrNi powder (gradation of 45μm) into a molten pool on the surface of the substrate[4]. Overlapping tracks were first made on the flat surface.

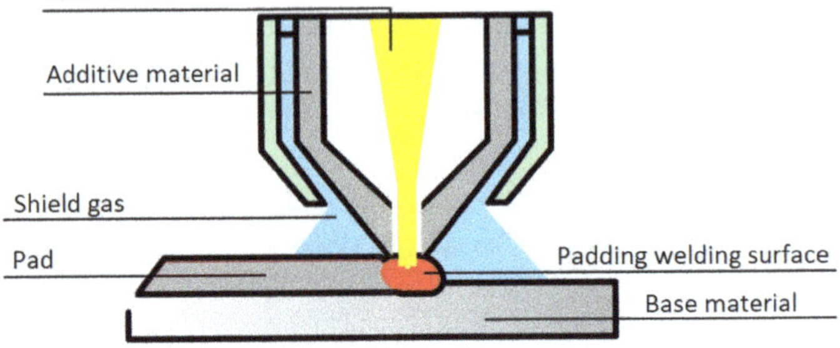

Fig 1. Laser padding welding process.

The chemical composition of the additional material was 65% of Nickel and 35% of Chromium. In order to study the effect of heat input in the component, laser padding welding was conducted at three different power levels to shown quantity of heat input during the laser process, while keeping other parameters constant[5]. The laser padding welding parameters are given in Table 1. Laser clad specimens were investigated by visual testing (VT) using a Hirox KH-8700 confocal digital optical microscopy, a JEOL JSM – 7100 F (SEM) scanning electron microscopy and microhardness measurement.

Table 1. Laser padding welding parameters.

Parameter:	Padding weld 1	Padding weld 2	Padding weld 3
Power	2.5 [kW]	3 [kW]	3.5 [kW]
Process velocity	1200[mm/min]	1200[mm/min]	1200[mm/min]
Shielded gas	Argon 15[l/min]	Argon 15[l/min]	Argon 15[l/min]
Carried gas	Helium	Helium	Helium

Testing results.

VT results, including padding welding profiles and macrostructure in cross section, were carried out. The results showed that a pad welded with the power of 3kW has good profile characteristics (Fig 2.) and less scatters than the other two samples. A pad made with those parameters was carried out for microscopic and hardness tests.

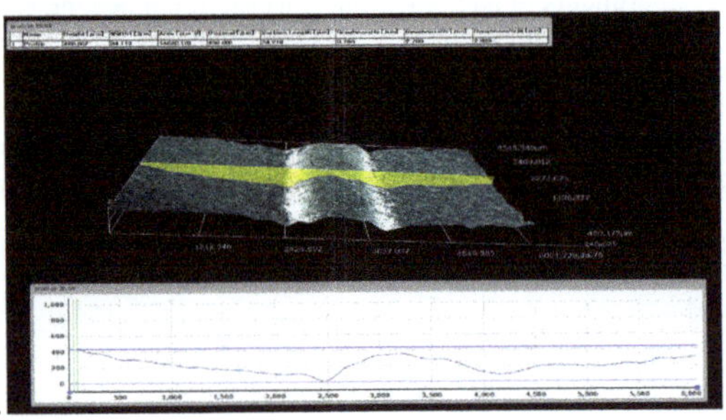

Fig. 2. Profile of laser padding welding sample, for the power of 3kW

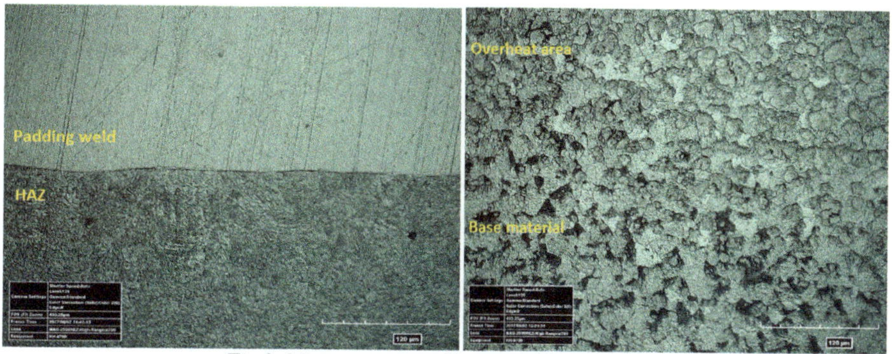

Fig 3. Microstructure of laser padding welding.

Figure 3 presents the results of microscopic testing. In the cross section of padding welding four characteristic areas were shown. Padding weld and the Heat Affected Zone (HAZ) were shown on the left and two similar microstructures of overheat area and base material (BM). The overheat area has little grain refining comparing to BM and its transition lattice between BM and HAZ.

Metallographic tests showed dissimilarity in the structure of the material. Chromium contained in the additive material improved hardenability in steels and might have caused structure transformation [6]. To further investigate this phenomenon, chemical distribution in cross section of padding welding (Fig 4.) would be conducted.

Sample	Al	Si	Cr	Mn	Fe	Ni	Nb	Mo
1		1.06	14.14	0.48	28.35	44.60	2.63	5.80
2				0.63	99.37			
3	0.23		0.08	0.47	99.22			
4		0.79	14.38		30.79	47.71	1.56	4.77

Fig 4. Measurement points of chemical composition, and amount of elements.

Chemical composition of selected points using a SEM microscope was carried out. For the achievement of padding welding characteristics, linear distribution of elements was conducted (Fig 5).

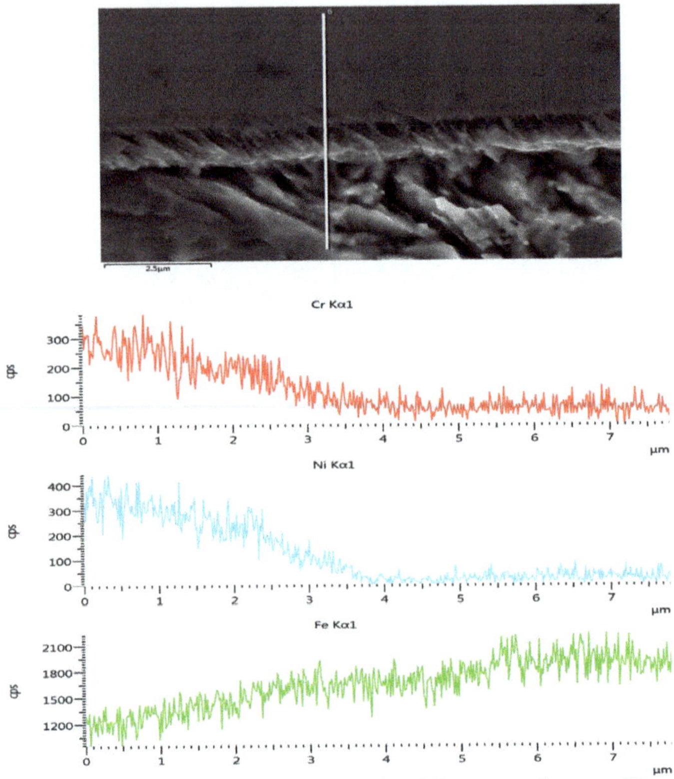

Fig 5. Linear distribution of Chromium, Nickel and Ferrite in a laser padding welding.

To investigate changes in mechanical properties of the padding welded material, microhardness tests were conducted. The results of the Vickers Hardness test with the load of 5kgF were shown below.

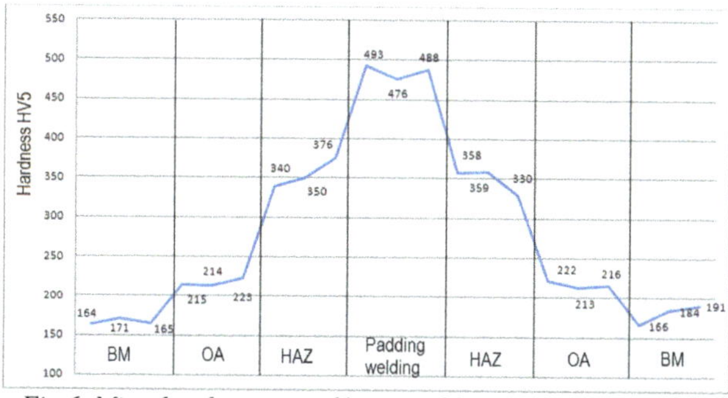

Fig 6. Microhardness test of laser padding welding results.

Laser padding welding rebuilding of connection rodsurface.

The experiment and results presented in the previous chapters were carried out to investigate proper padding welding parameters for the rebuilding of a connection rod surface. The wear surface was regenerated by metallic powder deposited on a cylindrical sphere of part (Fig 7.). After rebuilding, a sliding process to achieve nominal geometry is required.

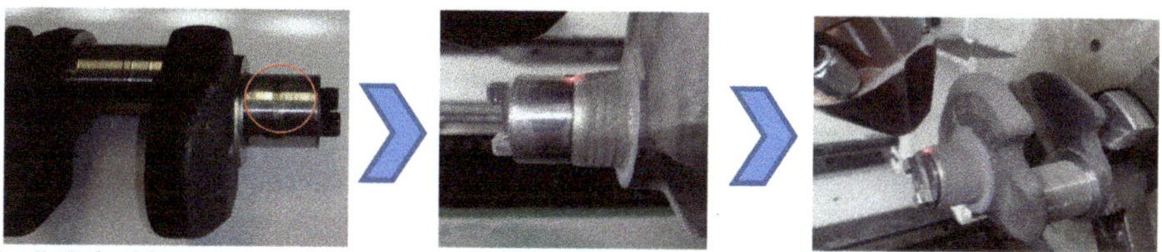

Fig 7. Rebuilding process of a connection rod surface using laser padding welding technology.

Summary

Rebuilding process using laser padding welding could be used to regenerate attrition surfaces. Using proper additive materials, surface properties could be modified. Laser padding welding characterized in the narrow heat affected zone in the material, and process power and ratio of velocity affect the metallographic structure of material by determining the amount of heat absorbed in the material. Improvement of material properties including hardness of padding welded surfaces affects the exploitation time of regenerated components [7,8].

References

[1] N. Radek, J. Pietraszek, B. Antoszewski, The average friction coefficient of laser textured surfaces of silicon carbide identified by RSM methodology, Advanced Materials Research, 874 (2014) 29-34. https://doi.org/10.4028/www.scientific.net/AMR.874.29

[2] N. Radek, E. Wajs, M. Luchka, The WC-Co electrospark alloying coatings modified by laser treatment, Powder Metallurgy and Metal Ceramics 47 (2008)197-201. https://doi.org/10.1007/s11106-008-9005-7

[3] J. Pietraszek, N. Radek, K. Bartkowiak, Advanced statistical refinement of surface layer's discretization in the case of electro-spark deposited carbide-ceramic coatings modified by a laser

beam, Solid State Phenom. 197 (2013) 198-202.
https://doi.org/10.4028/www.scientific.net/SSP.197.198

[4] Z. Bogdanowicz, K. Grzelak, Laser rebuilding of engine exhaust valves, Journal of KONES Powertrain and Transport, Vol. 18, No. 1 2011

[5] R. Kaul, P. Ganesh, S. K. Albert, A. Jaiswal, N. P. Lalla, A. Gupta, C. P. Paul and A. K. Nath, Laser cladding of austenitic stainless steel with nickel base hadrfacing alloy, Surface Engineering, Vol 19, No 3, 2003.

[6] J. Song, Q. Deng , C. Chen , D. Hu , Y. Li, Rebuilding of metal components with laser cladding forming, Applied Surface Science 252, p7934–7940, 2006.
https://doi.org/10.1016/j.apsusc.2005.10.025

[7] B. Graf, S. Ammer , A. Gumenyuk , M. Rethmeier, Design of experiments for laser metal deposition in maintenance, repair and overhaul applications, Procedia CIRP 11, 2013

[8] B. Graf , A. Gumenyuk , M. Rethmeier, Laser metal deposition as repair technology for stainless steel and titanium alloys, Physics Procedia 39, p376–381, 2012. https://doi.org/10.1016/j.phpro.2012.10.051

Terotechnology 2017 Materials Research Forum LLC
Materials Research Proceedings **5** (2018) 189-193 doi: http://dx.doi.org/10.21741/9781945291814-33

Sintered Mesh Layers for the Production of Efficient Phase – Change Heat Exchangers

ORMAN Łukasz J.[1, a*], RADEK Norbert[1,b] and BRONČEK Jozef[2,c]

[1]Kielce University of Technology, al. Tysiaclecia P.P. 7, 25-314 Kielce, Poland

[2]University of Zilina, Univerzitna 1, 01026 Zilina, Slovakia

[a]orman@tu.kielce.pl, [b]norrad@tu.kielce.pl, [c]jozef.broncek@fstroj.utc.sk

Keywords: Boiling, Heat Exchangers, Sintering

Abstract. The paper focuses on the production and thermal performance of metal meshed heaters. The sintering process of sample production is described and the boiling heat transfer performance of multilayer meshed surfaces is discussed. The experimental results of nucleate boiling are compared with selected models and correlations available in literature.

Introduction

Metal meshes are commonly used in the industry (e.g. as sieves). They are cheap and offered in a large variety of sizes. The number of materials from which they are made is also wide. The application of meshes can be favourable to increase heat fluxes dissipated from heaters during pool boiling on horizontal and vertical surfaces, on tubes or in heat pipes.

Smirnov et al. [1] considered boiling of water and ethanol on a horizontally located heater that was covered with copper and brass meshes. These meshes were attached mechanically. Generally, heat flux was found to be independent of the structure height (namely, the number of layers) and weakly dependent on the structural material from which it was made. Asakavičjus et al. [2] testes the boiling of R-113, ethanol and water inside a heat pipe with an internal coating composed of two, eight and twelve meshes made of copper and stainless steel. They were mechanically attached to the heating surface. It was found that mesh layers improve boiling heat transfer performance comparing to the smooth surface, but this effect diminished with the increase of the heat flux value. The heat transfer coefficient for water was 1.8 – 3.5 times higher than for ethanol and R-113 with the same other process parameters. Franco et al. [3] presented results of boiling heat transfer of the R141b dielectric refrigerant on meshed surfaces. Stainless steel, aluminum, copper and brass meshes were used for the experiments. The meshes were clamped to the heater in a special mounting which enabled the control of the channels in the screen. It was proven that if meshes of different aperture are used together, the most significant enhancement is obtained for coatings of finer meshes at the base on the heater and coarser meshes higher. Li et al. [4] considered water boiling on meshes heaters. The microstructures were produced using sintering at the temperature of 1030°C in the gas mixture of nitrogen and hydrogen. Up to nine meshes were used for the production of one sample layer. It was found out that the proper contact between the elements results in better heat transfer conditions and that all the meshed surfaces enhanced boiling in comparison with the smooth surface without such covering.

Material and method

The heating surfaces can be covered with meshes and, as a consequence, boiling heat transfer can typically be enhanced. However, there is a problem with the proper attachment of the mesh to the surface (and to other meshes if multilayer structures are to be used). Mechanical clamping is sometimes used, however, the clamps themselves may hamper the heat transfer. The application of epoxy or glues may not be effective either, due to the increased thermal resistance (low thermal conductivity of the joining material). Thus, sintering is the best technology for the production of meshed surfaces. The process occurs in the reduction of temperature to prevent oxidation of materials at the temperature which is slightly lower than the melting temperature. As a result, the materials are joined together to form one element and strong bonds are created between the mesh and the base surface (and the meshes with one another for the multilayer structures). The samples are sintered in the atmosphere of hydrogen and nitrogen and remain in this oxygen-free atmosphere until being cooled to ambient temperature.

Fig. 1a presents the sample consisting of a single copper mesh sintered onto a copper disc. Here, the wires are simply molten into the base surface. The problem, however, is the proper production of multilayer microstructures. Fig. 1b presents the SEM image of a multi-mesh microstructure. The wires in contact with the heater surface (top right-hand side of the picture below) are nearly molten into the surface. The majority of the wires in this microstructure of 4 layers of mesh are also in a very good contact, which minimises the thermal resistance and, thus, the structure is highly efficient for a heat transfer. It needs to be noted, however, that some elements underwent destruction (a crack is visible along the long wire at the bottom of Fig. 1b). This may be caused by too much pressing of the sample during sintering (the sample was located in the mechanical clamp). Such defects of the structure could result in the increased thermal resistance and reduced heat flux dissipated from the heater and, consequently, should be avoided.

a)

b)

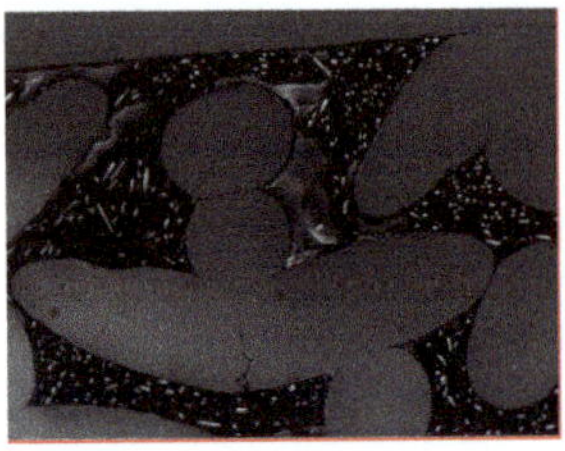

Fig. 1. Copper meshed surfaces on the copper base: a) photo of the single layer sample, b) SEM image of the cross – section of the multilayer sample

Results and discussion

The application of metal meshes on smooth surfaces can result in a significant increase in heat flux (q) dissipated from heat exchangers. Thus, such highly efficient heat exchangers can be smaller or transfer much higher heat loads at the same temperature differences (θ). Moreover, sintering produces stable bonds and such surfaces can be used in devices that undergo vibrations.

Fig. 2 shows the enhancement ratio, which is the ratio of heat flux transferred from the surface covered with the meshes (one to three) to the heat flux from the smooth surface. The data of heat flux changes with temperature difference was adopted from the paper by the co-author [5]. Three different values of temperature difference have been taken into consideration. As can be seen in Fig. 2, the use of a mesh layer can considerably improve heat transfer. Heat flux may be almost ten times higher than for the smooth surface without any additional covering. A typical phenomenon is that this highest enhancement occurs at lowest temperature differences. It needs to be noted that the application of more mesh layers favourably influences heat transfer. This may be linked with the increased surface area and higher density of active nucleation sites.

The description of physical phenomena which occur on the heater surface is another crucial problem for the proper design of heat exchangers. Models and correlations of boiling available in literature can provide equations for heat flux, but they are often inaccurate. Fig. 3 presents the comparison of the experimental data for a meshed surface with selected correlations from literature, namely those developed by Smirnov et al. [1, 6, 7], Nishikava et al. [8] as well as Xin and Chao [9]. Performing calculations according to the model presented by Xin and Chao was possible after introducing certain modifications as for the geometrical parameters used in their formulae. Namely, the width of a single cell had to be considered as the total of wire diameter and aperture and the width of the tunnel as aperture. The results of the calculations have been presented below for distilled water as the boiling liquid.

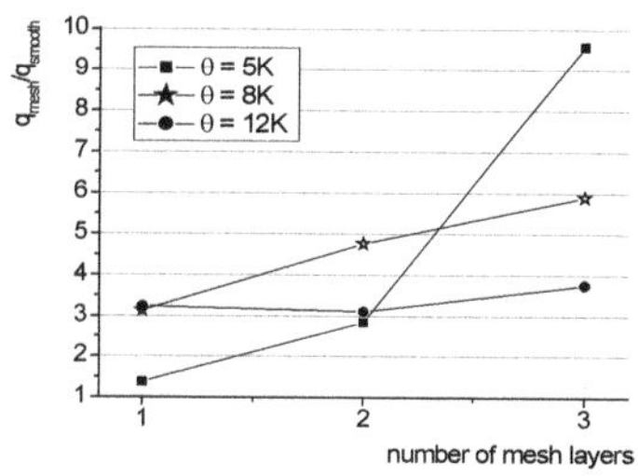

Fig. 2. Enhancement ratio for ethyl alcohol boiling

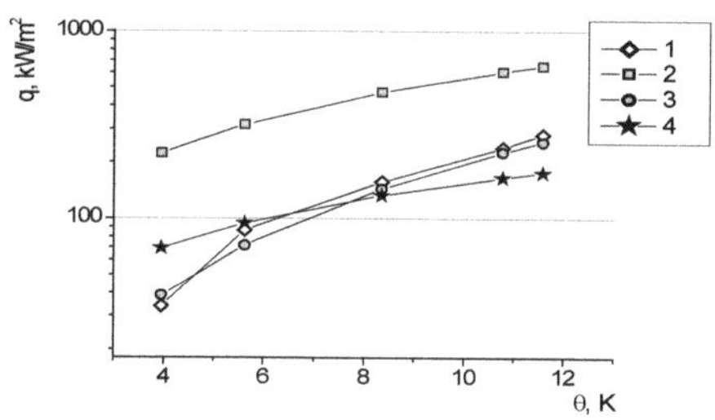

Fig. 3. Comparison of experimental data with the models; 1 - experimental results, 2 - calculation results according to Nishikava et al. correlation, 3 - calculation results according to Xin and Chao correlation, 4 - calculation results according to Smirnov et al. correlation

The model by Xin and Chao with modifications was the most accurate in this case, while the least effective – model by Nishikava et al., which is the simplest and does not consider many parameters of the microstructure.

Other surface treatment techniques are also possible and may include laser treatment of the surface as discussed by the co-author in [10, 11] and other surface treatment methods [12].

Conclusions

Sintered metal meshes can be highly effective in dissipating significant heat fluxes (even ten times higher than for the smooth surface) during pool boiling (although treated surfaces can also be very efficient in flow boiling [13]). The sintering production technology enables creating durable and strong bonds, which means that such heat exchangers can be used in vehicles or mechanical devices that undergo e.g. vibrations [14, 15]. Advanced materials science methods [16, 17], especially the image analysis of microstructures [18-21] and fuzzy assessment of uncertainty [22] seem to be useful in further investigations due to their previous usefulness in similar application in the electro-spark deposition and laser treatment of coatings [23].

Acknowledgement

Work was created within statutory research at Kielce University of Technology, no. 05.0.07.00/2.01.01.01.0033 MNSP.IKSI.16.003.

References

[1] G.F. Smirnov, A.L. Coba, B.A. Afanasiev, The heat transfer by boiling in splits, capilaries, wick structures, AIAA Paper (1978) 78 – 461.

[2] I.P. Asakavičjus, A.A. Žukauskas, V.A. Gajgalis, V.K. Eva, Teplootdača freona-113, etilovogo spirta i vody v setčatych fitiljach, Lietuvos TSR Moksly akademijos darbai, B serija, I, 104 (1978), 87-93.

[3] A. Franco, E.M. Latrofa, V.V. Yagov, Heat transfer enhancement in pool boiling of a refrigerant fluid with wire nets structures, Exp. Thermal and Fluid Science, 30 (2006), 263-275. https://doi.org/10.1016/j.expthermflusci.2005.07.002

[4] C. Li, G.P. Peterson, Y. Wang, Evaporation/boiling in thin capillary wicks (I) – wick thickness effects, J. of Heat Transfer, 128 (2006) 1312-1319. https://doi.org/10.1115/1.2349507

[5] Ł.J. Orman, T. Orzechowski, Analysis of boiling heat transfer on the non-isothermal surface covered with selected wire mesh structures, in: Proc of 7th World Conference on Experimental Heat Transfer ExHFT-7, 2009, 2159 – 2166.

[6] G.F. Smirnov, Približennaja teorija teploobmena pri kipenii na poverchnostjach pokrytych kapilljarno – poristymi strukturami, Teploenergetika, 9 (1977) 77-80.

[7] G.F. Smirnov, B.A. Afanasiev, Investigation of vaporisation in screen wick – capillary structures, in: Proc. VI Int. Heat Pipe Conf. "Advances in Heat Pipe Technology", London, 1982, 405-413.

[8] K. Nishikawa, T. Ito, K. Tanaka, Enhanced heat transfer by nucleate boiling on a sintered metal layer, Heat transfer – Japanese Research, 8 (1979) 65-81.

[9] M.-D. Xin, Y.-D. Chao, Analysis and experiment of boiling heat transfer on T-shaped finned surfaces, Chem. Eng, Comm., 50 (1987) 185-199.

[10] N. Radek, J. Konstanty, Cermet ESD coatings modified by laser treatment. Archives of Metallurgy and Materials, 57 (2012) 665-670. https://doi.org/10.2478/v10172-012-0071-y

[11] N. Radek, E. Wajs, M. Luchka, The WC-Co electrospark alloying coatings modified by laser treatment, Powder Metallurgy and Metal Ceramics 47 (2008)197-201. https://doi.org/10.1007/s11106-008-9005-7

[12] R. Ulewicz, Hardening of steel X155CrVMo12-1 surface layer, Journal of the Balkan Tribological Association, 21 (2015) 166-172.

[13] M. Piasecka, Heat transfer research on enhanced heating surfaces in flow boiling in a minichannel and pool boiling. Annals of Nuclear Energy, 73 (2014) 282-293. https://doi.org/10.1016/j.anucene.2014.06.041

[14] R. Ulewicz, F. Novy, J. Selejdak, Fatigue strength of ductile iron in ultra-high cycle regime, Advanced Materials Research 874 (2014) 43-48. https://doi.org/10.4028/www.scientific.net/AMR.874.43

[15] U. Ferdek, M.S. Kozien, Simulation of Application of FGM Piezoelectric Actuators for Active Reduction of Beam Vibrations, Acta Phys. Pol. A 123 (2013) (6) 1044-1047.

[16] A. Szczotok, B. Chmiela, Effect of Heat Treatment on Chemical Segregation in CMSX-4 Nickel-Base Superalloy, J. Mater. Eng. Perform. 23 (2014) (8) 2739-2747.

[17] K. Trzewiczek, A. Szczotok, A. Gadek-Moszczak, Evaluation of the State for The Material of the Live Steam Superheater Pipe Coils of V Degree, Adv. Mater. Res.-Switz. 874 (2014) 35-42.

[18] A. Szczotok, S. Roskosz, New possibilities of light microscopy research resulting from digital recording of images, Mater. Sci.-Poland 23 (2005) (2) 559-565.

[19] J. Korzekwa, A. Gadek-Moszczak, M. Bara, The Influence of Sample Preparation on SEM Measurements of Anodic Oxide Layers, Prakt. Metallogr.-Pr. M. 53 (2016) (1) 36-49.

[20] A. Gadek-Moszczak, History of Stereology, Image Anal. Stereol. 36 (2017) (3) 151-152.

[21] A. Gadek-Moszczak, P. Matusiewicz, Polish Stereology - a Historical Review, Image Anal Stereol 36 (2017) (3) 207-221.

[22] J. Pietraszek, M. Kolomycki, A. Szczotok, R. Dwornicka, The Fuzzy Approach to Assessment of ANOVA Results, Lect. Notes Artif. Int. 9875 (2016) 260-268. https://doi.org/10.1007/978-3-319-45243-2_24

[23] N. Radek, K. Bartkowiak, Laser Treatment of Cu-Mo Electro-Spark Deposited Coatings, Physcs. Proc. 12 (2011) 499-505. https://doi.org/10.1016/j.phpro.2011.03.061

[24] N. Radek, A. Sladek, J. Broncek, I. Bilska, A. Szczotok, Electrospark alloying of carbon steel with WC-Co-Al_2O_3: deposition technique and coating properties, Adv. Mater. Res.-Switz. 874 (2014) 101-106.

Terotechnology 2017 Materials Research Forum LLC
Materials Research Proceedings 5 (2018) 194-199 doi: http://dx.doi.org/10.21741/9781945291814-34

The Machines Operation Effectiveness Analysis in Polish industry enterprise

STASIAK-BETLEJEWSKA Renata

Czestochowa University of Technology, Faculty of Management, Department of Production Engineering and Safety, Al. Armii Krajowej 19B, 42-200 Częstochowa, Poland

renata.stasiak-betlejewska@wz.pcz.pl

Keywords: Geosynthetics, Quality, Machine Operation Effectiveness, ABS Method

Abstract. TPM (Total Productive Maintenance) is one of the main Total Quality Management components. It is an activity aimed at increasing the productivity and efficiency of processes related to maintaining traffic by increasing creative involvement of operational employees involved in these processes. The primary goal of TPM is zero failures and zero defects resulting from machines and devices operation. Machine operation effectiveness and machine modernity level analysis presented in the paper allows for identifying the machine effectiveness level that determines the product quality level in the selected textile industry enterprises.

Introduction

Geosynthetics has now become a group of materials that are attractive to designers, investors and contractors of the construction industry due to potential savings in comparison to traditional engineering solutions. However, it should be remembered that the quality of geosynthetics depends to a large extent on the level of modernity of production machines and the level of effectiveness of their work. According to a definition given by the International Geosynthetics Society, a *geosynthetic* is a flat material made of polymer, used in construction in contact with soil, rock or other geotechnical material [1]. *Geotextiles* (GTX) are the basic group of permeable geosynthetic products. According to [2], they are defined as flat, permeable, polymeric (synthetic or natural) textile products that can be woven, nonwoven or knitted, used in contact with soil and/or other materials in geotechnics and construction. This group includes geo-textiles, geotextiles and geeks. Geotextile (GTX-W), according to [3], is a textile product made of two (or more) systems of yarns, continuous fibers, tapes or other elements, usually interlaced at right angles, made with a classic weaving technique (warp has the longitudinal direction to the length of the fabric, the thread is made of threads perpendicular to the warp) [11,12].

The aim of the research findings presented in the paper is the analysis of the selected machine operation effectiveness and its modernity level with regard to the textile industry product quality that is determined mainly by the machine operation conditions and the machine use level. TPM indicators and the machine modernity level analysis method (ABC method) were used to indicate selected machine conditions responsible for the final textile product quality.

Research methodology

Operation of machines and devices in the textile industry is one of the key factors affecting the final product quality level and thus its functionality and durability. One of the main element of Total Quality Management is Total Productive Maintenance that is aimed at preventing deterioration of machines or defects and eliminating serious factors responsible for losses in the form of low quality products. The essence of the TPM concept is uncompromising prevention of damage, which is based on three assumptions [3, 4]:

1. Maintaining normal working conditions of technical facilities. In order to maintain normal machine working conditions, employees must prevent damage to technical objects by regular cleaning, inspecting, lubricating, tightening loosened components and checking accuracy.
2. Early detection of incorrect work of technical facilities. In doing so, operators should use their senses and measuring instruments available in order to detect the earliest signs of abnormal operation. In order to detect incorrect operation of technical objects, maintenance workers should also carry out periodic diagnostic checks with the help of specialist equipment.
3. Immediate reaction to incorrect work of technical facilities. Operators and maintenance staff cannot afford to delay in responding to malfunction of technical objects, as this could seriously undermine the achieved level of productivity of technical objects.

The TPM concept covers the full life cycle of devices and such issues in the cycle as: operation, inspection, maintenance, repair, diagnostics, design and manufacture of machinery and equipment. Statistical data show that 80% of the full equipment life cycle costs are set at the stage starting from design to commissioning of the machine, while costs related to operation (maintenance, repairs) often far exceed the purchase price of the device [9,10]. The TPM system is a system which serves maximizing the effectiveness of the equipment and establishing a general pro-active service system covering the entire equipment lifetime. It covers all departments (including planning department, departments using equipment, departments supporting equipment), all personnel from top management to workshop staff who participate in it, promotes handling oriented to the manufacturing process thanks to incentive management, directs enterprise activities to the elimination of the so-called six great losses [5].

In order to implement the TPM system in an enterprise, PM (Preventive Maintenance), that is prophylactic maintenance of technical facilities should be introduced first. The implementation of TPM brings many benefits. According to [6], the main benefits include reduction of failures by 90%, reduction of customer complaints by 75%, reduction of the maintenance services need, increase of technical facilities readiness for immediate production by 40%, increase of technical facilities work rate by 10%, increase of productivity by 150%, reduction of maintenance costs and technical objects maintenance with a 30% reduction in manufacturing costs by 30%, reduced inventory levels by 50% and a serious decrease in the number of the work accidents.

To measure the effectiveness of TPM implementation, the OEE (Overall Equipment Effectiveness) parameter describing overall efficiency of equipment is used. It is calculated as a result of the mentioned indicator which calculates the device availability indicator, the use of working time indicator, the efficiency indicator and production quality indicator. In general, it is calculated as [6]:

$$\text{OEE} = \text{availability} \times \text{use of working time} \times \text{efficiency} \times \text{quality}. \qquad (1)$$

Equipment availability (availability) is the percentage of time in which the station is available for production. The use of working time (operating efficiency) is the percentage share of the actual operating time in the time available for production. The use of working time is affected by periods of inactivity of the device due to the lack of product, lack of operator, break for meals and breaks at meetings of production staff.

Efficiency, as the efficiency ratio (efficiency rate) is the ratio between the capacity of the actual and theoretical maximum efficiency of the device. *Quality*, as the quality index (rate of quality) indicates the percentage of good products in total production made on the device. Products that are eligible for improvement are also considered to be defective [7].

Machine components modernity level is the next element which is crucial for the product quality level identification. The level of modernity of the machine can be assessed by the ABC method using the Parker scale with the following levels [5, 6]:

- level 1 - simple parts, it is possible to create them with the help of craft techniques, e.g. cover,
- level 2 - parts for which unchanged and well-known technologies have been used for many years, e.g. standard engine cooling system,
- level 3 - parts produced using a controlled technology, requiring appropriate technical knowledge, e.g. a standard electric motor,
- level 4 - parts that have been produced using modern market technologies, e.g. displaying information on the monitor of the control panel,
- level 5 - parts resulting from the production of a combined use of modern technologies, patented and existing only in a specific company's machine.

The use of this modern categorization allows for the alignment of component parts at the level of relevance of their development and investment, e.g. decision on which of them need to be moved to a higher level. Technologies often create constellations that use rules of scientific approach to a problem. However, the ABC analysis of the technology always allows for the specification of significant technological capabilities of the company. The ABC analysis of technology is based on the principle that each statistical group can be distinguished by several of its segments, conventionally marked with the A letter, which determines the main part of the effects. At the other end of distribution there are many members of the group marked with the C symbol, with a small contribution to the results of its activity. However, their work cannot be completely disregarded. Others are an indirect group members marked with the B letter [5, 6, 8]. The ABC analysis consists in classifying technologies used in a given machine or device. Basic technologies are technologies of A category, which include fundamental components of machines that give a product special attributes. The B category is assistive technology, often of a general nature. Side technologies, the C category, are mainly technologies that are not subject to any innovative activity of the user and are of little importance when buying new machines.

Research object and research findings analysis

The subject of the analysis in the article is a selected machine that plays a key role in the process of weaving a geotextile in the form of a flat loom. Weaving involves the interlacing warp thread resulting in a fabric. Two main mechanisms of the loom are involved in the weaving operation: the feeding mechanism which provides the warp and the metastatic mechanism which introduces the weft. The employee is responsible for the operation of a machine and controls a visually resultant fabric. Characteristics of a flat loom has been presented in the context of the analysis of the results using the ABC method.

The analyzed machine, named Sulzer Power Leno loom, was manufactured in 2007. It is used for the production of polypropylene fabrics. It consists of two main elements: the loom and winder. It has two types of sensors spurt: the sensor warp and sensor weft. Specifications loom flat are as follows: 8 kW power output, yield 265 rev / min, voltage of 400V, the noise level approx. 75 dB. The level of modernity of the loom was assessed based on a comparison of this machine with a loom manufactured by Textima, a market leader. Using a scale and the method of Parker ABC technology, results were obtained which are shown in Table 1.

Materials Research Proceedings **5** (2018) 194-199 doi: http://dx.doi.org/10.21741/9781945291814-34

Table 1. Evaluation of the Sulzer Power Leno loom parts modernity level

Part category	No	Machine subassembly	Technology level
A	A1	The control system	5
	A2	Control panel	5
	A3	Programming system	4
	A4	Measuring system	4
	A5	Repair system	4
	A6	Winding the finished product	4
	A7	Automatic change in threading density	4
	A8	Fabric regulator	4
B	B1	Main propulsion system	4
	B2	Adjustment system	4
	B3	Pneumatic system	5
	B4	Compressor (air network)	4
	B5	Attaching the workpiece	4
	B6	An innovative system of changing the weaving isthmus	5
C	C1	Machine construction (housing)	4
	C2	Lubrication system	4
	C3	Ribbon guide elements	5

Source: own elaboration.

The ABC analysis of the technology of individual components of the Power Leno flat bar components using the Parker five-degree scale showed that 71% of its components are at the 4th level of modernity of this scale, while the remaining 29% are components classified at the 5th level of modernity according to the Parker scale. The occurrence of parts at the modern level 5 indicates that the machine uses the latest technology. It is a guarantee of producing high quality products.

The evaluation of the effectiveness of the use of the Power Leno flat loom with the use of TPM coefficients were carried out for a period of 22 weeks. The individual elements of the OEE calculation sheet have been presented in Table 2.

The value of the overall efficiency of the OEE device allows for assessing the overall efficiency of the flat loom. The analyzed machine achieves results in the range of 48.42% ÷ 59.25%. The UCD indicator reaches values in the range of 68.27% ÷ 83.75%. The machine also achieves a high level of WJ index that is over 99% (which guarantees the production of high-quality geotextiles).

Conclusions

The article presents the results of the efficiency factor analysis for 22 weeks. The value of OEE (desirable over 60%, satisfying over 80%) allows for the assessment of the efficiency of the use of machines, and consequently of the whole process from the point of view of machines and devices.

The analysis of TPM coefficients for machines involved in the production of geotextiles made it possible to state that the efficiency of these machines is not satisfactory. The desired value of OEE should be at the level of 60%, and satisfactory at above 80%.

Table 2. Analysis of the efficiency of the Power Leno flat loom during the research period of 22 weeks

The research period [week]	Shift work time fund TZ [h]	Downtime TP [h]	Machine operation time (TZ-TP) [h]	Operating coefficient WE [%]	Perfect unit time ICJ [min/rot.]	Real unit time RCJ [[min/rot.]	Operating speed factor WPD [%]	Production P [rot.]	Usable time of action UCD [%]	The efficiency coefficient WW [%]	The quality level PJ [%]	The quality coefficient WJ [%]	Total machine efficiency OEE [%]
1	168	8	160	95.24	0.004	0.0055	72.73	2371200	81.51	59.28	0.05	99.95	56.43
2	168	6.5	161.5	96.13	0.004	0.0055	72.73	2335300	79.53	57.84	0.05	99.95	55.57
3	168	5	163	97.02	0.004	0.0055	72.73	2318200	78.22	56.89	0.10	99.90	55.14
4	168	3.5	164.5	97.92	0.004	0.0055	72.73	2489900	83.25	60.54	0.05	99.95	59.25
5	168	5.5	162.5	96.73	0.004	0.0055	72.73	2474500	83.75	60.91	0.10	99.90	58.86
6	168	3.5	164.5	97.92	0.004	0.0055	72.73	2255000	75.40	54.83	0.05	99.95	53.66
7	168	9	159	94.64	0.004	0.0055	72.73	2337050	80.84	58.79	0.05	99.95	55.62
8	168	7	161	95.83	0.004	0.0055	72.73	2265500	77.39	56.29	0.05	99.95	53.91
9	168	4	164	97.62	0.004	0.0055	72.73	2035700	68.27	49.65	0.10	99.90	48.42
10	168	9.5	158.5	94.35	0.004	0.0055	72.73	2376500	82.47	59.97	0.10	99.90	56.53
11	168	3.5	164.5	97.92	0.004	0.0055	72.73	2277600	76.15	55.38	0.05	99.95	54.20
12	168	2	166	98.81	0.004	0.0055	72.73	2134050	70.71	51.42	0.10	99.90	50.76
13	168	4.5	163.5	97.32	0.004	0.0055	72.73	2385000	80.23	58.35	0.25	99.75	56.64
14	168	8	160	95.24	0.004	0.0055	72.73	2175500	74.78	54.39	0.15	99.85	51.72
15	168	4	164	97.62	0.004	0.0055	72.73	2253000	75.56	54.95	0.05	99.95	53.62
16	168	4.5	163.5	97.32	0.004	0.0055	72.73	2346500	78.93	57.41	0.05	99.95	55.84
17	168	10	158	94.05	0.004	0.0055	72.73	2354600	81.96	59.61	0.05	99.95	56.03
18	168	6.5	161.5	96.13	0.004	0.0055	72.73	2355300	80.21	58.34	0.10	99.90	56.02
19	168	4	164	97.6	0.004	0.0055	72.73	2375500	79.67	57.94	0.25	99.75	56.42
20	168	7	161	95.83	0.004	0.0055	72.73	2379500	81.29	59.12	0.10	99.90	56.60
21	168	3.5	164.5	97.92	0.004	0.0055	72.73	2365400	79.09	57.52	0.05	99.95	56.29
22	168	7	161	95.83	0.004	0.0055	72.73	2374750	81.13	59.00	0.10	99.90	56.49

Source: own elaboration.

References

[1] Recommended descriptions of geosynthetics functions, geosynthetics terminology, mathematical and graphical symbols. International Geosynthetics Society, 2000.

Materials Research Proceedings **5** (2018) 194-199
doi: http://dx.doi.org/10.21741/9781945291814-34

[2] PN-EN ISO 10318:2007 Geosyntetyki. Terminy i definicje.

[3] J. Łunarski J. Zarządzanie jakością. Standardy i zasady, Wydawnictwo Naukowo – Techniczne, Warszawa, 2008.

[4] S. Borkowski, R. Ulewicz, J. Selejdak, M. Konstanciak, D. Klimecka-Tatar, The Use of 3x3 Matrix to Evaluation of Ribbed Wire Manufacturing Technology. METAL 2012. 21st International Conference on Metallurgy and Materials. Conference Proceedings. TANGER. May 23rd - 25th, Brno, Czech Republic. 2012.

[5] S. Borkowski, J. Selejdak, S. Salamon. Efektywność eksploatacji maszyn i urządzeń, Wydawnictwo Wydziału Zarządzania Politechniki Częstochowskiej, Częstochowa 2006.

[6] J. Pietraszek, M. Kolomycki, A. Szczotok, R. Dwornicka, The Fuzzy Approach to Assessment of ANOVA Results, Computational Collective Intelligence, ICCCI 2016, PT I, Lecture Notes in Artificial Intelligence 9875 (2016), pp. 260-268

[7] M. Urbaniak. Zarządzanie jakością. Teoria i praktyka, Wydawnictwo Difin, Warszawa. 2004.

[8] R. Ulewicz, D. Jelonek, M. Mazur, Implementation of logic flow in planning and production control, Management and Production Engineering Review 7/1 (2016), pp. 89-94. https://doi.org/10.1515/mper-2016-0010

[9] Maszke A., The analysis of machine operation and equipment loss in ironworks and steelworks, Production Engineering Archives 17(2017), pp. 45-48. https://doi.org/10.30657/pea.2017.17.10

[10] S. Stasiak-Betlejewska, R. Ulewicz, The effectiveness of selected machinery and equipment in the woodworking joinery, Path forward for wood products: a global perspective, Proceedings of scientific papers, 2016 , pp. 149-156.

[11] T. Ylipaa, A. Skoogh, J. Bokrantz, Jon; et al., Identification of maintenance improvement potential using OEE assessment, International journal of productivity and performance management 66(1)2017. pp.126-143. https://doi.org/10.1108/IJPPM-01-2016-0028

[12] A. Tifan, Directions for improving management accounting in the textile industry enterprises, Industria Textila 65(2) 2014, pp.101-106

Terotechnology 2017 Materials Research Forum LLC
Materials Research Proceedings **5** (2018) 200-205 doi: http://dx.doi.org/10.21741/9781945291814-35

Tests of New Methods of Manufacturing Elements for Water Hydraulics

Justyna KRAWCZYK[1,a], Andrzej SOBCZYK[2,c], Jarosław STRYCZEK[1,b], Paweł WALCZAK[2,d*]

[1]Wrocław University of Science and Technology, Department of Fundamentals of Machine design and Tribology, Fluid Power Research Group, ul. Łukasiewicza 5, 50-371 Wrocław, Poland

[2]Cracow University of Technology, Faculty of Mechanical Engineering, Laboratory of Techno - Climatic Research and Heavy - Duty Machines, al. Jana Pawła II 37, 31-864 Kraków, Poland

[a]justyna.krawczyk@pwr.edu.pl, [b]jaroslaw.stryczek@pwr.wroc.pl, [c]andrzej.sobczyk@mech.pk.edu.pl, [d]pawel.walczak@mech.pk.edu.pl

Keywords: Water Hydraulics, Hydraulic Cylinders, Hydraulic Pumps, Directional Control Valve, Coating, Plastic

Abstract. The article presents a review of modern methods of producing hydraulic components that can work with water as a working medium. The presented technology of coating and production of plastic elements is an alternative to the currently used stainless steel construction. Large-scale testing of cylinders and valves made in coating technology has been carried out. Innovative methods for designing pumps made of plastic have been developed, and were confirmed by laboratory tests.

Introduction

Research conducted on the use of water as a working medium not only in power hydraulics, but also in other branches of industry (fire extinguishing, refrigeration, etc.) confirm the existence of problems with its use [1, 2, 3]. However, the advantages far outweigh its disadvantages. That is why it is so important to use water hydraulics in natural environment friendly applications [4]. Especially if their price will not be a big obstacle. Currently, the cost of components made of stainless steel 3 ÷ 4 times exceeds the price of standard oil hydraulic elements [5, 6]. The first solution to its reduction is the technology of chemical-physical coatings integrating the Diamond Like Carbon coating technology (DLC) Coating with innovative technology with self-lubricating properties of the Columnar Nanostructured Coating (CNC). The research of these coatings was one of the topics of the EU project called "Novel high-pressure water hydraulic equipment for application in the mining and mining sector" with the working name "Hydrocoat". Tests were carried out in the fluid power laboratory which currently is a part of the Laboratory of Techno-Climatic Research and Heavy-Duty Machines at the Cracow University of Technology. The second method is the use of plastics. The research of hydraulic pumps and other elements made of plastic bodies is carried out by the Fluid Power Research Group at the Wrocław University of Science and Technology.

Materials

The parts produced for oil hydraulics were specially treated. The metal-contacting parts were covered with special patented protective layers. This technology is treated as a solution enabling an adaptation of the already existing valves and cylinders to work in a water hydraulic system. These layers are the following: the Diamond Like Carbon (DLC) and Columnar Nanostructured Coating (CNC) with self-lubricating properties. Their use contributes to the lowering of the friction force and also provides a layer protecting steel from corrosion [2, 7]. In addition, using the experience gained

Materials Research Proceedings **5** (2018) 200-205 doi: http://dx.doi.org/10.21741/9781945291814-35

in the research of water hydraulic systems, the seals were replaced in the tested cylinders. The new seals were made of materials that could cooperate with water, such as: ULTRALEN 90 and KEFLOY 22 [1].

In turn, the use of plastics in the construction of pumps brings design, technological, operational and economic benefits such as reduction of mass, simplification of construction, self-sealing ability, improvement of tribological properties between cooperating elements, increase of resistance to contaminants in the working fluid, increase of the ability to damp vibrations and reduce noise [8]. And above all, pumps made of plastics can work with various working fluids, such as hydraulic oils, water, emulsions, nanofluids and chemical fluids [2, 9]. As a plastics easily available on the market, cheap and easy to process polyoxymethylene POM was chosen [10].

Cylinder tests

The aim of the research was to confirm the effectiveness of the developed cover technologies. The Polish Standard PN - 72 / M - 73202 concerning tests of oil hydraulic cylinders was used during the tests of a water hydraulic cylinder. No standards have yet been developed for testing elements using water as a working fluid. According to it, external leak tests and internal leak tests were conducted, the friction force, volumetric efficiency, hydraulic - mechanical efficiency and total efficiency were determined [11].

Tests with a static pressure load did not show measurable external or internal leaks. Therefore, the determination of hydraulic - mechanical efficiency was reduced to determine the equivalent, total efficiency of the cylinder as a function of piston velocity and working pressure. A special stand was designed for the tests, where in the welded frame of C - profiles, cylinders connected with a trolley were mounted. The wheels of a trolley were run along the inner surface of the frame. One of the cylinders was used for the drive, while the other was used to generate the load. The displacement of the piston L, the pressure on the piston side p_{a1} and the pressure on the rod side p_{a2} of the cylinder were measured. During efficiency tests, the temperature of the water in the hydraulic system was T = + 40 °C. They were repeated, for comparison, for other values of temperature. To determine the impact of pressure on the total efficiency, tests were carried out at the piston velocity of v = 0,2 m/s and the working pressure p_Z changing from the minimum to the nominal value. To determine the influence of velocity on the total efficiency of the cylinder, tests were carried out at the nominal load pressure and the speed changing from 0,05 to 0,2 m/s.

Figure 1 shows three - dimensional characteristics of the determined total cylinder η efficiency for the extension and retraction depending on the velocity of the v piston and the theoretical F force. The force was calculated as a product of pressure and surface area on the piston side of the cylinder during extension and on the rod side of the cylinder during retraction.

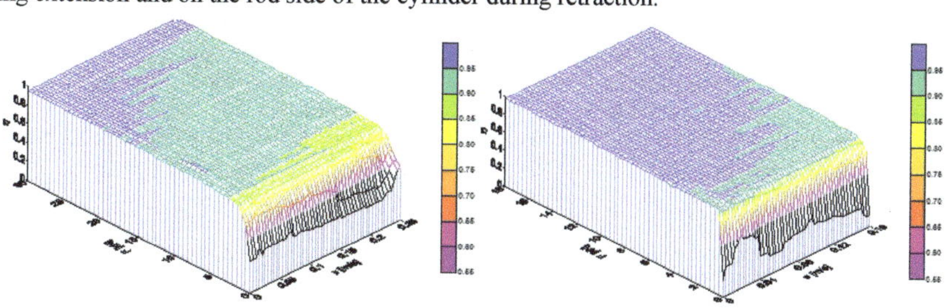

Fig. 1. Characteristics of the total cylinder efficiency during extension and retracting [7]

Materials Research Proceedings **5** (2018) 200-205 doi: http://dx.doi.org/10.21741/9781945291814-35

Valves tests

The problem of a proper design of the gap between the movable elements of the spool directional control valve is particularly difficult when water is used as a working medium [12]. The low value of its viscosity coefficient causes the increase of leakages and a friction force. For these reasons, a directional control valve was constructed of four two-way ON/OFF valves. They were chosen because they are durable and relatively uncomplicated. Four separate ON/OFF valves were covered with a CNC coating and mounted in a specially designed aluminium block (Fig. 2). To control the coils of these valves, a programmable electronic module was designed and made. It enabled an independent activation and deactivation of each ON/OFF valve. That is why the distributor could work in any configuration of ways. Tests of the directional control valve consisted of functional tests in a hydraulic system with a cylinder and determination of flow characteristics for individual way of valve (Fig. 3) [13].

Fig. 2. View of a four - way, three - position manifold body with four ON/OFF valves [13].

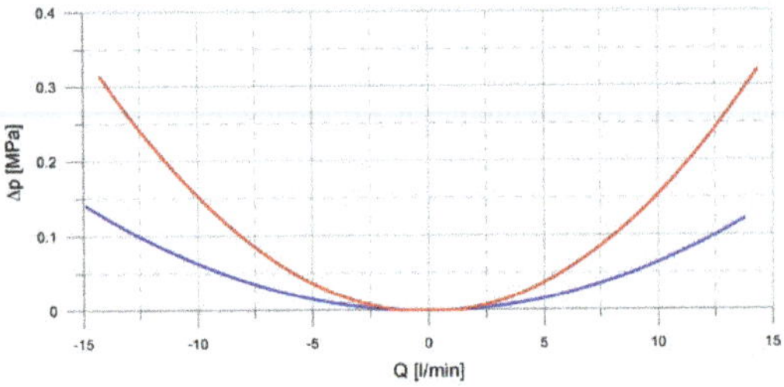

Fig. 3. Flow characteristics for roads P - A (red graph, negative values Q), P - B (red graph, positive values Q), A - T (blue graph, positive values Q), B - T (blue graph, negative values Q)

Pump tests

The tests of pumps and rotary motors with the use of covers did not bring the expected results. The exact analysis of the problem revealed that the use of coatings presents difficulties in keeping the strength and clearances between the elements. Therefore, it was considered reasonable to make such an element from scratch, ideally, from an easily accessible and easily machinable material such as plastics [14, 15]. It was assumed that the general shape of a pump body should be formed as a prism with a square base. By removing the material from this general shape and applying the principles of global and local modification, the final shape of the pump body was obtained (Fig. 4). In addition, using the generally known methods of designing hydraulic machines and systems and the Finite Element Method (FEM), the author's methodology for shaping the bodies of hydraulic machines from plastics was developed [14].

Materials Research Proceedings **5** (2018) 200-205 doi: http://dx.doi.org/10.21741/9781945291814-35

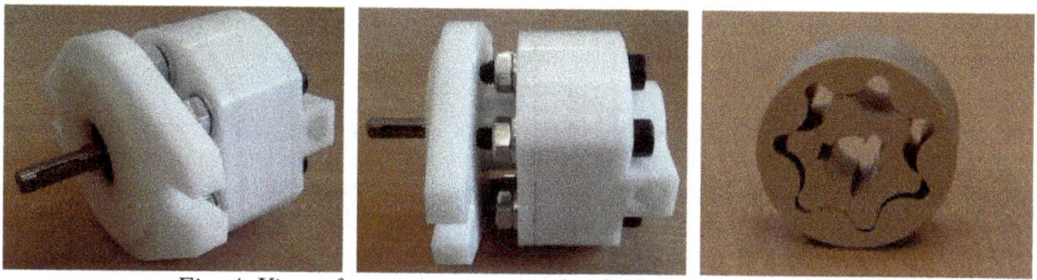

Fig. 4. View of a gerotor pump with a body made of plastics

The assembled gerotor pump with a plastic body has been subjected to experimental research on a test stand in order to determine its basic hydraulic characteristics. The test stand enabled the measurement of hydraulic parameters, such as flow and pressure at the inlet and outlet of the pump, and measurements of mechanical parameters, such as torque and rotational speed. Signals from measuring instruments were sent to the computer using a laboratory signal amplifier. The CatmanEasy computer program was used to archive and analyze the measurements on the computer. The tested pump was driven by a DC electric motor with the power of 30 kW and the maximum speed of n = 3000 rpm. A throttle valve was used to load the pump. To protect the pump and electric motor from overload, an overflow valve was used.

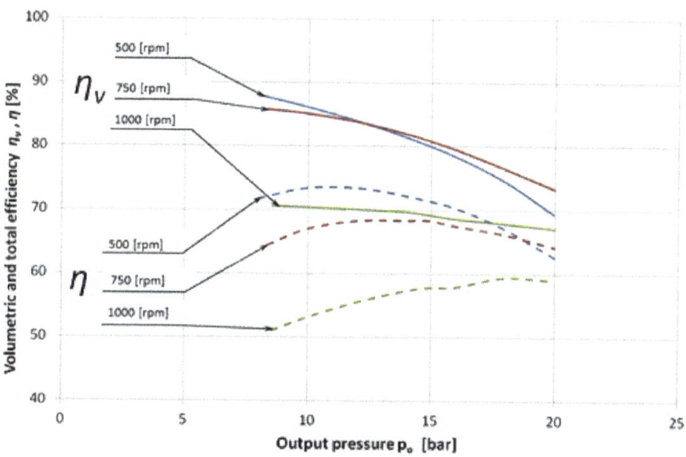

Fig. 5. The characteristics of volumetric and total efficiency, depending on the output pressure η_v, $\eta = f(p_{out})$, for different rotational speeds n = 500 ÷ 1000 rpm

Figure 5 shows that the pump was working in the range of working pressure of p = 0 - 20 bar and rotational speed of n = 500 - 1000 rpm. The η_v volumetric efficiency varied in the range of η_v = 89 - 70%, and the total η efficiency varied in the range of η = 73 - 50%. The relatively low efficiency prevented further loading of the pump, which could lead to its seizure. Figure 4 shows that the volumetric η_v and total η efficiency decreases with the increase of the p output pressure and the n rotational speed. This is explained by the fact that as the p pressure and n rotational speed increase, the deformation of the plastics pump body increases, which causes internal leakages in the pump, which in turn decreases the efficiency of the pump.

Summary

The results of tests of cylinders, directional control valves and pumps presented in this paper confirmed the possibility of using water in hydraulic systems. Working parameters of the tested hydraulic components are at satisfactory levels [16]. High efficiency and no leakages in the cylinders and low resistance of flow in the directional control valve decrease hydraulic, mechanical and volumetric losses in the entire system. Energy consumption is reduced. The whole system becomes more efficient and more economical. Covering the inner surface of the cylinder and the surface of the piston and piston rod confirms the protective properties of the coatings. They protect steel elements against corrosion. The lack of observable internal and external leakages during the static pressure test, testify proper cooperation between the sealing material, cylinder surface and piston rod, despite extremely low viscosity of water in relation to viscosity of hydraulic oil.

Two main objectives have been achieved in relation to hydraulic pumps: the methodology of designing bodies of hydraulic machines made of plastics was developed and verified by designing, making and experimental testing of a gerotor pump with a plastic body. The pump with the POM body works properly in the pressure range of $p = 0 \div 20$ bar and the speed of $n \leq 1000$ rpm. At the same time in future, plastics with higher strength and greater dimensional stability should be chosen, so it can be used for medium pressure pump bodies ($p = 60$ bar).

Using the developed methodology of designing bodies of hydraulic machines made of plastics, the construction solution can be improved from the point of view of various criteria, including material, strength, geometrics and technology. The principles of global and local modifications of the pump body are open and in the future can be expanded and improved. By using the modification process, the structural solution of pump, valves and motor bodies can be corrected at the design stage.

Promising results were obtained, giving a basis for further work on a new generation of relatively cheap components for water hydraulics. Currently, the Fluid Power Research Group from Wrocław University of Science and Technology is working on adapting the plastic pump to work on water as a working medium. The tests will be carried out at a stand prepared by the Water Hydraulics Laboratory at the Cracow University of Technology. Considering extensive experience of scientists from both universities, the success of the project is expected.

References

[1] W. Okularczyk, A. Sobczyk, Experimental investigations of guide rings made of UHMWPE and PTFE composite in water hydraulic systems, Proceedings of 5th FPNI PhD Symposium, Krakow (2008)

[2] J. Pobędza, A. Sobczyk, Modern coating used in high pressure water hydraulic components, Key Engineering Materials, Vol. 542, Advanced Materials in Machine Design, Trans Tech Publications Ltd., Zurich-Durnten (2013) 143-155.

[3] Kędzia Krzysztof: An algorithm for the determination of the control parameters of a multisource drive system, Technical Transactions, R. 114, vol. 3 (2017) 173 - 182.

[4] F. Conrad, J. Pobędza, A. Sobczyk, IT-Tools Concept for Simulation and Design of Water Hydraulic Mechatronic Test Facilities for Motion Control and Operation in Environmentally Sensitive Application Areas, Proceedings of ASME Intl Mechanical Engineering Congress IMECE´04: Fluid Power Systems and Technology. Vol. 2, New York (2004) 1 - 9. https://doi.org/10.1115/IMECE2004-62411

[5] A. Sobczyk, Water Hydraulics - the Industrial Perspective. Water Hydraulics. The Natural Choice, Bergen University College, Bergen, 2004, pp 34 – 67.

[6] A. Sobczyk, P. Walczak P, Selection of materials to fabricate pressure control valve components for water hydraulics systems application - selected aspects, Proceedings of 2011 International Salon of Hydraulics and Pneumatics - HERVEX, Calimanesti-Caciulata, (2011) 158 – 166.

[7] J. Pobedza, A. Sobczyk, Properties of high pressure water hydraulic components with modern coatings. Advanced Materials Research. Trans Tech Publications Ltd (2014)

[8] J. Stryczek, K. Biernacki, J. Krawczyk, J. Wołodźko, Application of plastics in the building of fluid power elements, Vìsnik Nacìonal'nogo Tehnìčnogo Unìversitetu Ukraïni "Kiïvs'kij Polìtehnìčnij Ìnstitut", Journal of Mechanical Engineering of the National Technical University of Ukraine, nr 2 (2014) 5 - 11.

[9] K. Nering, K. Rup, Unsteady natural convection in micropolar nanofluids, Archives of Thermodynamics, Vol. 35, No. 3, (2014) 155 - 170.

[10] P. J. Gamez-Montero, P. Antoniak, R. Castilla, J. Freire, J. Krawczyk, J. Stryczek, E. Codina, Magnet-sleeve-sealed mini trochoidal-gear pump prototype with polymer composite gear, Energies, vol. 10, nr 10 (2017) 1 - 18. https://doi.org/10.3390/en10101458

[11] A. Sobczyk, P. Walczak P, New Approach in Simulation of Water Turbine Guide Apparatus Control System, Proceedings of the 7th FPNI PhD Symposium, Reggio Emilia, (2012) 981 – 991.

[12] P. Śliwiński, Influence of Water and Mineral Oil on the Leaks in Satellite Motor Commutation Unit Clearances. POLISH MARITIME RESEARCH 3 (95) Vol. 24 (2017) 58 - 67.

[13] A. Gawlik, J. Pobędza, A. Sobczyk, P. Walczak, Tests of directional poppet valves with DLC and CNC coatings for water hydraulic, International Scientific-Technical Conference: Hydraulics and Pneumatics, Wroclaw (2012) 519 - 531.

[14] J. Krawczyk, J. Stryczek, Designing of the gerotor pump body made of plastics Proceedings of ASME 2016 9th FPNI Ph.D. Symposium on Fluid Power [electronic paper], Florianópolis (2016). ASME, New York (2016) 1 - 6.

[15] J. Krawczyk, J. Stryczek, Construction and experimental research on plastic cycloidal gears used in gerotor pumps, 8th FPNI Ph.D Symposium on Fluid Power [electronic paper], Lappeenranta (2014). ASME,New York (2014) 1 - 6.

[16] A. Sobczyk, P. Walczak P, Simulation of Water Hydraulic Control System of Francis Turbine, ASME Proceedings of the 8th FPNI PhD Symposium, Lappeenranta (2014).

Materials Research Proceedings **5** (2018) 206-209 doi: http://dx.doi.org/10.21741/9781945291814-36

Microstructure and Tribological Properties of ESD Coatings after Laser Processing

RADEK Norbert [1, a *], PIETRASZEK Jacek[2,b] and SZCZOTOK Agnieszka[3,c]

[1]Kielce University of Technology, Faculty of Mechatronics and Mechanical Engineering, Al. 1000-lecia P.P. 7, 25-314 Kielce, Poland

[2]Cracow University of Technology, Institute of Applied Informatics, Faculty of Mechanical Engineering, Al. Jana Pawła II 37, 31-864 Cracow, Poland

[3]Silesian University of Technology, Institute of Materials Science, str. Krasinskiego 8, 40-019 Katowice, Poland

[a]norrad@tu.kielce.pl, [b]pmpietra@gmail.com, [c]agnieszka.szczotok@polsl.pl

Keywords: Electro-Spark Deposition, Laser Processing, Coatings, Properties

Abstract. The paper is concerned with determining the influence of the laser treatment process on the properties of electro spark coatings. The properties were assessed after laser treatment by analysing the microstructure, measuring the microhardness and friction coefficient. The tests were carried out for Mo and Cu coatings (the anode) electro-spark deposited over the C45 steel substrate (the cathode) and molten with a laser beam. The coatings were deposited by means of an ELFA-541. The laser treatment was performed with an Nd:YAG, BLS 720 laser. The coatings are desirable in sliding friction pairs.

Introduction

The processes of coating formation on metal parts including electro-spark deposition involve mass and energy transport accompanied by chemical, electrochemical and electrothermal reactions [1]. Today, different electro-spark deposition techniques are used; they are suitable for coating formation and surface microgeometry formation [2-4].

Electro-spark alloying is becoming more and more popular as a surface processing technology. Electro-spark deposited coatings are frequently applied in industry, for example, to produce implants or cutting tool inserts. The coatings are deposited with manually operated equipment or robotized systems.

Coatings produced by electro-spark deposition are applied:

1. to protect new elements,
2. to recover the properties of worn elements.

As electro-spark coatings are reported to be resistant to wear and corrosion, they can be applied, for example, to the following:

- ship propeller components,
- casting moulds,
- fuel supply system components,
- exhaust system components.

Electro-spark deposited coatings are not free from disadvantages but these can be easily eliminated. One of the methods is laser processing; a laser beam is used for surface polishing, surface geometry formation, surface sealing or for homogenizing the chemical composition of the coatings deposited [5-8].

The work discusses the properties of electro-spark deposited Cu-Mo coatings subjected to laser treatment. The properties were established based on the results of a microstructure analysis, microhardness tests and tribological studies.

There are many alternative technologies for producing coatings and material properties improvements in relation to ESD technology [9-11].

Experimental

The tests were conducted for Cu-Mo coatings produced by electro-spark deposition, which involved applying Cu and Mo electrodes with a diameter of 1 mm (the anode) on the C45 steel substrate (the cathode). Here, copper constitutes the core coating material in the formation of low-friction surface layers; it also compensates for the occurrence of residual stresses. Molybdenum acts as the reinforcing constituents. The coating materials, i.e. molybdenum (99.8% Mo) and copper (99.2% Cu) in the form of a wire (ϕ = 1 mm) were purchased from BIBUS Metals Sp. z o.o. (certificate included).

The heterogeneous coatings were electro-spark deposited on the C45 steel substrate by means of the ELFA-541 made by a Bulgarian manufacturer. Based on the analyses of the current characteristics as well as the manufacturer's recommendations, it was assumed that the parameters of the ESD operation should be as follows: current intensity I = 16 A (for Cu I = 8A); table shift rate v = 0.5 mm/s; rotational speed of the head with electrode n = 4200 rev/min; number of coating passes L = 2; capacity of condenser system C = 0.47 μF; pulse duration T_i = 8 μs; interpulse period T_p = 32 μs; frequency f = 25 kHz.

The subsequent laser treatment was performed with the aid of a BLS 720 laser system employing the Nd:YAG type laser operating in the pulse mode. The following parameters were assumed for the laser treatment: laser spot diameter d = 0.7 mm; laser power P = 20 W; beam shift rate v = 250 mm/min; nozzle-sample distance h = 1 mm; pulse duration t_i = 0.4 ms; frequency f_i = 50 Hz.

Results and discussion

A Joel JSM-5400 scanning microscope equipped with an Oxford Instruments ISIS-300 X-ray microanalyzer was used to test the coating microstructure. Figure 1 shows the microstructure of electro-spark deposited on two-layer Cu-Mo coatings. The layer thickness is approximately 8÷10 μm, and the range of the heat affected zone (HAZ) inside the (underlying) substrate material is about 10÷15 μm. In the photograph, the boundary line between the two-layer coating and the substrate is clear. The melting and solidifying processes during laser treatment resulted in the migration of elements across the coating-substrate interface. Laser radiation caused intensive convective flow of the liquid material in the pool and, in consequence, the homogenization of the chemical composition (Fig. 2). In the heat affected zone (HAZ), which was 20÷50 μm thick, there was an increase in the content of carbon.

The microhardness of the specimens with Cu-Mo coatings was analyzed by applying a load of 0.4 N and using the Vickers method. The indentations were made in perpendicular microsections in three zones: the white homogeneous difficult-to-etch coating, the heat affected zone (HAZ) and the substrate. The microhardness of the substrate after electro-spark deposition equalled, on average, 280 HV0.4; the same value was reported for the substrate before the process. There was a considerable increase in microhardness after depositing the heterogeneous Cu-Mo coating. The microhardness of the Cu-Mo coating was approx. 587 HV0.4 - a rise of 110 %. The microhardness of the Cu-Mo coating in the heat affected zone (HAZ) after electro-spark treatment was 51 % higher than that of the substrate material. Laser treatment had a favorable effect on the changes in the microhardness of the electro-spark deposited on the Cu-Mo coating. There was an increase of 161 % in the microhardness of the Cu-Mo coating.

Materials Research Proceedings **5** (2018) 205-209 doi: http://dx.doi.org/10.21741/9781945291814-36

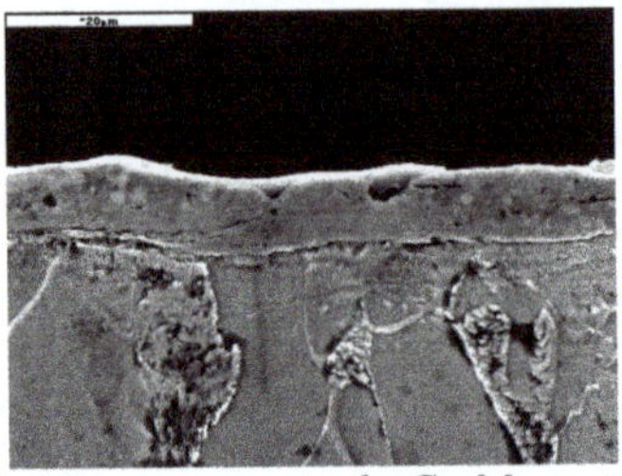

Fig. 1. Microstructure in the Cu-Mo coating

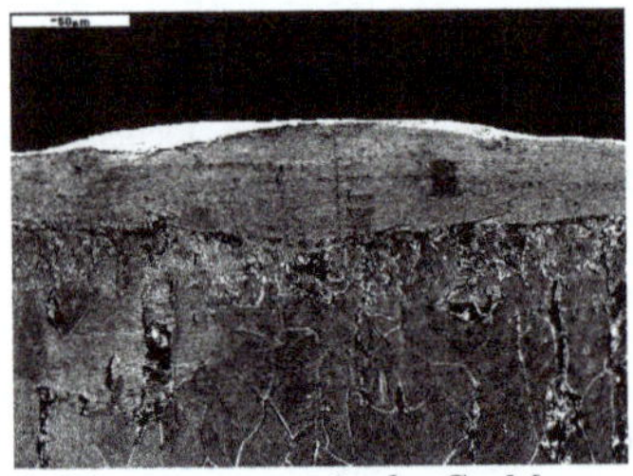

Fig. 2. Microstructure in the Cu-Mo coating after laser treatment

The coefficient of friction for the Cu-Mo coating before and after laser treatment was determined using a T-01- pin on disc type tribological tester, whose principle of operation is shown in Fig. 3. The tester enables continuous measurement of friction force at a set load. The pin of ϕ4 x 20 mm was made of medium-carbon steel with a hardness of 27 HRC. The testing was performed at the following parameters: load $Q = 10$ N, rotational speed $n = 382$ rpm, test duration $t = 500$ s.

Figure 4 shows the friction coefficient in the function of time at a load of 10N. This diagram illustrates the Cu-Mo coating before and after modification with a laser beam. Dry friction observed in the case of the coatings resulted in the transformation of the outer layer into a surface layer. This was mainly due to the sliding stresses and speed, and the interaction with the medium. The state stabilization of the anti-wear surface layer was observed.

In Fig. 4, one can see stabilization of the friction coefficient after 80 sec., its value fluctuating at 0.16÷0.18. In the case of the laser modified Cu-Mo coating, the friction coefficient stabilizes after 240 sec., and its value fluctuates at 0.35÷0.37. The average friction coefficient of the Cu-Mo coating is lower than that of the laser-modified Cu-Mo coating (at the moment of stabilization).

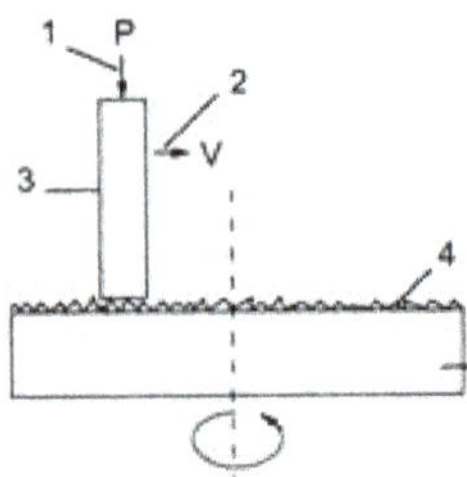

Fig. 3. Principle of operation of the pin on disc type tester: 1-applied force, 2-linear movement, 3-substrate,4-abrasive surface, 5-rotating disc

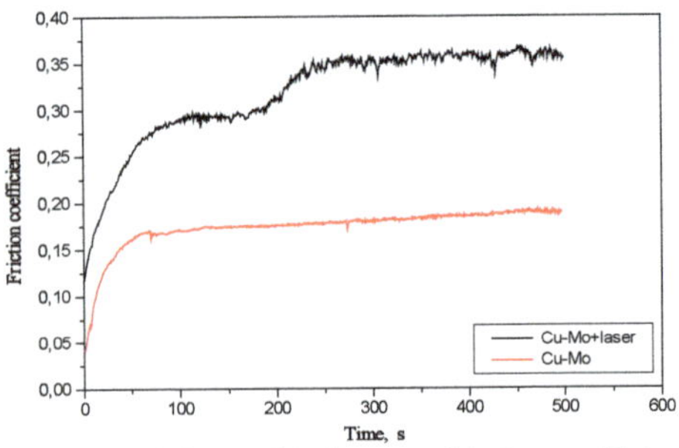

Fig. 4. Relationship between friction coefficient and time

Summary

1. There is no change in the chemical composition of electro-spark deposited coatings after laser treatment in spite of their melting and solidification. The results of laser radiation are the

homogenization of the chemical composition, structure refinement and the healing of microcracks and pores.
2. The average value of the friction coefficient (at the moment of stabilization) obtained during the tribological tests for the Cu-Mo coating is approximately 54% lower than that obtained for the same coating after laser modification.
3. Laser treatment caused a 20% increase in the microhardness of the electrospark deposited Cu-Mo coatings.

References

[1] I.V. Galinov, R.B. Luban, Mass transfer trends during electrospark alloying, Surface & Coatings Technology 79 (1996) 9-18. https://doi.org/10.1016/0257-8972(95)02434-4

[2] T. Chang-bin, L. Dao-xin, W. Zhan, G. Yang, Electro-spark alloying using graphite electrode on titanium alloy surface for biomedical applications, Applied Surface Science 257 (2011) 6364-6371. https://doi.org/10.1016/j.apsusc.2011.01.120

[3] B. Antoszewski, E. Evin, J. Audy, Study of the effect of electro-spark coatings on friction in pin-on-disc testing, Journal of Tribology-Transactions of the ASME 3 (2008) 253-262.

[4] N. Radek, A. Sladek, J. Broncek, I. Bilska, A. Szczotok, Electrospark alloying of carbon steel with WC-Co-Al_2O_3: deposition technique and coating properties, Advanced Materials Research 874 (2014) 101-106.

[5] N. Radek, E. Wajs, M. Luchka, The WC-Co electrospark alloying coatings modified by laser treatment, Powder Metallurgy and Metal Ceramics 47 (2008)197-201. https://doi.org/10.1007/s11106-008-9005-7

[6] N. Radek, J. Konstanty, Cermet ESD coatings modified by laser treatment. Archives of Metallurgy and Materials 57 (2012) 665-670. https://doi.org/10.2478/v10172-012-0071-y

[7] J. Pietraszek, N. Radek, K. Bartkowiak, Advanced statistical refinement of surface layer's discretization in the case of electro-spark deposited carbide-ceramic coatings modified by a laser beam, Solid State Phenom. 197 (2013) 198-202. https://doi.org/10.4028/www.scientific.net/SSP.197.198

[8] N. Radek, J. Pietraszek, B. Antoszewski, The average friction coefficient of laser textured surfaces of silicon carbide identified by RSM methodology, Advanced Materials Research 874 (2014) 29-34. https://doi.org/10.4028/www.scientific.net/AMR.874.29

[9] R. Ulewicz, Hardening of steel X155CrVMo12-1 surface layer, Journal of the Balkan Tribological Association 21 (2015) 166-172.

[10] A. Dudek, L. Adamczyk, Properties of hydroxyapatite layers used for implant coatings, Optica Applicata 43 (2013) 143-151.

[11] A. Dudek, A. Wronska, L. Adamczyk, Surface remelting of 316 L+434 L sintered steel: microstructure and corrosion resistance, Journal Of Solid State Electrochemistry 18 (2014) 2973-2981. https://doi.org/10.1007/s10008-014-2483-2

Materials Research Proceedings **5** (2018) 210-215 doi: http://dx.doi.org/10.21741/9781945291814-37

The Influence of Reinforcing Layers and Varnish Coatings on the Smoke Properties of Laminates Based on Selected Vinyl Ester and Polyester Resins

RADZISZEWSKA-WOLIŃSKA Jolanta Maria [a*], TARKA Izabela[b]

Instytut Kolejnictwa, Materials & Structure Laboratory, ul.Chłopickiego 50, 04-275 Warsaw, Poland

[a] jradziszewska-wolinska@ikolej.pl, [b] itarka@ikolej.pl

Keywords: Flame Retardation of Unsaturated Polyester Resins, Smoke Properties, Polyester-glass Laminates

Abstract. The smoke properties of non-flammable polyester-glass laminates with reinforcing layers and laminates with paint coatings in relation to base laminates, unmodified, were compared. The obtained parameters were referred to the requirements applicable in rolling stock. These results were compared with the results of smoke tests for polycarbonate, which as a construction and lining material does not meet the requirements of the EN 45545-2 European rolling stock [1].

Introduction

The wide use of polyester-glass laminates in rail transport results from their high resistance to environmental factors and chemical compounds, ease of processing, aesthetics and good mechanical properties at a relatively low price. Unfortunately, polyester resins are also characterized by high flammability and intense smoking during combustion [2]. It is caused by, inter alia, high content of styrene used for their crosslinking. Therefore, in order to increase the safety of their use, they are modified with the use of different flame retardants [3-6]. Traditional micron flame retardants widely used for polyester resins include aluminum hydroxide (ATH), which was also used in the tested laminates. It is a mineral compound capable of releasing water molecules during the decomposition process. At temperatures above 200^0C aluminium hydroxide is split into aluminium oxide and water. The water cools the source of fire and thins the resulting fumes. Aluminium oxide forms a ceramic protective layer [3]. The flame retardants used must not only increase fire resistance but also limit the emission of dangerous fumes and toxic gases when burning of the composite material. The biggest smoke is generated by polymers, the decomposition of which is accompanied by the separation of condensed aromatic structures [4-9]. In rail transport, smoke emitted in the initial phase of fire is more dangerous than fire at this stage. Smoke makes it difficult to orientate passengers and causes panic, which creates problems with efficient evacuation. In addition, toxic decomposition products irritate the respiratory system and cause poisoning [2]. Therefore, in order to increase the safety of passengers in case of fire, all non-metallic materials used in European rolling stock must meet the requirements of EN 45545-2 [1]. This requirement also applies to the intensity of smoking and the content of selected toxic gases. A number of polyester resins have already been developed, which in laminates with a flame-retardant gel coat meet these requirements. Laminates based on these resins are, however, subjected to modifications in order to improve the stiffness of the products or reduce their weight and resin consumption. Laminates are also coated with paint coatings - this is aimed not only at changing the aesthetic properties of the products but also at improving their resistance to water.

Laboratory tests

The purpose of the described work was to evaluate the effect of modification of vinyl ester-glass and polyester-glass laminates intended for rolling stock on their smoke properties during combustion. In the tested laminates, layers of rigid PVC foams, cork biocomposites and honeycomb layers were used as 'spacers'. Surface modification was also made by applying varnish coatings: Nuvovern WR Emaillack and MIPA coating.

The tests were carried out for laminates with a thickness of 10 mm made on the basis of two following types of resins:

-VE11M vinyl ester resin,

-Firestop 8175-W-1 polyester resin and Firestop S285 gelcoat.

Table 1 Marking and composition of the tested materials

Type of resin	Laminate composition	Marking of samples
vinyl ester resin VE11	9 x S600 – four-way glass fabric 600 [g/m^2], vinyl ester resin VE11M	VE
	2 x gelcoat Firestop GC S714-2720, 9 x S600 – four-way glass fabric 600 [g/m^2], vinyl ester resin VE11M	VE+2xG
	2 x gelcoat Firestop GC S714-2720 + 1 x S + CSM 900 – glass mat 300 [g/m^2] + four-way glass fabric 600 [g/m^2]+ 1 x S600 – four-way glass fabric + Soric XF [6mm] – polyester synthetic layer + 2 x S600 – four-way glass fabric + vinyl ester resin VE11M	VE+2xG+S
	2 x gelcoat Firestop GC S714-2720 + 1 x S + CSM 900 – glass mat 300 [g/m^2] + four-way glass fabric 600 [g/m^2] + 1 x S600 – four-way glass fabric + Corecork 6 [mm] – cork biocomposite + 2 x S600 – four-way glass fabric + vinyl ester resin VE11M	VE+2xG+C
	2 x gelcoat Firestop GC S714-2720 + 1 x S + CSM 900 – glass mat 300 [g/m^2] + four-way glass fabric 600 [g/m^2]+ 1 x S600 – four-way glass fabric +DIV 6 mm – PVC foam layer + 2 x S600 – four-way glass fabric + vinyl ester resin VE11M	VE+2xG+D
Polyester resin Firestop 8175-W-1	polyester resin Bufa Firestop 8175-W-1, gelcoat Bufa Firestop S270/, glass mat 150 [g/m^2], glass fabric WR 600	R+G
	polyester resin Bufa Firestop 8175-W-1, gelcoat Bufa Firestop S285, glass mat (1x [150 g/m^2], 2x 450 [g/m^2]), topcoat Bufa Firestop S285.	R+G+T
	polyester resin Bufa Firestop 8175-W-1, 2 x gelcoat Bufa Firestop GC S714, 9 x S600 – glass fabric 600 [g/m^2]	R+2xG
	polyester resin Bufa Firestop 8175-W-1, gelcoat Bufa Firestop S285, glass mat (1x [150 g/m^2], 2x 450 [g/m^2]), topcoat Nuvovern WR Emaillack.	R+G+N
	polyester resin Bufa Firestop 8175-W-1, gelcoat Bufa Firestop S285, glass mat (1x [150 g/m^2], 2x 450 [g/m^2]), foam 25 [mm], topcoat Nuvovern WR Emaillack.	R+G+F+N
	polyester resin Bufa Firestop 8175-W-1, gelcoat Bufa Firestop S285, glass mat (1x [150 g/m^2], 2x 450 [g/m^2]), Coremat 5 [mm] topcoat Nuvovern WR Emaillack.	R+G+C+N
Polycarbonate	Lexan F2000	PC

Materials Research Proceedings **5** (2018) 210-215 doi: http://dx.doi.org/10.21741/9781945291814-37

For comparison, tests of smoke properties were also made for polycarbonate, which has high impact strength - it is difficult to break, and at the same time relatively light and could be successfully used for structural elements. However, it is currently impossible to use in rolling stock due to too high smoke emission during combustion.

The composition of individual layers of the tested laminates is shown in Table 1.

The measurements were carried out using a test stand for determining smoke intensity according to ISO 5659-2 [11] (Fig.1). For each individual specimen, the percentage value of light transmission was determined and the appropriate specific optical density was calculated on this basis. The results were reported in terms of D_s4 (optical density in first 4 min), and VOF4 (cumulative value of optical density in the first 4 mins - integral for the D_s4 curve in 4 mins).

Fig. 1. Smoke chamber according to ISO 5659-2 [11].

The results of the tests carried out for vinyl ester and polyester laminates modified with sandwich layers and paint coatings are presented in Tables 2 and 3 and in Figs. 2 and 3.

Table 2 Results of smoke intensity determination for vinyl ester based laminates and the values for polycarbonate.

Marking of samples (vinyl ester resin)	D_s4	VOF4
VE	629.17	955.4
VE+2xG	179.92	407.8
VE+2xG+S	234.6	478.1
VE+2xG+C	335.9	596.1
VE+2xG+D	254.6	584.4
PC	1070.1	1720

The introduction of modifications to the laminate structure based on vinyl ester resin resulted in a reduction of the amount of smoke emitted. The unmodified VE11M vinyl ester resin is characterized by high optical density values in the fourth minute of the D_s4 test. This parameter significantly exceeds the value of 300 which is, according to EN 45545-2 [1], the maximum acceptable limit for materials that can be used as lining elements (walls, ceilings) in European rolling stock. The VOF4 parameter is even more exceeded, where the permissible value according to EN 45545-2 [1] is 600. The most effective in terms of reducing the intensity of smoking turned out to cover the laminate with a double layer of the Firestop GC S714-2720 gel coat. In this version it meets the requirements of EN 45545-2 [1] in the field of smoke properties. Covering the gel coat resulted in the reduction of the Ds4 parameter over three times and the VOF4 parameter more than twice in relation to the unmodified laminate. A thick, airtight carbonized layer, fused with metal oxides of flame retardants, was formed on the surface of these samples. The formation of a protective layer on the surface of the sample can also change the release dynamics of the degradation products [5]. This process reduced the number of particles emitted in the gas phase [6]. Then the same laminate system was tested with three types of the following spacers:

- 6 mm synthetic spacer Soric XF synthetic polyester layer reducing weight, and through its structure optimizing the flow of resin between the layers - deterioration of the Ds4 parameter by 30% and VOF4 by 17% - the material meets the requirements of EN 45545-2 [1].
- 6 mm Corecork 6 mm spacer (it is a chipped, glued and then pressed cork oak bark layer, which has excellent insulating properties and resistance to acids and oils) - significant deterioration of smoke properties - Ds4 parameter increased by 87% and VOF4 46% - material no longer meets the requirements of EN 45545-2 [1],
- 6 mm foam spacer - PVC DIV (acoustic seals, anti-vibration primers, thermal insulation) - deterioration of the Ds4 parameter by 41% and VOF4 by 43% - nevertheless, the material meets the requirements of EN 45545-2 [1].

Table 2 Results of smoke intensity determination for polyester based laminates

Marking of samples (polyester resin)	D_s4	VOF 4
R+G	58.5	124
R+G+T	98.4	184.4
R+2xG	141.7	313.3
R+G+N	188.3	313.3
R+G+F+N	203.5	374.5
R+G+C+N	250.2	411.2

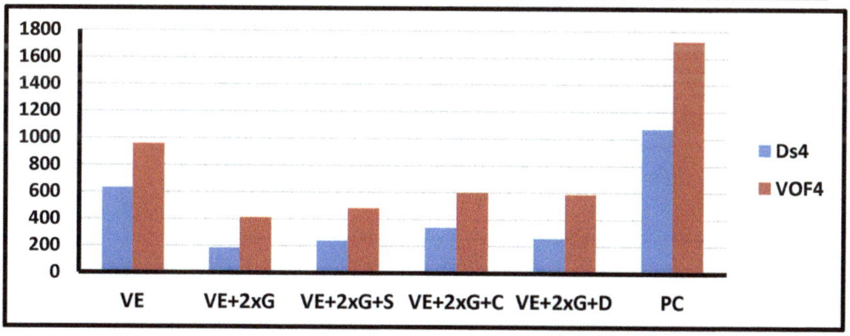

Fig.2 Values of the Ds4 and VOF4 after 4 mins. of tests for vinyl ester resin laminates and polycarbonate.

However, polycarbonate tested in the same conditions on samples with the thickness of 2 mm exceeded the upper limits of smoke parameters according to EN 45545-2 [1] in the following ranges:

- three times the value of parameter Ds4,
- almost three times the value of the VOF4 parameter.

Materials Research Proceedings **5** (2018) 210-215
doi: http://dx.doi.org/10.21741/9781945291814-37

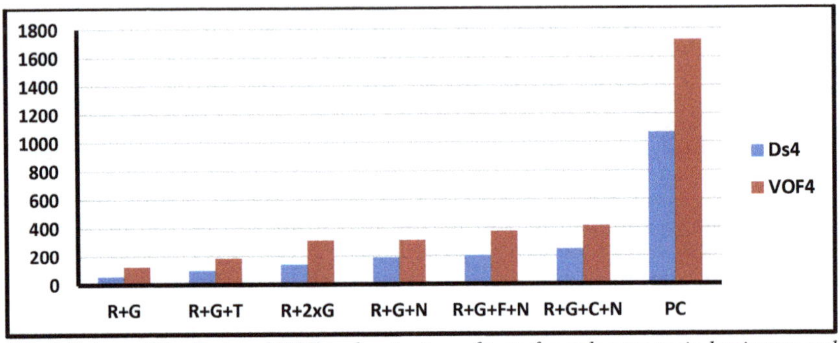

Fig.3 Values of the Ds4 and VOF4 after 4 mins. of tests for polyester resin laminates and polycarbonate.

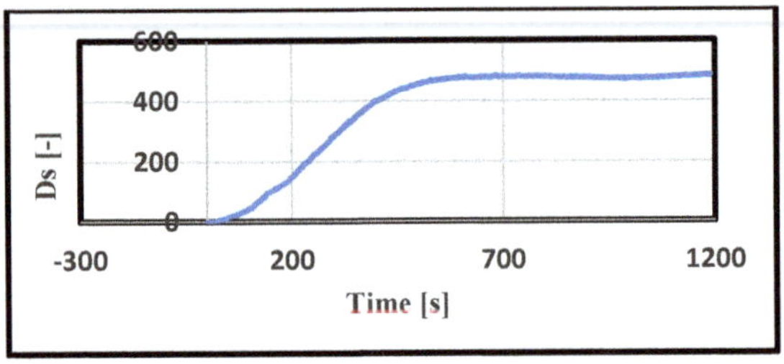

Fig. 4 Optical density in time for the GRP (resin Firestop 8175-W-1, with the Coremat layer)

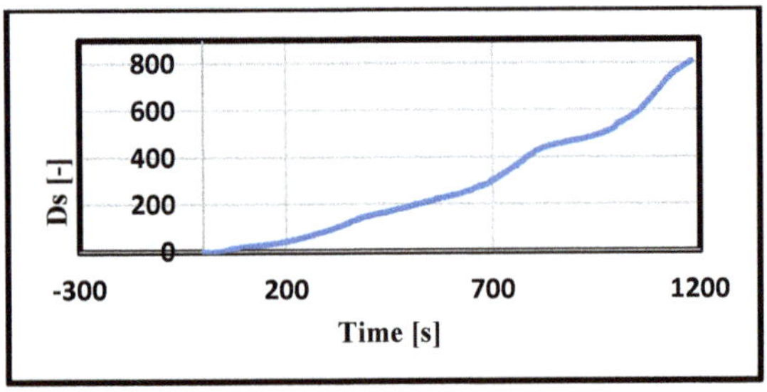

Fig. 5 Optical density in time for the GRP (resin Firestop 8175-W-1, no modification)

Based on the value of the VOF4 parameter, it is possible to assess how much smoke the burning material emits within the first 4 minutes of fire which are decisive for effective evacuation. For laminates based on the Firestop 8175-W-1 resins, the value of this parameter remains low, regardless of the method of laminate modification. The lowest value of the Ds4 and VOF4 parameters was recorded for the gel coat resin without modification. Further modifications by covering the laminate with a double gel coat, the Nuvovern WR Emaillack painting coating, adding a foam layer or the Coremat mat (improving the stiffness of the laminate, reducing its weight and resin consumption) have drastically reduced the smoke properties. The maximum increase in parameters took place for the laminate reinforced with the Coremat mat. The Ds4 parameter has increased by a factor of four and the VOF4 increased by more than threefold in relation to the unmodified base laminate. Nevertheless, the material still meets the requirements of EN 45545-2 [1]. The graph (Fig.4) shows an example of an optical density curve during the test for a laminate with the Coremat mat, which was characterized by the most intense smoke emission in the group of modified laminates based on the Firestop 8175-W-1 polyester resin. For comparison, in the next graph (Fig. 5) the optical density curve for the unmodified laminate, whose smoke parameters were at the lowest level in this group, is presented.

Summary

Smoke properties of composites based on polyester and vinyl ester resin in the base version and after the application of modifying layers were investigated. These modifications included covering with external coatings improving aesthetics, as well as introducing into the structure of composites materials that improve insulation properties, chemical resistance, and weight reduction. It was determined to what extent the improvement of functional properties influences the smoke parameters of these composites important from the fire safety perspective.

It has been shown that for laminates based on vinyl, ester resin coating with a double layer of flame-retardant gelcoat greatly reduces the amount of smoke emitted. On the other hand, other tested modifications worsened smoke properties. However, the introduction of the Soric XF layer or the PVC foam DIV increases the amount of smoke generated in such a range that the obtained values of the Ds4 and VOF4 parameters meet the requirements of EN 45545-2 [1]. On the other hand, in the case of laminates based on polyester resin, it has been shown that for each modification there is only a slight deterioration of the smoke properties and it is possible to use these composites in the rolling stock.

References

[1] PN-EN 45545-2:2013+A1:2015: Railway applications – Fire protection on railway vehicles – Part 2: Requirements for fire behaviour of materials and components.

[2] J. Radziszewska-Wolińska, D. Milczarek, Uniepalnienie materiałów niemetalowych a ich właściwości funkcjonalne, TTS 11-12, 2012, (in Polish) 56-59.

[3] S. Bourbigot, S. Duquesne, Fire retardant polymers, recent developments and opportunities, Journal of Materials Chemistry, 17 (2007) 2283-2300. https://doi.org/10.1039/b702511d

[4] S. Boryniec, W. Przygocki, Procesy spalania polimerów. Cz III Opóźnianie spalania materiałów polimerowych, Polimery 10 (1999) 656-665 (in Polish).

[5] E. Kocko-Walczak, Nowe bezhalogenowe antypireny-uniepalnianie nienasyconych zywic poliestrowych z zastosowaniem związku boru, Polimery 2 (2008) (in Polish).

[6] Zatorski W., Sałasińska K., Analiza palności nienasyconych żywic poliestrowych modyfikowanych nanocząstkami, Polimery, 61 (2016) 11-12 (in Polish).

[7] E. Konecki M. Półka M, Analiza zasięgu widzialności w dymie powstałym w czasie spalania materiałów poliestrowych, Polimery, 51 (2006), (in Polish).

[8] D. Riegert, Sposoby modyfikowania właściwości palnych tworzyw sztucznych, Bezpieczeństwo i Technika Pożarnicza, 30 (2013) (in Polish).

[9] D. Hohenwarter Experience Gained from Fire Testes According to EN 45545-2 and DIN 5510-2 for Testing of Seats, Railway Reports, 6 (2016) 27-38.

[10] ISO 5659-2:2012: Plastics - Smoke generation - Part 2: Determination of optical density by a single-chamber test.

Surface Treatment Technologies for Boiling Heat Transfer augmentation

ORMAN Łukasz J.[1, a *], RADEK Norbert[1,b] and KAPJOR Andrej[2,c]

[1]Kielce University of Technology, al. Tysiaclecia P.P. 7, 25-314 Kielce, Poland

[2]University of Zilina, Univerzitna 1, 01026 Zilina, Slovakia

[a]orman@tu.kielce.pl, [b]norrad@tu.kielce.pl, [c]andrej.kapjor@fstroj.utc.sk

Keywords: Boiling, Laser Treatment, Sintering

Abstract. The paper discusses different methods of surface treatment for the production of efficient phase – change heat exchangers. Special focus is given to sintered metal mesh microstructures and laser treatment. The augmentation of boiling heat transfer is described for meshed surfaces due to their best performance characteristics.

Introduction

There are many techniques of augmenting boiling heat transfer. They can be classified as active and passive methods, depending on whether direct input of external power is necessary. Surface treatment with various methods is a passive technique and generally involves modification of the surface either with mechanical or thermal technologies or application of various types on surface coatings (microstructures). Poniewski [1] gave an overview of microstructures used for boiling heat transfer augmentation. According to the author, heat flux enhancement is caused by such a change in the heater structure that increases a number of active nucleation sites and ensures their stabilised action. Industrial microstructures, sintered or thermally sprayed powder structures as well as metal fibrous, wire mesh and combines structures are, among others, boiling heat transfer enhancing structures. Industrially produced surfaces could be made of sintered metal particles, differently shaped microfins, microstructures with internal tunnels and others. Sintered layers are produced with metal particles of a given diameter and enable obtaining designed porosity and height. Thermal spraying includes plasma, electric arc and flame spraying. In the case of plasma and flame spraying, the coating material is supplied to the burner and sprayed onto the base surface, while the structural material to form porous layers using the electric arc technique comes from the electrodes themselves. Metal fibrous structures are sintered using fine metal fibers. Mesh structures are made of layers of mesh and joined together mechanically or thermally [1]. It is worth noting that even producing roughness on the heat exchanger surface with emery paper can augment boiling heat transfer. This phenomenon is linked with increasing the number of active nucleation sites (locations where vapour bubbles are created and grown) due to increased surface roughness, because such nucleation sites are typically located at cracks or irregularities on the heater surface. The same principle can be behind surface treatment with the laser beam.

The present paper deals with mesh structures, which are often used for heat pipe production and easily available, commercial meshes made of different metals as well as laser treatment of the heater surface.

Sintered metal meshes

The use of metal meshes for boiling heat transfer augmentation is well documented. However, much literature data is given for meshes which were mechanically attached to the heater surface. For example, Rannenberg and Beer [2] presented the results of boiling heat transfer of R-11 and R-113 at atmospheric pressure on a horizontal copper surface covered with 2-9 layers of stainless steel and

Materials Research Proceedings **5** (2018) 216-219 doi: http://dx.doi.org/10.21741/9781945291814-38

bronze meshes. In this case, the meshes were mechanically attached to the heater surface and improved heat transfer in comparison to the surface without the meshes.

Mechanical attachment may cause high thermal resistance to heat transfer. Consequently, sintering is another technique. It offers the possibility of producing strong bonds due to local melting of the mesh into the base surface of the heater. Li et al. [3] considered water boiling on a horizontal copper heater covered with mesh layers. The structure was sintered at 1030°C in the gas mixture of 75% N_2 and 25% H_2. It was reported that sintering minimised the contact resistance between the meshes and the heater. It was proven that a proper contact results in better heat transfer conditions. Five kinds of meshes were tested, while the number of layers ranged from 2 to 9. All the meshed surfaces augmented boiling in comparison with the smooth surface.

The experiments of the authors were carried out on samples that were prepared in such a way that the meshes were sintered to the copper discs of 3 cm in diameter in the reduction atmosphere of hydrogen and nitrogen to prevent oxidation. The sintering process took place at a temperature that is slightly lower than the melting temperature in order to produce strong bonds between the heater disc and the mesh layers, while preventing melting of the sample altogether. In the analysed case of the copper sample, the sintering temperature amounted to about 920°C. Fig. 1 presents the SEM image of a multi-mesh microstructure. The wires in contact with the heater surface are nearly molten into the base surface (bottom part of Fig. 1a). The wires comprising the microstructure are also typically in very good contact with one another. This minimises thermal resistance and increases conduction through the structure of the layer. However, careful analysis of the figures below indicates that some of the wires comprising the layer are not in proper contact, which leads to increased thermal resistance and lower heat fluxes transferred from such a surface to the boiling liquid. This can be eliminated with a careful production of the samples, which are pressed in a special mechanical clamp, in which they are kept during sintering in the oven.

a) b)

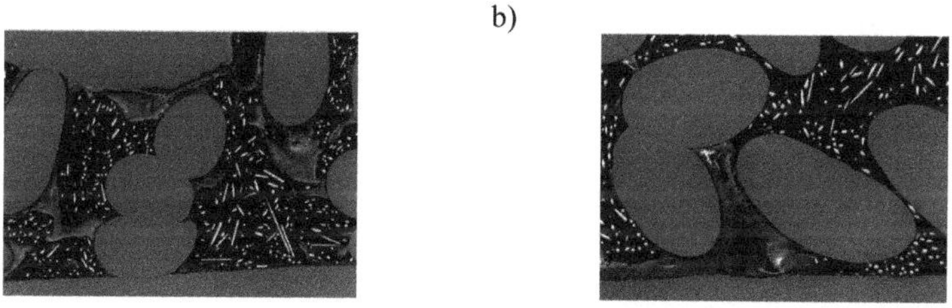

Fig. 1. SEM image of the multilayer copper mesh surface: a) magnification 75x, b) magnification 100x

The application of the multilayer copper meshes augmented boiling heat transfer in comparison to the smooth reference surface. Heat flux (q) can be even several times higher for such treated surfaces than for the surface without any coating at the same temperature difference (θ). Fig. 2 presents the enhancement ratio expressed as the ration of the value of heat flux dissipated from the meshed surface to the heat flux from the smooth surface without such coating. The data has been adopted from the paper by the co-author [4]. Here, distilled water boiling under ambient pressure is considered on the copper heater surface covered with sintered layers of copper meshes. Three different values of the temperature difference have been taken into account for the analyses.

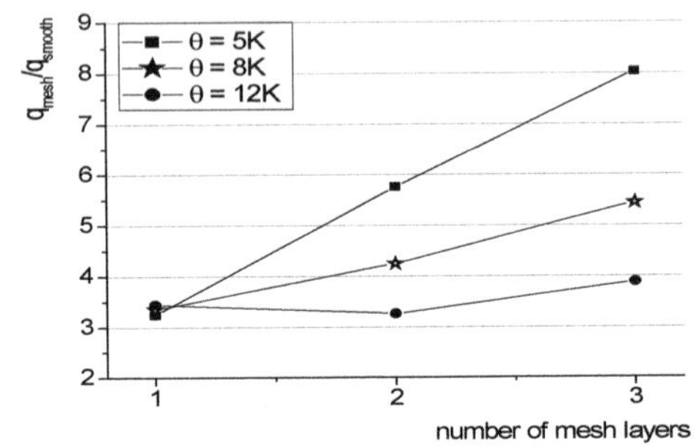

Fig. 2. Enhancement ratio for distilled water boiling

As can be seen in the figure above, the heat flux dissipated from the meshed surface (made of three sintered meshes) can be over eight times higher than for the smooth reference surface without any coating. It is worth noting that heat flux is typically larger as the number of mesh layers is increased and as the temperature difference is lower.

Laser treatment for boiling augmentation

The modification of the surface with the use of the laser beam is another method of surface treatment experimentally analysed by the authors.. This has been discussed by the co-author in [5, 6]. The treatment of the heater surface was conducted with a Nd:YAG type laser working in the pulse mode. The laser spot diameter was 0.7 mm, the beam shift rate: 1200 mm/min, the nozzle-sample distance: 6 mm, while the pulse duration equalled 0.45 ms. Fig. 3 presents the surface morphology of the laser treated heater.

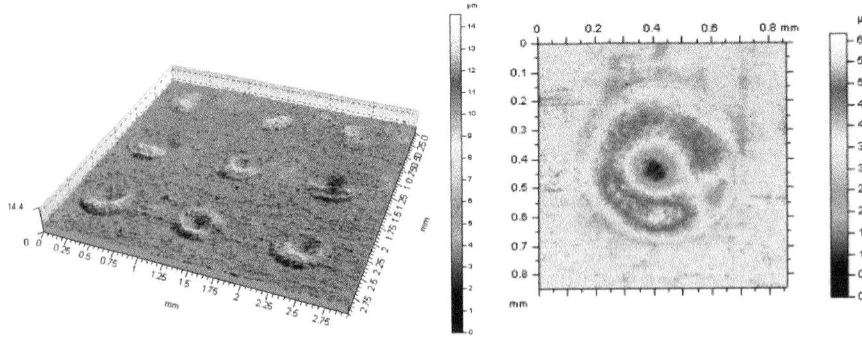

Fig. 3. Surface morphology of the laser treated sample (cavity depth: 2.6 μm, cavity diameter: 0.17 mm)

The application of the laser treatment can lead to increased heat fluxes being dissipated from the heaters. According to the authors' tests a laser treated surface dissipated even twice as much heat flux as the non-treated surface.

Summary

The use of special surface treatment techniques can result in higher heat fluxes that can be dissipated from heaters during boiling heat transfer. Thus, heat exchangers can be smaller and still transfer the same heat fluxes or can dissipate more heat if the surface area remains the same. It offers significant possibilities in many areas of engineering such as for the production of refrigeration devices or electronic cooling units, where phase – change processes of different agents are used. Although the present paper deals with pool boiling, treated surfaces can also be highly effective in the flow boiling heat transfer mode as indicated in [7] and other systems [8].

Acknowledgement

This work was created within statutory research at Kielce University of Technology, no. 05.0.07.00/2.01.01.01.0033 MNSP.IKSI.16.003.

References

[1] M.E. Poniewski, Wrzenie pęcherzykowe na rozwiniętych mikropowierzchniach, Wydawnictwo Politechniki Świętokrzyskiej, Kielce, 2001.

[2] M. Rannenberg, H. Beer, Heat transfer by evaporation in capillary porous wire mesh structures, Letters in Heat and Mass Transfer, 7 (1980) 425-436. https://doi.org/10.1016/0094-4548(80)90019-3

[3] C. Li, G.P. Peterson, Y. Wang, Evaporation/boiling in thin capillary wicks (I) – wick thickness effects, J. of Heat Transfer, 128 (2006) 1312-1319. https://doi.org/10.1115/1.2349507

[4] Ł.J. Orman, T. Orzechowski, Analysis of boiling heat transfer on the non-isothermal surface covered with selected wire mesh structures, in: Proc of 7th World Conference on Experimental Heat Transfer ExHFT-7, 2009, 2159-2166.

[5] J. Pietraszek, N. Radek, K. Bartkowiak, Advanced statistical refinement of surface layer's discretization in the case of electro-spark deposited carbide-ceramic coatings modified by a laser beam, Solid State Phenom. 197 (2013) 198-202. https://doi.org/10.4028/www.scientific.net/SSP.197.198

[6] N. Radek, J. Pietraszek, B. Antoszewski, The average friction coefficient of laser textured surfaces of silicon carbide identified by RSM methodology, Advanced Materials Research, 874 (2014) 29-34. https://doi.org/10.4028/www.scientific.net/AMR.874.29

[7] M. Piasecka, Flow boiling heat transfer in a minichannel with enhanced heating surface, Heat Transfer Engineering, 35, 10 (2014) 903-912. https://doi.org/10.1080/01457632.2014.859505

[8] L. Radziszewski, Intrusive effect of a contact transducer on testing results, Metrology and Measurement Systems, 11 (2004) 31-43.

Materials Research Proceedings 5 (2018) 220-224 doi: http://dx.doi.org/10.21741/9781945291814-39

Distribution of Forces and Analysis of Friction Coefficient during Envelope Rolling of Tapered Tubes

KRUZEL Robert

Czestochowa University of Technology, Akademicka 3, 42-201 Czestochowa, POLAND

kruzel@bud.pcz.pl

Keywords: Conical Pipe, Envelope Formation Process

Abstract. Conical pipes belong to a group of economical profiles which find broad application in various metal products. Manufacturing of conical pipes belong to complicated engineering processes, and the existing methods do not allow for the demand for conical pipes to be met– especially the long ones with large crosswise dimensions, e.g. pipes for lamp posts. This situation makes it reasonable to look for new processes with wider engineering possibilities when compared to the existing methods. It seems that the process of conical pipe envelope formation will allow for a wide conical pipes assortment.

Introduction

Laboratory testing of conical pipes formed by means of the envelope rolling method – a test stand. The test stand (Fig. 1) to form conical pipes was installed on a lathe. The stand base, including the device body was screwed down to the lathe slide.

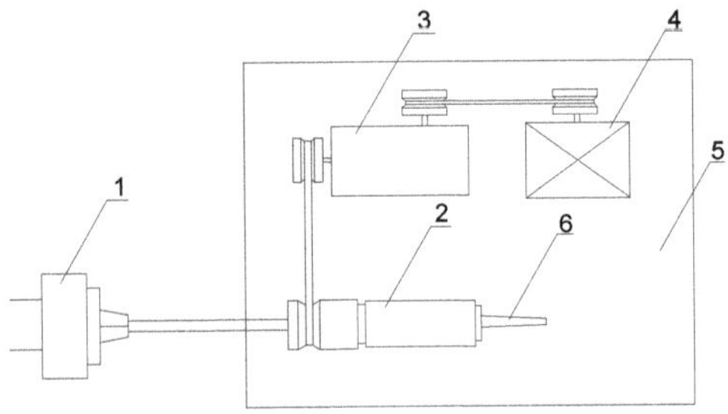

Fig. 1. Diagram of a test stand to roll conical pipes; 1 – lathe slide, 2 – rolling device, 3 – worm gear, 4 – gear drive motor, 5 – base plate, 6 – rolled pipe

In the process, a pipe is put in rotations, and its radial straining takes place through four rolls symmetrically arranged in relation to the deformed pipe (Fig. 2). The location of the rolls in relation to the pipe is determined by the conical surface of the *1* fixing sleeve connected with body *2* by a thread, so setting the sleeve in rotations causes its axial displacement, and simultaneously starts radial shifting of rolls *4*, and that motion makes it possible to receive a conical pipe. Conical pipe formation may occur with increasing or decreasing draft, and the pipe taper depends on process kinematics conditions, i.e. lengthwise displacement speed of forming rolls and their radial shifting speed.

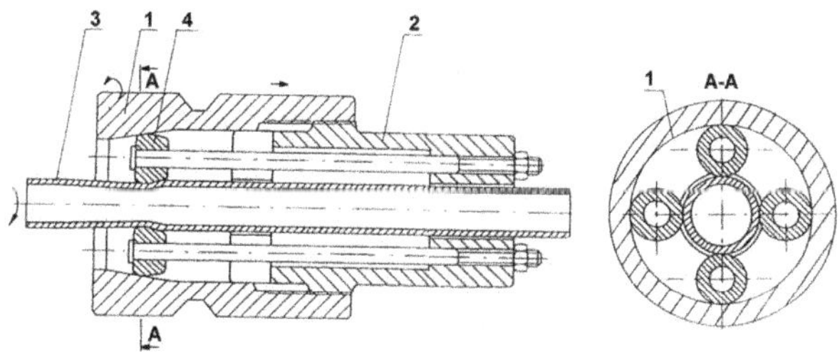

Fig. 2. Cross-section of the conical pipe envelope rolling device; 1 – fixing sleeve, 2 – body, 3 – pipe, 4 – roll, 5 – mandrel, → - designation of particular element motion during the process

Process characteristics and experimental

In the process of conical pipe envelope forming, an input cylindrical pipe put in rotations, is drawn between four rolls, symmetrically arranged in relation to the deformed pipe, and the roll generator is declined at some angle in relation to the deformed pipe, so a gradual decrease of the outer diameter takes place. During the process, a transition of a given point through the plastic strain valley is connected with multiple entering in a zone of direct roll impact, and the total strain is a sum of unit strains resulting from each roll impact on the deformed pipe. Pipe plastic strains that are expressed by diminishing the outer pipe diameter and by lengthwise and radial strains, take place as a result of direct forming roll impact and zones of non-contacted interaction on the boundary of zones between forming rolls, in which elastic state prevails, as well as zones of plastic state. The enveloped formed of tapered tubes involves drawing the cylindrical part of a starting tube set in the rotary motion between radially moving rolls, whose generating lines are inclined at a certain angle relative to the axis of the tube being formed [1, 2]. During the production process, working rolls move together with the entire device along the rotating tube, and owing to a decreasing distance between the rolls, a tapered tube is obtained. The length and taper angle of the tapered surface of tubes are dependent on the peripheral speed imparted to the tube and the speed of displacement of the device along the tube. The displacement of the formed tube in relation to the rolls occurs by the action of axial force applied to the already deformed part of the tube. A feature specific to envelope forming of tapered tubes is that the resultant deformation representing changes of the overall dimensions (e.g. diameter, wall thickness and tube length) is a result of summation of component deformations occurring during multiple partial deformation in the plastic deformation area. In this respect, the envelope rolling process is similar to the process of forging on lever swaging machines, where unit deformation is small, however, large resultant deformation is obtained as a result of summation of partial deformations, and the correct shape of the forged material after the process is assured. It is a common view that forging on lever swaging machines assures uniform deformation, and the forged product is characterized by high accuracy of shape. During forging on lever swaging machines, anvils perform active motion when exerting pressure and idle motion when returning. In the envelope rolling process, forming rolls rotate with uniform motion, and no idle phase of the process occurs. Under the radial action of rolls and longitudinal stresses associated with drawing stresses, the radial flow of a tube material occurs in the range of the direct action of rolls. The surfaces of contact of the deformed tube with a particular roll are characterized by a small width in relation to the length. On a certain segment adjacent to the arc of contact of the tube with the roll, a zone of contactless plastic forming occurs, and then these segments pass into a zone of the elastic state of the deformed tube. When tapered tubes are formed by the discussed method, a given material point enters the plastic forming zone multiple times, then exits this zone to remain for some time in the elastic state in the zone between the forming rolls, and then again enters the zone of contactless plastic deformation. The shape of the plastic deformation region is characterized by the overall dimensions of the plastic deformation region, i.e. the outer diameter of the stock tube, D_p; the outer diameter of the tube leaving the plastic deformation region; the plastic deformation region length; the angle of generating line inclination relative to the tube axis; the shape of the surface of contact of deformed tube with the roll; and the shape of the contactless deformation zones. Having in mind the necessity of assuring a stable deformation of the tube, the deformation zone should be covered by a large possible number of rolls, although instances are known where plastic deformation is concentrated within a small volume in relation to the volume of material being deformed, e.g. spinning which is performed using a single roll. On the other hand, however, small diameter rolls may create difficulties in assuring their proper hardness. With the above-mentioned aspects in mind, the number of rolls seem to be from three to five. A decision to use four rolls was made. The angle of the taper of the roll tapered surface was chosen based on the criterion saying that the taper angle should be within the angles for which the force during sink drawing of round tubes reaches the smallest values (in practice, most often 12÷16°) [3, 4]. A tapered surface inclination angle of 14° was taken. It seems that this angle may be either larger or smaller as well, because, in the range of optimum angles, the effect of tapered surface inclination angle on the value of longitudinal force is relatively small. Rolls, together with the device, move along the axis of the rotating tube. At the same time, their mutual distance gradually decreases, thus the pressure of rolls on metal increases. By the combination of this effect with the displacement of the entire device along the axis of the rotating tube, a tapered surface of the tube is obtained.

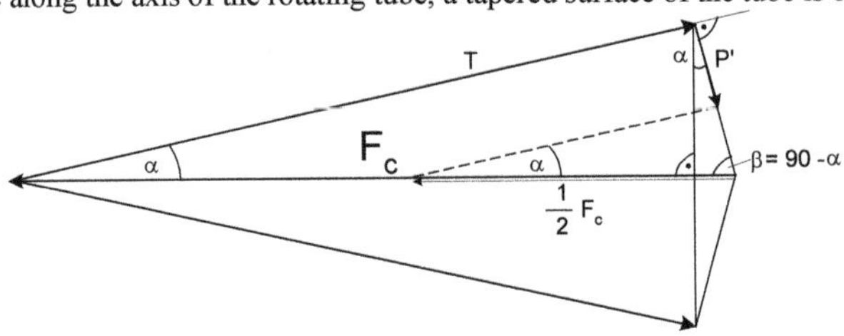

Fig. 3. Distribution of forces during the process of envelope rolling of conical pipes

In this process, a system of balance of forces acting in the rolling head on the tapered tube can be written for four rolls and a single longitudinal force, F_c, as measured extensometrically.

$$F_c = 4P' \sin\alpha \tag{1}$$

From the above balance system, the force of roll pressure on the tube, P', has been determined:

$$P' = \frac{F_c}{4\sin\alpha} \tag{2}$$

In the process of envelope forming of tapered tubes, the displacement of a formed tube in relation to the rolls occurs by the action of axial force applied to the already deformed part of the tube. In addition to the longitudinal force, torque moment acts on the deformed tube segment resulting in tangential stresses, $\tau_{r\theta}$, which, together with the drawing stress, σ_l, cause some effort of the material. The reduced stress, according to the hypothesis of deformation energy, is equal to:

$$\sigma_H = \sqrt{\sigma_l^2 + 3 \cdot \tau_{\theta r}^2} \tag{3}$$

and should be smaller than the yield stress for a given material condition.

Beyond the plastic deformation region, the material should be in the elastic state, and the coefficient of effort defined by the ratio of reduced stress expressed by Formula (3) to yield stress should be less that a unit, and practically should be in the range of $0.6 \div 0.9$. The smaller the coefficient of material effort, the higher the likelihood of a trouble-free completion of the process. The coefficient of material effort is defined by the following equation:

$$n = \frac{\sigma_H}{\sigma_p} \tag{4}$$

Example results obtained from the calculation of σ_H and n are given in Table 1. Figure 4 shows the relationship of the coefficient of material effort versus the tapered tube deformation.

Table 1. Summary of the results obtained from the calculation of material effort and stress intensity

No.	Tube outer diameter	Tube inner diameter	Tube deformation	Reduced stress	Coefficient of material effort	Yield stress
	D_z *[mm]*	D_w *[mm]*	$-\ln\frac{D_k}{D_p}$ *[-]*	σ_H *[MPa]*	n *[-]*	σ_p *[MPa]*
0	13.0	11.50	0.00	-	-	183.5
1	12.5	10.96	0.04	19.5	0.10	185.0
2	12.0	10.40	0.08	21.5	0.12	186.0
3	11.5	9.82	0.12	27.6	0.15	187.5
4	11.0	9.22	0.17	37.4	0.20	189.0
5	10.5	8.62	0.21	44.9	0.24	190.5
6	10.0	8.08	0.26	50.7	0.26	193.0

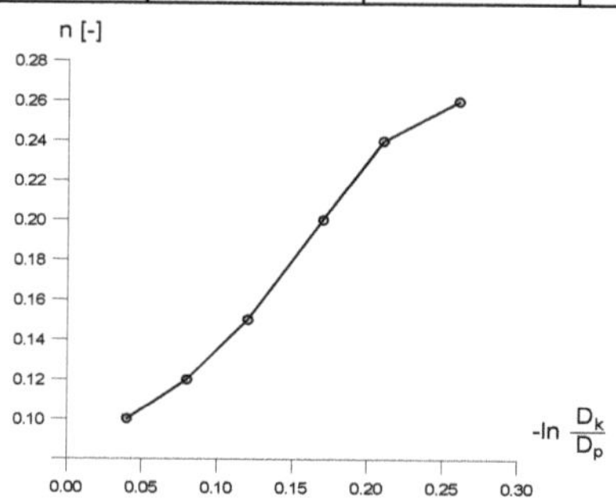

Fig. 4. Relationship of the coefficient of material effort vs. deformation

Another factor limiting the value of deformation in a single pass of a tube through the deformation head is the risk of losing stability during plastic forming.

External friction occurring between the working surface of the tool and plastically deformed metal has a substantial effect on the course of plastic deformation, and thus on the service properties of the product (i.e. tapered tube) and the lifetime of the tools. Friction forces, i.e. the tangential components of the reactions of working roll surfaces on the surface of plastically formed tapered tube have substantial influence on the field and state of stresses in the deformed material, particularly in relation to working tool surfaces. The stress field has, in turn, an effect on the process of metal deformation, and thus on the movement of the metal surface in relation to the working tool surfaces, so also on friction forces. Thus, a certain kind of feedback occurs here. The performed microscopic measurements of tapered tube sections clearly indicate that the final wall thickness of the tube increases during the process, which means that the condition $\varepsilon_r = 0$, or generally $\sigma_r \neq \frac{\sigma_l + \sigma_\theta}{2}$, has not been satisfied, which depends on parameters affecting the value of stress state components. These parameters are the following: tool geometry, tube dimensions, friction coefficient, properties of rolled material, temperature, etc.

Summary

Envelope rolling of tapered tubes is used chiefly to reduce the outer tube diameter without a controlled change of the final tube wall thickness. The dependence of friction coefficient, μ, on the velocity of relative translational motion, W, of bodies in friction is complex and very difficult to establish, because the relative velocity of the displacement of a rolled metal in the rolling head is different at different points of contact of the tube with rolls, depending on the roll angle, α, rolled metal diameter, as well as on the magnitude of the reduction used. According to a suggestion contained in some works, including [5], it can be assumed that $\mu = f(W) = const.$ This assumption greatly simplifies theoretical considerations and at the same time does not lead to any significant discrepancies between theoretically obtained results and experimental test results. Assuming that the unit force, t, of the friction of metal with the roll has the μp value and its direction is coincident with the direction of the resultant velocity, V, then by changing the direction of the resultant velocity a change in the position of the vector of t friction force will be obtained. The position of the V vector of resultant velocity at a given point, and thus also the vector of t, is defined by the β angle.

References

[1] R. McCulloch, Sumitomo unveils new seamless pipe technology, Metall Bull. Monthly, 304 (1996) 18-23.

[2] W. Dobrucki, Podstawy konstrukcji i eksploatacji walcowni / Basics of construction and operation of a rolling mill, Wyd. Śląsk, Katowice, Poland, 1981 [in Polish].

[3]. F. Knap, R. Kruzel, Walcowanie obwiedniowe rur stożkowych, Rudy i Metale Nieżelazne R42 (1997) 338-340.

[4] F. Knap, Stan naprężenia i odkształcenia w procesie obwiedniowego kształtowania rur stożkowych. Rudy i Metale Nieżelazne R43 (1998) 81-86.

[5] O.C. Fuller, Cylindryczna rura stożkowa, Patent-USA: 4565950. Cab nr: 5/86, s. 46-47.

Properties of BO100-AGR Clearcoat Anti-Graffiti Coating Systems used in the Railway Industry

PASIECZYŃSKI Łukasz[1, a *], RADEK Norbert[2,b], KŁONICA Mariusz[3,c], KAMIŃSKI Janusz[4,d] and ŚWIDERSKI Jacek[5,e]

[1]Firma Handlowa Barwa, ul. Warkocz 3-5, 25-253 Kielce, Poland

[2]Kielce University of Technology, Faculty of Mechatronics and Mechanical Engineering, Al. 1000-lecia P.P. 7, 25-314 Kielce, Poland

[3]Lublin University of Technology, Faculty of Mechanical Engineering, ul. Nadbystrzycka 38 D, 20-618 Lublin, Poland

[4]Warsaw University of Technology, Faculty of Materials Science and Engineering, ul. Wołoska 141, 02-507 Warsaw, Poland

[5]Kielce University of Technology, Faculty of Mechatronics and Mechanical Engineering, Al. 1000-lecia P.P. 7, 25-314 Kielce, Poland

[a]lpasieczynski@barwa.kielce.pl, [b] norrad@tu.kielce.pl,, [c]m.klonica@pollub.pl, [d]jmkaminsk@gmail.com, [e]swiderski@tu.kielce.pl

Keywords: Anti-Graffiti Coatings, Free Surface Energy, Rail, Anti-Graffiti Corrosion

Abstract. The paper presents tests results of the selected properties of anti-graffiti paint system for rolling stock industry. The system consists of high solid corrosion protection primer, putty, filler, basecoat and anti-graffiti clearcoat. The results of the research focus on the analysis of corrosion resistance and free surface energy of anti-graffiti clearcoat.

Introduction

In the globalization era, companies' competitiveness is improved by manufacturing processes, focusing attention on new technologies. The fact that the introduction of new products is usually accompanied by the use of new high-strength materials leads to a significant increase in the demand for unconventional, high-performance machining processes using a concentrated energy beam [1-5]. In addition to the optimization of already known and used processes, the effects are consistently implemented by new technologies and materials [6-9].

We are currently observing the development of paint coatings used in various industries [10- 12].

Application of various types of paints (graffiti) on a vehicle may lead to shortening durability of the coating surface. Spray paints should be treated as dangerous substances for coatings because they contain various solvents and other substances that might soften or migrate into protective coating and cause delamination of the coating system. The study of wettability and free surface energy is of great practical importance in industrial processes, especially for paints with special properties like anti-graffiti for rolling stock industry. The paper presents results of free surface energy for anti-graffiti BO100-AGR coating developed in the laboratory of Barwa Company.

To minimize the corrosion problem, the industry uses anticorrosion systems which allow a vehicle to operate without any repairs until the planned term. At the next stage of the study, measurements of the corrosion resistance of the anti-graffiti coating system were performed. Potentiodynamic studies were performed in the water solution of Na_2SO_4 and NaOH at 25°C.

At the end, roughness measurement of the coating system was performed. A test was made to determine the influence of roughness to adhesion properties.

Materials

Coatings were applied with a SATA spray gun on S355 carbon steel and before the application the surface of steel was polished with 80-grit sandpaper. A coating system consisting of the following layers: anti-corrosion epoxy primer, putty, filler, basecoat and an anti-graffiti BO100-AGR clearcoat. Each layer is applied and dried in accordance with the requirements of technological cards. The prepared samples were conditioned at 23 °C and 50 % humidity for minimum 14 days in order to perform tests on dry coating.

Measurement of contact angle and free surface energy

The value of free surface energy of the construction materials is determined indirectly by measuring the contact angles of the selected measuring fluids. Distilled water and diododomethane (DIM) are used to measure the contact angle. A stereoscopic microscope with a camera and a MicroScan v 1.3 software were used for droplet observation and contact angle measurement. The following values of free surface energy constants of the measuring fluids and their polar and dispersion components were assumed as follows: γ_w=72.8 [mJ/m^2], γ^p_w=51.0 [mJ/m^2], γ^d_w=21.8 [mJ/m^2], γ_d=50.8 [mJ/m^2], γ^p_d=2.3 [mJ/m^2], γ^d_d=48.5 [mJ/m^2]. The measuring liquid was applied to the test surface with a 5 µl constant volume micropipette.

a) b)

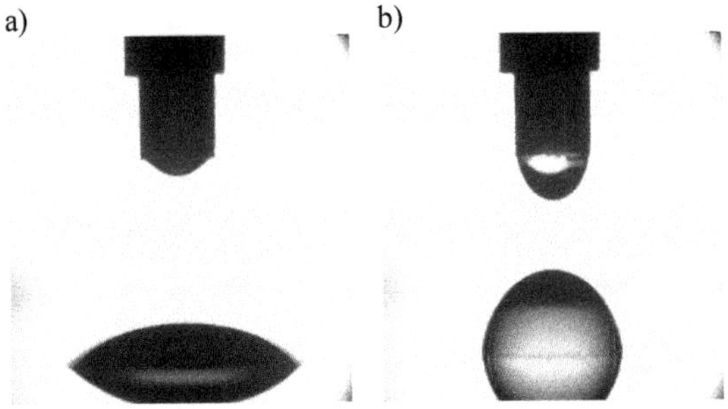

Fig. 1. View of measurement of the contact angle of diododomethane (a) and of distilled water (b)

Free surface energy (FSE) values were determined by measuring the contact angle. The FSE estimation was done at least six times for each surface. Table 5 summarizes the averaged wetting angle and free surface energy measurements of the BO100-AGR anti-graffiti coating systems. Analyzing the obtained results, we can observe high repeatability of the measurements, as evidenced by small values of standard deviations (Table 1).

One of the most commonly used methods for determining free surface energy is the Owens-Wendt method in which it is assumed that the free surface energy is the sum of two components, dispersion and polar:

$$\gamma_s = \gamma_s^d + \gamma_s^p \qquad (1)$$

where:

γ^d_S - dispersion component of free surface energy,

γ^p_S - polar component of free surface energy.

Materials Research Proceedings **5** (2018) 225-230 doi: http://dx.doi.org/10.21741/9781945291814-40

Table 1. Results of contact angle measurements and free surface energy of the BO100-AGR anti-graffiti coating systems

Measuring series number	Average angle - water	Average angle - DIM	E_p [mJ/m^2]	γ^d [mJ/m^2]	γ^p [mJ/m^2]
1	78.4	40.2	49.2	40.1	9.1
2	80.3	40.0	48.5	40.3	8.3
3	80.9	40.5	48.0	39.9	8.1
4	83.6	39.7	47.2	40.3	6.9
5	79.8	41.1	48.3	39.7	8.6
6	80.8	41.4	47.7	39.5	8.2
arithmetic average	80.63	40.48	48.15	39.97	8.20
standard deviation	1.71	0.66	0.69	0.33	0.73

Corrosion resistance tests

Corrosion resistance examinations were carried out by means of potentiodynamic methods using an Autolab PGSTAT100 potentiostat/galwanostat with the FRA2 module, in non-deaerated solution at 25 °C, which consisted of the following:

- 0.1M Na_2SO_4
- 0.1M NaOH

Prior to electrochemical studies, the samples were exposed to a corrosive solution in current-free conditions for 120 minutes. Potentiodynamic test were conducted in a three-electrode setup: the test electrode; reference electrode (saturated calomel electrode (SCE)); auxiliary electrode (platinum), up to the potential of 1800 mV, or to the current above 1mA/cm^2. The test material was polarized with a potential sweep rate of 0,2 mV/s. After the tests (after 24h), potentialadynamic tests were repeated. The extension "corrosion2" has been added to the names of the samples.

In both studied solution coating systems through it protective properties lower corrosion current, increasing corrosion resistance and vehicle lifetime. Also, the isolation steel from the electrolyte causes increasing polarity resistance. The results are shown in Tables 1 and 2 and also in Fig. 1 and Fig. 2.

Table 2. Results of electrochemical studies, potentiodynamic test conducted in 0.1M Na_2SO_4

Na_2SO_4	Rt [kΩcm^2]	E_{kor} [mV]	I_{kor} [μA/cm^2]
Steel	2.39	-735	7.80
Corrosion	798	-765	0.017
Corrosion 2	343	-730	0.038

Table 3. Results of electrochemical studies, potentiodynamic test conducted in 0.1M NaOH

NaOH	Rt [kΩcm^2]	E_{kor} [mV]	I_{kor} [μA/cm^2]
Steel	62.2	-350	0.523
Corrosion	5330	-305	0.0056
Corrosion 2	129800	-180	0.00019

Materials Research Proceedings **5** (2018) 225-230 doi: http://dx.doi.org/10.21741/9781945291814-40

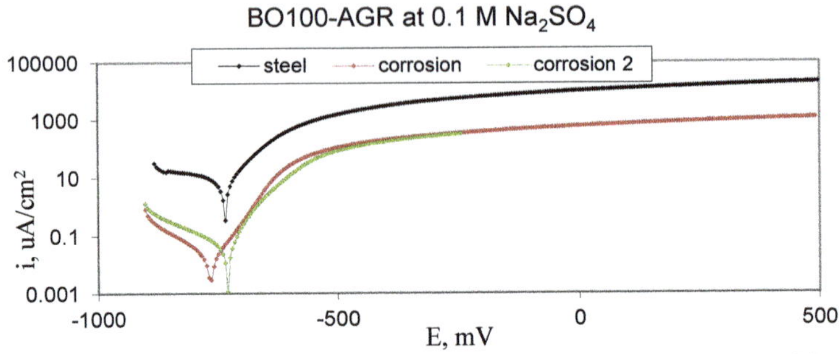

Fig. 2. Steel polarization curve with the BO100-AGR coating system in 0.1 M Na_2SO_4 solution

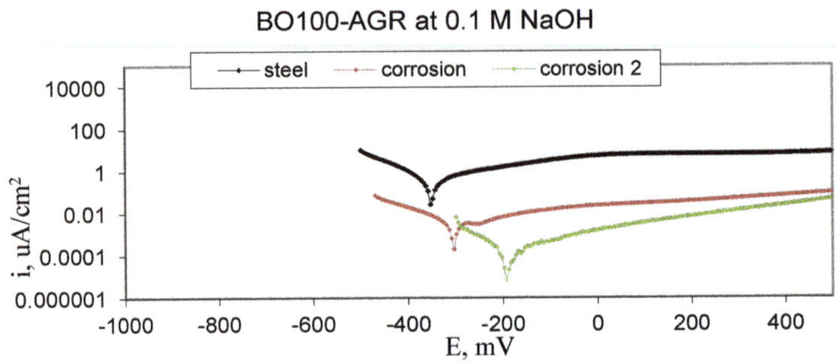

Fig. 3. Steel polarization curve with the BO100-AGR coating system in 0.1 M NaOH solution

Roughness measurements

The roughness of the BO100-AGR anti-graffiti coating systems was measured at the Laboratory for Measurement of Geometric Quantities of the Kielce University of Technology using TALYSURF CCI equipment.

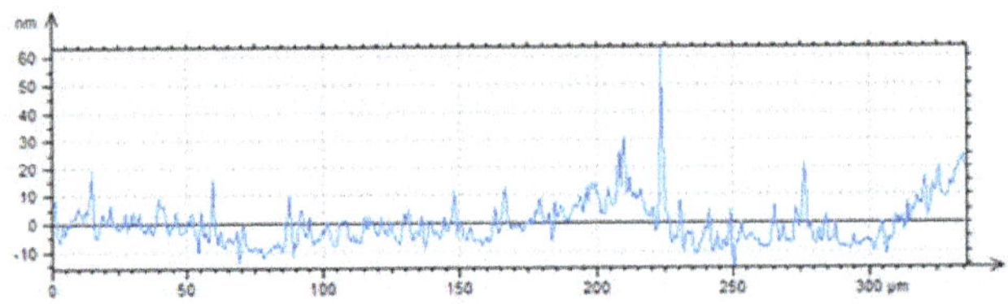

Fig. 4. View of roughness profile of the BO100-AGR anti-graffiti clearcoat

The roughness was measured in two directions perpendicular to each other. Then, the following average value was calculated: Ra = 3.92÷4.04 nm. Samples of S355 steel after grinding with P80 abrasive paper, on which coatings were applied, had roughness from 0.74 to 0.79 µm.

Fig. 4 presents an example of two-dimensional surface microgeometry measurement of the BO100-AGR anti-graffiti coating system. Table 4 presents the most important average roughness parameters of the tested coating system. The low value of roughness parameters also has influence on adhesion properties.

Table 4. Results of roughness profile according to ISO 4287

Parameters of roughness	Anti-graffiti coating system
Rp [nm]	37.21
Rv [nm]	9.20
Rz [nm]	46.42
Rc [nm]	16.36
Rt [nm]	71.37
Ra [nm]	3.98
Rq [nm]	5.70
Rsk	2.08
Rku	12.94

Summary

The study revealed that BO100-AGR anti-graffiti coating systems are characterized by low value of free surface energy and BO100-AGR anti-adhesive properties can be significant in potential applications on rail vehicles. Possible further research, with deeper investigation on the impact of particular factors, should include statistical multivariate analyzes [13, 14], including CA (cluster analysis) [15, 16] and PCA (principal components analysis) [17, 18], as well as alternative – in relation to probabilistic – uncertainty of results e.g. fuzzy approach [19, 20]. It may also be interesting to compare subjective and objective – by means of image analysis methods [21-25] – visual assessments of protected surfaces.

References

[1] N. Radek, E. Wajs, M. Luchka, The WC-Co electrospark alloying coatings modified by laser treatment, Powder Metallurgy and Metal Ceramics 47 (2008) 197-201. https://doi.org/10.1007/s11106-008-9005-7

[2] N. Radek, J. Konstanty, Cermet ESD coatings modified by laser treatment. Archives of Metallurgy and Materials 57 (2012) 665-670. https://doi.org/10.2478/v10172-012-0071-y

[3] J. Pietraszek, N. Radek, K. Bartkowiak, Advanced statistical refinement of surface layer's discretization in the case of electro-spark deposited carbide-ceramic coatings modified by a laser beam, Solid State Phenomena 197 (2013) 198-202. https://doi.org/10.4028/www.scientific.net/SSP.197.198

[4] N. Radek, J. Pietraszek, B. Antoszewski, The average friction coefficient of laser textured surfaces of silicon carbide identified by RSM methodology, Advanced Materials Research 874 (2014) 29-34. https://doi.org/10.4028/www.scientific.net/AMR.874.29

[5] M. Scendo, N. Radek, J. Trela, Influence of laser treatment on the corrosive resistance of WC-Cu electrospark coatings, International Journal of Electrochemical Science 8 (2013) 9264-9277.

[6] R. Ulewicz, F. Novy, J. Selejdak, Fatigue Strength of Ductile Iron in Ultra-High Cycle Regime, Advanced Materials Research 874 (2014) 43-48. https://doi.org/10.4028/www.scientific.net/AMR.874.43

[7] R. Ulewicz, Hardening of steel X155CrVMo12-1 surface layer, Journal of the Balkan Tribological Association 21 (2015) 166-172.

[8] A. Dudek, C. Kolan, Assessments of shrinkage degree in bioceramic sinters HA+ZrO_2, Solid State Phenomena 165 (2010) 25-30. https://doi.org/10.4028/www.scientific.net/SSP.165.25

[9] A. Dudek, R. Włodarczyk, Structure and properties of bioceramics layers used for implant coatings, Solid State Phenomena 165 (2010) 31-36. https://doi.org/10.4028/www.scientific.net/SSP.165.31

[10] J. Radziszewska-Wolińska, Fire properties of anticorrosion coatings to rolling stock, Technical Transactions 3-M (2016) 79-86.

[11] A. Gergely, E. Pfeifer, I. Bertóti, T. Török, E. Kálmán, Corrosion protection of cold-rolled steel by zinc-rich epoxy paint coatings loaded with nano-size alumina supported polypyrrole, Corrosion Science 53 (2011) 3486-3499. https://doi.org/10.1016/j.corsci.2011.06.014

[12] M.F. Montemor, Functional and smart coatings for corrosion protection: A review of recent advances, Surface & Coatings Technology 258 (2014) 17-37. https://doi.org/10.1016/j.surfcoat.2014.06.031

[13] A.J. Izenman, Modern Multivariate Statistical Techniques. Regression, Classification, and Manifold Learning. Springer, New York, 2008.

[14] J. Pietraszek, Response surface methodology at irregular grids based on Voronoi scheme with neural network approximator, Adv. Soft Comp. (2003) 250-255.

[15] B.S. Everitt, S. Landau, M. Leese, D. Stahl, Cluster Analysis, Wiley, Hoboken, 2012.

[16] E. Skrzypczak-Pietraszek, J. Pietraszek, Seasonal Changes of Flavonoid Content in Melittis melissophyllum L. (Lamiaceae), Chem. Biodivers. 11 (2014) (4) 562-570.

[17] I.T. Jolliffe, Principal Component Analysis, Springer, New York, 2010.

[18] E. Skrzypczak-Pietraszek, J. Pietraszek, Chemical profile and seasonal variation of phenolic acid content in bastard balm (Melittis melissophyllum L., Lamiaceae), J. Pharmaceut. Biomed. 66 (2012) 154-161. https://doi.org/10.1016/j.jpba.2012.03.037

[19] J. Pietraszek, A. Gadek-Moszczak, The Smooth Bootstrap Approach to the Distribution of a Shape in the Ferritic Stainless Steel AISI 434L Powders, Solid State Phenom. 197 (2013) 162-167. https://doi.org/10.4028/www.scientific.net/SSP.197.162

[20] J. Pietraszek, E. Skrzypczak-Pietraszek, The Optimization of the Technological Process with the Fuzzy Regression, Adv. Mater. Res.-Switz. 874 (2014) 151-155. https://doi.org/10.4028/www.scientific.net/AMR.874.151

[21] A. Szczotok, M. Sozanska, A Comparison of Grain Quantitative Evaluation Performed with Standard Method of Imaging with Light Microscopy and EBSD Analysis, Prakt. Metallogr.-Pr. M. 46 (2009) (9) 454-468.

[22] A. Gadek-Moszczak, L. Wojnar, Objective, Quantitative and Automatic X-Ray Image. Analysis of the Bone Regenerate in the Ilizarov Method, ECS10: The10th European Congress of Stereology and Image Analysis (2009) 453-458.

[23] J. Korzekwa, A. Gadek-Moszczak, M. Bara, The Influence of Sample Preparation on SEM Measurements of Anodic Oxide Layers, Prakt. Metallogr.-Pr. M. 53 (2016) (1) 36-49.

[24] A. Gadek-Moszczak, History of Stereology, Image Anal. Stereol. 36 (2017) (3) 151-152.

[25] A. Gadek-Moszczak, P. Matusiewicz, Polish Stereology – a Historical Review, Image Anal. Stereol. 36 (2017) (3) 207-221.

Terotechnology 2017 Materials Research Forum LLC
Materials Research Proceedings **5** (2018) 231-236 doi: http://dx.doi.org/10.21741/9781945291814-41

Powder Surface Modification as a Method of Corrosion Rate Limitation of the Magnetic RE-M-B Composite in an Acid Medium with Different pH

KLIMECKA-TATAR Dorota

Department of Production Engineering and Safety, Faculty of Management, Czestochowa University of Technology, Al. Armii Krajowej 19b, 42-218 Czestochowa, Poland

dorota.klimecka-tatar@wz.pcz.pl

Keywords: Bonded Magnets, Nd-Fe-B Alloys, Polarization Resistance

Abstract. This paper presents the results of research on the effects of chemical treatment of the magnetic powder particles surface on the corrosion resistance of composites based on $Nd_{12}Fe_{77}Co_5B_6$ powder bonded with an epoxy resin. In the studies, a magnetic material formed of a commercial MQP-B powder has been used and subjected to chemical modification in such aqueous solutions as acetic acid, EDTA, citric acid and oxalic acid. Chemical treatment of the powder particles surface was applied for etching the oxide paramagnetic phases formed on the surface as an effect of spontaneous oxidation of the alloy elements (Nd and Fe) in the presence of oxygen and moisture. On the basis of changes in the polarization resistance value, which were recorded for a series of samples in 0.5M acidic sulphate solutions (pH = 2, 3 and 4), it has been found that the proposed pre-treatment has beneficial effect on the stability of the material in acid environments.

Introduction

Magnetic permanent RE-M-B materials contain rare earth metals (RE, e.g. Nd, Dy) and transition group metals (M, e.g. Fe, Co, Ni) in their composition, which ensures excellent magnetic properties. Due to a wide range of possibilities to control functional properties by production technologies, this kind of magnetic materials are successfully used in various technical fields (elements of hard drives, equipment devices in cars, components of qualified equipment for magnetic resonance imaging, retention systems in dentistry, orthodontics, prosthetics, etc.) [1-3]. The major limitation to their use is unsatisfactory stability under the use conditions. A multi-phase structure, segregation of chemical components, large differences in electrochemical potential of individual phases and considerable porosity are the main causes of poor resistance to corrosion and a significant degradation of the structure [4-7]. The unique properties of the RE-M-B magnetic material are determined by the presence of the $RE_2M_{14}B$ ferromagnetic phase (2:14:1, e.g. $Nd_2Fe_{14}B$). On the other hand, phases enriched with boron (B-rich) and neodymium (Nd-rich) are also present in a typical structure of the RE-M-B sintered materials [1, 3-9]. Segregation of B and Nd in the structure is the reason why the phase which contains approx. 95% of neodymium is considered to be the one that is most easily degraded - the most active phase, whereas the phase in which boron is the predominant component is considered as the most noble. The structure modeling is possible by using alloying additives, which together with neodymium create more noble phases. Co, Ni and Cu [3, 5, 8] have to be among the most commonly used alloying additives (which replace the alloy composition of the main component, i.e. Fe). Other methods of reducing corrosion susceptibility of the RE-M-B materials are focused on the use of various production technologies. Among them, magnetic powders consolidation using a metal of low melting point or of an organic binder (epoxy resin, polyester resin, nylon) are worth mentioning. Magnetic powder bonding with an organic binder (caring about sufficient homogenization of the components), makes it possible to isolate individual powder particles, which is particularly important for powders of an alloy containing a large amount of Nd [4,

6, 9]. It is known that alloys based on rare earth metals easily oxidize even in the presence of trace amounts of oxygen and moisture [10,11], and the presence of paramagnetic oxide phases on the powder particles surface adversely affects the adhesive connections stability, and thus resistance to corrosion of bonded magnets [6,10]. Increasing the adhesion of the binder to the surface of the particles of magnetic powder has beneficial effects on many levels as it reduces porosity structure and limits the penetration of the corrosive medium into the material. The use of appropriate chemical modification of the magnetic powder particle surface prior to the consolidation process can help to improve not only corrosion resistance but also mechanical properties and magnetic material. It should be noted that reducing the corrosion rate of materials based on rare earths is becoming an increasingly important aspect. Technologies for the recovery of rare earths from used equipment are already being sought [12, 13].

The aim of this study is to determine the effectiveness of powder particle surface modification (chemical modification) and its effect on the corrosion rate. To evaluate the effect of a bonded magnetic material on corrosion properties, a series of tests in the solution of 0.5M Na_2SO_4 acidified to pH = 2, 3 and 4 were carried out.

Experimental

The Nd-M-B magnetic powder-based composite with the MQP-B manufacturer symbol (Magnequench) and $Nd_{12}Fe_{77}Co_5B_6$ chemical formula (subscripts indicate atomic percentages) has been used as the experimental material. According to the manufacturer information, the powder was prepared in three main steps: rapid solidification of the alloy from a molten state, thermal treatment of the received tapes (short-term annealing at the temperature of 600 °C) and mechanical crushing. Heat treatment allows for obtaining a nanocrystalline structure of the intermetallic compound, single-phase $Nd_2M_{14}B$ structure (M - transition metal) and mechanical fragmentation results in the "lamellar" powder particles shape. As the presence of oxides on the powder particles surface may considerably limit the adhesion of the binder to the surface, the powder was subjected to preliminary etching. Digestion of the powder was carried out in aqueous solutions and this process can be seen as a type of chemical modification of the powder. As the etching media were selected aqueous solutions of the compounds strongly complexing metal ions or weak acid anions of the complexing properties, as follows: acetic acid, sodium edetate (EDTA), citric acid, oxalic acid.

The etching of the $Nd_{12}Fe_{77}Co_5B_6$ powder aimed at removing paramagnetic oxide phases, which are formed on the powder surface during storage under exposure to air and humidity. The powder was digested in a solution of acetic, citric and oxalic acid until the appearance of the first bubbles of gas (H_2) at the surface of the powder particles. Exposure times were various for different applied solutions. For solutions of acetic, citric and oxalic acid it was approx. 20, 15 and 5 minutes (at 20 °C).

The prepared magnetic powder has been bonded by two kinds of epoxy resins in the form of acetone solutions. In both cases the resin content in the magnetic composite was 2.5 wt. %: epidian 100 (Method B) - thermosetting resin, (Method C) – chemosetting resin. Resins used as a binder have been prepared as an acetone solution, to cover the surface of the powder particles with a thin adhesive layer (microencapsulation), thereby increasing the homogenization of the composition. For epoxy resin cure, the samples made with the B method have been annealed at 180°C for 2 h and the samples prepared with the C method have been allowed to stand for 48 hours in a dessicator. The effect of the powder surface etching has been determined on the basis of potentiokinetic tests and characteristic parameters of magnetic materials in sulphate solutions having three different pH values. The corrosion current density has been appointed by two methods: the method of extrapolation of Tafel's straight sections and the polarization resistance method based on linear overvoltage changes at potentials close to the corrosion potential - accordingly to the Stern-Hoar equation.

Results and discussion

The corrosion current density values (i_{corr}) determined by the anode Tafel plot extrapolation, and by polarization resistance are shown in Tables 1 and 2, while Figure 1 illustrates the logi_{corr} relation in pH function. In all the presented cases (Figure 1), the increase of pH leads to a reduction in corrosion rate and the slope of a linear relation (logi_{corr} = f(*pH*)) for i_{corr} values appointed by the Tafel plot extrapolation is approx. −0.67 ÷ −0.40, which is in analogy with literature data for pure iron (at pH = 0 ÷ 2). In contrast, the slope of a linear relation (logi_{corr} = f(*pH*)) for i_{corr} values appointed by polarization resistance has a value of −0.30 ÷ −0.17.

Table 1. The values of the parameters determining the corrosion resistance of the materials from the $Nd_{12}Fe_{77}Co_5B_6$ powder in sulphate solutions of pH 2, 3 and 4. The powder $Nd_{12}Fe_{77}Co_5B_6$ priory bonding coated with a film of thermosetting epoxy resin – the B method

0.5 M SO_4^{2-}	modifying solution	E_{corr} [V]	i_{corr} *extrapolation* [mA·cm^{-2}]	R_p [Ω·cm^2]	i_{corr} *resistance polarization* [mA·cm^{-2}]
pH = 2	without modification	-0.58	2.5	3.9	3.3
	acetic acid	-0.56	2.4	4.0	3.3
	EDTA	-0.66	2.8	5.0	2.6
	citric acid	-0.58	2.1	5.4	2.4
	oxalic acid	-0.58	2.0	5.4	2.4
pH = 3	without modification	-0.60	0.73	7.6	1.7
	acetic acid	-0.62	0.71	9.1	1.4
	EDTA	-0.67	0.99	7.2	1.8
	citric acid	-0.62	0.89	9.7	1.3
	oxalic acid	-0.60	0.65	9.9	1.3
pH = 4	without modification	-0.66	0.29	19	0.91
	acetic acid	-0.64	0.18	20	0.87
	EDTA	-0.68	0.28	18	0.97
	citric acid	-0.64	0.10	22	0.79
	oxalic acid	-0.65	0.087	24	0.73

As it can be seen, the differences in inclination are quite significant. It seems that the main cause lies in the accepted method of determining i_{corr} value, not taking into account the course of curves in the cathodic field. The fact that the hydrogen evolution (consumption of H^+ ions) in the cathode area must lead to an increase in pH near the surface, the stronger then pH of the solution (with stronger buffer capacity of the solution). It can therefore be expected that the actual value of the pH at the surface (pH_S) varies in time (during a reference potential). In the highly cathodic potential range, the value of pH_S may be, for example, 1-2 units higher than in the solution volume (pH_V). With changes in the potential (*E*) through the anode range, due to rapid mixing forces probably $pH_V \approx pH_S$. Because the polarization resistance method takes into account the cathode/anode transition, one can guess that this method is more subject to errors caused by the increase of the pH near the surface. On the other hand, it should be noted that the value of the $\partial \log i_{corr} / \partial pH$ partial derivative of iron decreases with the increasing pH. The literature values of the derivative at pH = 2 are close to 0.4, hence it is not entirely clear whether measurements using the polarization resistance of the $Nd_{12}Fe_{77}Co_5B_6$ (−0.3 ÷−0.17) alloy are less reliable. The lowest corrosion rate values in the sulfate solution (Table 1 and 2) were observed for the bonded magnetic material based on powder whose

Materials Research Proceedings **5** (2018) 231-236 doi: http://dx.doi.org/10.21741/9781945291814-41

surface was pre-etched in an aqueous solution of citric acid and oxalic acid. The $Nd_{12}Fe_{77}Co_5B_6$ powder etching in a solution of acetic acid has little effect on the corrosion inhibition of the fabricated composite, while the corrosion current density of the composite based on the powder whose surface was modified with EDTA has comparable or higher values than the magnetic powder in the supply condition.

Table 2. The values of the parameters determining the corrosion resistance of the materials from the $Nd_{12}Fe_{77}Co_5B_6$ powder in sulphate solutions of pH 2, 3 and 4. The powder $Nd_{12}Fe_{77}Co_5B_6$ priory bonding coated with a film of chemosetting epoxy resin – the C method

0.5 M SO_4^{2-}	**modifying solution**	E_{corr} **[V]**	i_{corr} *extrapolation* **[mA·cm⁻²]**	R_p **[Ω·cm²]**	i_{corr} *resistance polarization* **[mA·cm⁻²]**
pH = 2	**without modification**	-0.57	2.4	4.7	2.8
	acetic acid	-0.57	2.1	5.2	2.5
	EDTA	-0.61	2.5	4.8	2.7
	citric acid	-0.57	1.9	6.5	2.0
	oxalic acid	-0.57	1.7	6.5	2.0
pH = 3	**without modification**	-0.63	0.82	8.8	1.5
	acetic acid	-0.64	0.79	10	1.3
	EDTA	-0.70	0.69	13	1.0
	citric acid	-0.63	0.55	16	0.81
	oxalic acid	-0.62	0.51	18	0.72
pH = 4	**without modification**	-0.66	0.15	17	1.0
	acetic acid	-0.65	0.12	21	0.83
	EDTA	-0.69	0.18	16	1.1
	citric acid	-0.65	0.091	23	0.76
	oxalic acid	-0.64	0.077	24	0.73

Among the used solutions, EDTA has the strongest complexing facilities and acetic acid the weakest. It should be noted that EDTA (edentate sodium salt) is almost neutral (pH ~ 7), while the other solutions of organic acids have a pH ≈ 2 (out of the three selected acids, the oxalic acid has relatively maximum power).

The most beneficial powder etching effect of the oxalic acid should therefore be attributed to both considerable acidity (pH <2) and good complexing ability of metal cations. The process of oxide phases removing from the surface of the powder particles involves acid digestion, wherein the complexing anions stimulate this process and do not allow for passivating the surface of the particles.

Materials Research Proceedings **5** (2018) 231-236 doi: http://dx.doi.org/10.21741/9781945291814-41

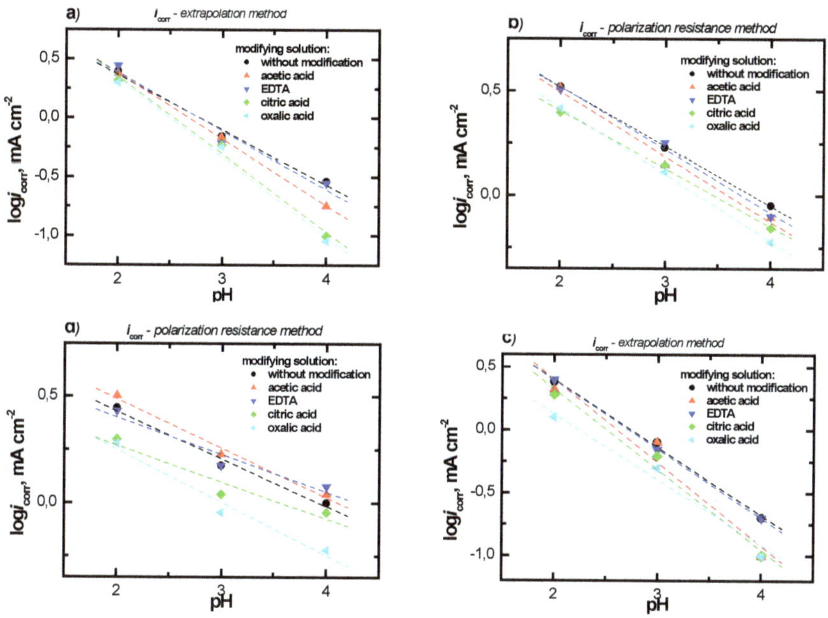

Fig. 1. The corrosion rate of magnetic materials as a function of 0,5 M SO_4^{2-} solution pH (i_{corr} value determined by: a, c) extrapolation, b, d) polarization resistance): a, b) - the B method; c, d) - the C method

Conclusions

The presented results show that chemical treatment of a powder particles surface in a complexing acid solution quite significantly improves corrosion characteristics of magnetic materials based on $Nd_{12}Fe_{77}Co_5B_6$, as evidenced is the reduction in the corrosion current density and polarization resistance increase in the sulphate environments. Removal of oxide phases spontaneously arising during storage in air from the surface of a powder ensures good adhesion of the binder directly to a metal substrate, effectively reducing the contact of individual particles in the composite structure. The elimination of oxide layers prevents the formation of gaps between particles and the binder in the conditions of acidic environment, which obviously inhibits corrosion processes.

References

[1] R. Ulewicz, E. Wysłocka, Magnets: History, the current state and the future, METAL 2015 - 24th International Conference on Metallurgy and Materials, Conference Proceedings, 1680-16862, Brno, Czech Republic

[2] P. Ceruti, S. Ross Bryant, Lee Jun-Ho, M.I. MacEntee, Magnet-Retained Implant-Supported Overdentures: Review and 1-Year Clinical Report, J Can Dent Assoc. 76 (2010) 52

[3] John J. Croat, Rapidly Solidified Neodymium-Iron-Boron Permanent Magnets, Elsevier Science, Woodhead Publishing Series in Electronic and Optical Materials, 2017

[4] K. Loewe, D. Benke, C. Kübel, T. Lienig, K.P. Skokov, O. Gutfleisch, Grain boundary diffusion of different rare earth elements in Nd-Fe-B sintered magnets by experiment and FEM simulation, Acta Materialia, 124 (2017) 421–429. https://doi.org/10.1016/j.actamat.2016.11.034

[5] T. T. Sasaki, T. Ohkubo, K. Hono, Structure and chemical compositions of the grain boundary phase in Nd-Fe-B sintered magnets, Acta Materialia, 115 (2016) 269–277. https://doi.org/10.1016/j.actamat.2016.05.035

[6] D. Klimecka-Tatar, H. Bala, B. Ślusarek, K.Jagielska-Wiaderek, The effect of consolidation method on electrochemical corrosion of polymer bonded Nd-Fe-B type magnetic material, Arch Metall Mater, 54/1 (2009)247-256

[7] G.A.Zickler, P. Toson, A. Asali, J. Fidler, Nanoanalytical TEM Studies and Micromagnetic Modelling of Nd-Fe-B Magnets, Physics Procedia, 75 (2015) 1442–1449. https://doi.org/10.1016/j.phpro.2015.12.164

[8] L. Z. Zhao, Q. Zhou, J. S. Zhang, D. L. Jiao, Z. W. Liu, J. M. Greneche, A nanocomposite structure in directly cast NdFeB based alloy with low Nd content for potential anisotropic permanent magnets, Materials & Design, 117 (2017) 326–331. https://doi.org/10.1016/j.matdes.2017.01.008

[9] K. Radomska, G. Pawłowska, D. Klimecka-Tatar, Profilometry surface analysis of bonded magnets based on biencapsulated Nd-(Fe,Co)-B powder, Ochrona przed Korozją 4 (2014) 142-146

[10] D. Klimecka-Tatar, G. Pawłowska, M. Sozańska, The Effect of Powder Particle Biencapsulation with Ni-P Layer on Local Corrosion of Bonded Nd-(Fe,Co)-B Magnetic Material, Archives of Metallurgy and Materials, 60/1 (2015) 153-157. https://doi.org/10.1515/amm-2015-0024

[11] A. A. El-Moneima, O. Gutfleish, A. Plotnikov, A. Gebert, The Influence of Co and Ga on the Corrosion Behavior of Nanocrystalline NdFeB-Based, Magnets J. Magn. Magn. Mater. 248 (2002) 121-133. https://doi.org/10.1016/S0304-8853(02)00286-X

[12] K. Kapustka, G. Ziegmann, D. Klimecka-Tatar, Technological and ecological safety in aspect of chemical properties of recycled neodymium magnets – electric motors and hard disk, Production Engineering Archives, 17/ 4 (2017) 36-39. https://doi.org/10.30657/pea.2017.17.08

[13] K. Kapustka, G. Ziegmann, S. Sdrenka, T. Elwert, D. Klimecka-Tatar, Microscopic characteristic of Nd-Fe-B magnets structure - magnets recovered from electric motors, Archiwum Wiedzy Inżynierskiej, 2/1 (2017) 39-41

Surface Improvement by WC-Cu Electro-Spark Coatings with Laser Modification

PLISZKA Izabela[1,a*], RADEK Norbert[1,b], GĄDEK-MOSZCZAK Aneta[2,c], FABIAN Peter[3,d] and PARASKA Olga[4,e]

[1]Kielce University of Technology, Faculty of Mechatronics and Mechanical Engineering, Al. 1000-lecia Państwa Polskiego 7, 25-314 Kielce, Poland

[2]Cracow University of Technology, Institute of Applied Informatics, Faculty of Mechanical Engineering, Al. Jana Pawła II 37, 31-864 Cracow, Poland

[3]University of Zilina, Faculty of Mechanical Engineering, Univerzitna 1, 01026 Zilina, Slovakia

[4]Khmelnitskiy National University, Department of Chemical Technology, st. Instytutska 11, 29016, Khmelnickiy, Ukraine

[a]ipliszka@tu.kielce.pl, [b]norrad@tu.kielce.pl, [c]aneta.moszczak@gmail.com, [d]fabianp@fstroj.uniza.sk, [e]olgaparaska@gmail.com

Keywords: Electro-Spark Deposition, Coatings, WC-Cu Coatings, Laser Treatment, Surface

Abstract. The article presents the possibilities of using laser surface modification on the way ESD to better tribological properties. The question presented in the paper can be used to expand the knowledge in the field of application of electrospark deposition. Surface treatment by applying a coating by electro spark deposition has many advantages (eg. local interface or applying thin layers), and therefore this technology is used in the industry. A concentrated stream of laser beam can effectively modify the state of WC-Cu electrospark coatings and improve their performance. The aim of the study is to evaluate the influence of laser treatment on the properties of electrospark coatings. The evaluation of the properties of coatings after laser treatment was carried out by surface geometry analysis.

Introduction

High performance characteristics of modern equipment components are obtained, inter alia, by the use of new materials of construction and the application of protective layers, ensuring the best resistance to wear, erosion, or high fatigue strength [1-3]. Manufacturing new layers on machine parts is economically viable for use fragments or surface layer and where the surface layer must have different characteristics from the core. Electro-spark deposition (ESD) is a cheap and efficient way to improve performance properties of metal [4-6]. Developed in the post-war period, the technology has been frequently modified. The main advantages are related to the ability of a precise choice of the area to be modified and the ability of coating thickness selection. This method makes it possible to deposit ultra-thin and slightly thicker coatings made of different metal [7].

Analysis of properties of coatings requires many methods [8-10]. There are many alternative technologies for producing coatings and material properties improvements in relation to ESD technology [11, 12].

Materials

Based on literature and our own experiences, we developed 2 powder blends which have been produced by powder metallurgy methods for ESD coating electrodes. The tests were performed on the WC-Cu (50%-50%) and WC-Cu (25%-75%) coatings produced on the normalized C45 grade steel specimens by electrospark deposition. The coatings were deposited in the argon atmosphere with the use of an EIL-8A pulse generator for triggering spark gaps, with manual electrode displacement [13]. The following parameters were established in compliance with the manufacturer's guidelines and previous experience of the authors: voltage U = 230 V; capacitor volume C = 150 μF; current intensity I = 0.7 A; deposition time τ = 2min/cm^2. The coatings were subjected to laser treatment at the Centre for Laser Technology of Metals. A BLS 720 Nd:YAG laser capable of generating 150 W maximum average power, operating in the pulse mode and manufactured by Baasel Lasertechnik was used. The laser treatment was performed in the ambient air atmosphere. The tests used a focusing head. The TEM_{00} beam defined the radiation energy distribution. The parameters used were as follows: spot diameter d = 0.7mm; laser power P = 50W, 60W, 70W; specimen movement rate v = 250mm/min; nozzle-workpiece distance Δl = 1mm; pulse duration t_i = 0.4ms; pulse repetition frequency f = 50Hz; beam shift jump S = 0.4mm.

Methodology

Images of the microstructure of a selected sample were obtained by means of a JSM-5400 scanning electron microscope, which is owned by the Kielce University of Technology. It is a kind of electron microscope that allows observation of the topography of the material being examined. It is used for observation and characterization of organic and inorganic materials on a scale from nanometric to micrometric. The electron beam is the primary beam in this research method.

The geometric structure of the surface has a significant impact on many processes occurring in the surface layer. Measurements of the geometric structure of the surface were carried out in the Laboratory of Computer Measurements of Geometric Size of the Kielce University of Technology. The tests were carried out using a Talysurf CCI optical profilometer using the Taylor Hobson patented coherence correlation algorithm, enabling measurements with an axis resolution of less than 0.8 nm. The measurement result is recorded in a matrix 1024x1024 measuring points which, with the x10 lens used, gives a measured area of 1.65 mm x 1.65 mm and a horizontal resolution of 1.65 μm x 1.65 μm.

Results

From the microstructure images presented in Fig.1, it can be concluded that the laser modification significantly homogenized the state of the coating. As a result of applying a coating, the pores and microcracks are eliminated and the distribution of elements (Fig.2) is more uniform at a positive effect on the modified surface. The spectral analysis for selected points is given in Fig.3.

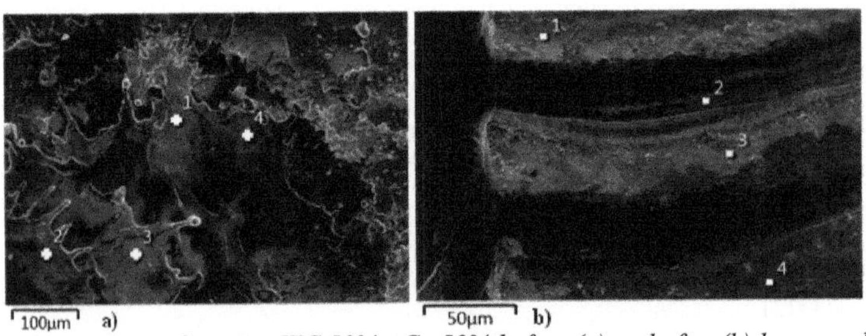

Fig. 1. Microstructure of coating WC 50% - Cu 50% before (a) and after (b) laser modification with the power of P=60W. Magnification x1000

Three-dimensional surfaces and their analysis using the TalyMap Platinium software allowed for a thorough understanding of the geometrical structure of the tested surfaces.

Materials Research Proceedings **5** (2018) 237-242 doi: http://dx.doi.org/10.21741/9781945291814-42

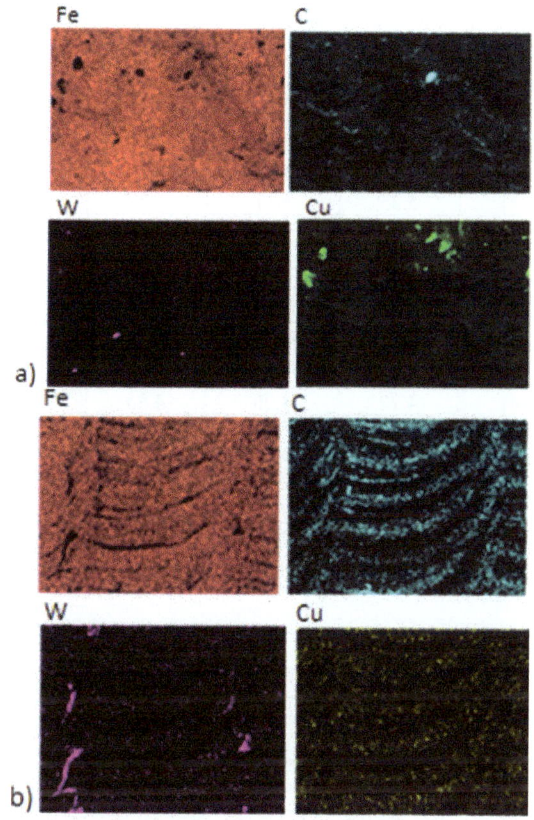

Fig. 2. Surface distribution of elements a) before laser modification b) after modification with the laser beam of P=60W

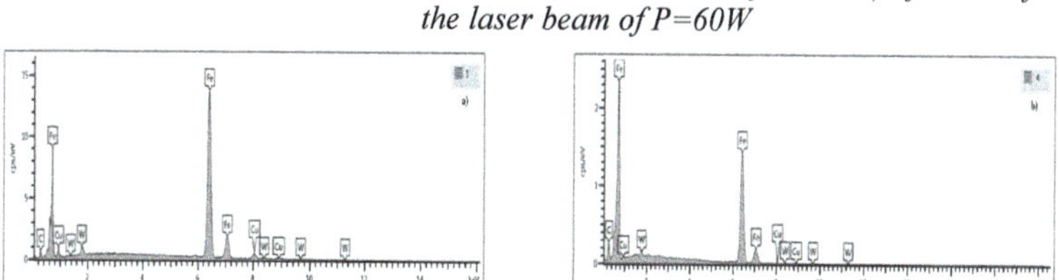

Fig. 3. Spectrum of EDS coating WC50%-Cu50% :a) before laser treatment in point 1, b) after laser treatment P=60W in point 4

The analysis of the results contained in Table 1 and the image of the surface topography of the samples indicates that, depending on the coating applied, different laser power used to modify the substrate improved the parameters of the geometrical structure.

Table 1. Parameters of the surface geometry of coatings before and after laser treatment

Parameters of the surface geometry	Laser power			
	ESD	P=50 [W]	P=60 [W]	P=70 [W]
WC50%-Cu50%				
Sa [μm]	3.98	4.08	3.80	5.36
Sq [μm]	5.25	5.11	4.69	6.75
Ssk	-0.22	-0.08	-0.28	0.01
Sku	3.87	2.98	2.70	2.93
Sp [μm]	18.20	22.37	12.89	24.97
Sv [μm]	21.52	19.44	18.96	25.05
Sz [μm]	39.73	41.08	31.86	59.03
WC25%-Cu75%				
Sa [μm]	4.22	4.06	4.35	4.53
Sq [μm]	5.35	5.12	5.44	5.66
Ssk	0.44	0.07	-0.28	-0.52
Sku	3.36	2.98	3.09	3.09
Sp [μm]	28.12	16.51	17.49	17.74
Sv [μm]	17.24	25.64	23.02	24.26
Sz [μm]	45.35	42.16	40.51	41.99

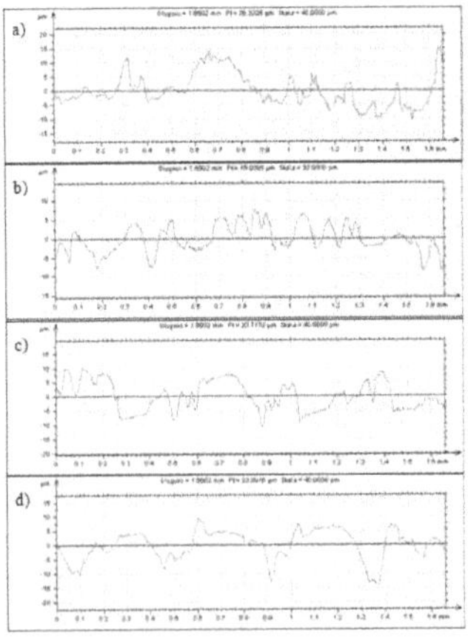

Fig. 4. Results of microscopy measurements of WC25%-Cu75%: a) without laser treatment b) laser-modified P=50W; c) laser modified P=60W; d) laser-modified P=70W

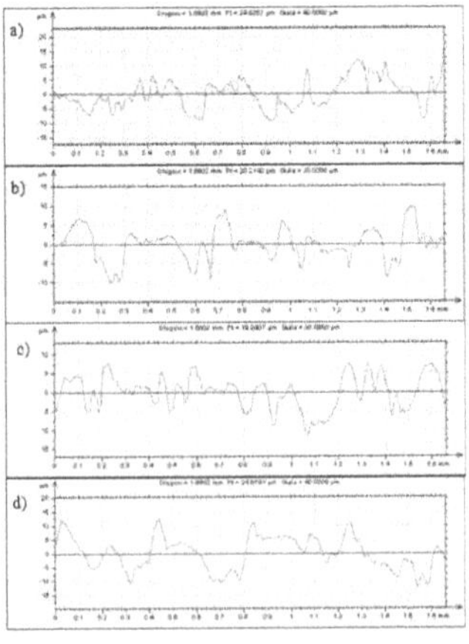

Fig. 5. Results of microscopy measurements of WC50%-Cu50%:a) without laser treatment b) laser-modified P=50W; c) laser-modified P=60W; d) laser-modified P=70W

For the WC50%-Cu50% coating, the best parameters are obtained when modifying the laser with P = 60W. For each analyzed parameter of the geometrical structure, the value has been improved. However, for the WC25%-Cu75% coating, the best parameters are obtained when modifying the laser with P =50W. Unfortunately, in this case one parameter, i.e. Sv, the maximum depth of the valley, has not been improved. In the discussed cases, laser treatment caused a decrease in the main values of geometrical parameters of the surface of the electroless coatings in relation to the SGP parameters of the coatings before the laser treatment. The same relationship can be seen in the description of the roughness profile amplitude parameters (Figs. 4 and 5). For the WC50%-Cu50% coating at the P-60W laser power, the Ra parameter, the arithmetic mean profile deviation, equalled 3.85 μm, while after laser treatment with the P = 60W power it decreased to 3.01 μm Rz. The maximum height of the profile was also reduced by 3.14 μm in relation to the value before modification of the laser beam. Only the Rv parameter had a worse value, the maximum valley depth of the roughness profile from 8.02 μm to 8.8 μm. The WC25%-Cu75% coating, when modified with the P-50W laser, also improved its performance. The mean of the arithmetic deviation changed its value from the initial Ra = 3.57 μm to Ra = 2.78 μm.

Summary

On the basis of the conducted tests, it can be stated that the surface of carbon steel can be modified by the ESD method using WC-Cu electrodes with different percentage of both components. The concentrated laser beam stream can be used to effectively modify the state of the ESD coating layer, WC-Cu and to improve their performance. The use of laser treatment for the modification of WC-Cu coatings applied with electro-discharge coating improved in some cases the geometrical structure of these coatings. Possible further research, with deeper investigation on the impact of particular factors, should include statistical multivariate analyzes [14, 15], including RSM (the response surface methodology) [16] and a factorial modeling [17], CA (the cluster analysis) [18, 19] and modified adjustment calculus [20]. It may also be interesting to compare subjective (by human) and objective (by software) – by means of image analysis methods [21-24] – visual assessments of observed surfaces.

References

[1] N. Radek, E. Wajs, M. Luchka, The WC-Co electrospark alloying coatings modified by laser treatment, Powder. Metall. Met. C. 47 (2008)197-201. https://doi.org/10.1007/s11106-008-9005-7

[2] N. Radek, J. Konstanty, Cermet ESD coatings modified by laser treatment, Arch. Metall. Mater. 57 (2012) 665-670. https://doi.org/10.2478/v10172-012-0071-y

[3] J. Pietraszek, N. Radek, K. Bartkowiak, Advanced statistical refinement of surface layer's discretization in the case of electro-spark deposited carbide-ceramic coatings modified by a laser beam, Solid State Phenom. 197 (2013) 198-202. https://doi.org/10.4028/www.scientific.net/SSP.197.198

[4] N. Radek, J. Pietraszek, B. Antoszewski, The average friction coefficient of laser texturedsurfaces of silicon carbide identified by RSM methodology, Adv. Mater. Res.-Switz. 874 (2014) 29-34. https://doi.org/10.4028/www.scientific.net/AMR.874.29

[5] A.D. Thamer, M.H. Hafiz, B.S. Mahdi, Mechanism of building-up deposited layer during elektro-spark deposition, J. Surf. Eng. Mater. Adv. Technol. 2 (2012) 258-263. https://doi.org/10.1016/j.ijmachtools.2011.08.016

[6] M.P. Jahan, M. Rahman, Y.S. Wong, A review on the convetional and micro-electrodischarge machining of tungsten carbide, Int. J. Mach. Tools Manuf. 51(2011) 837-858. https://doi.org/10.1177/0954407011411388

[7] N. Radek, A. Sladek, J. Brončck, I. Bilska, A. Szczotok, Electrospark alloying of carbon steel with WC-Co-Al_2O_3: deposition technique and coating properties, Adv. Mater. Re.-Switz. 874 (2014) 101-106. https://doi.org/10.1016/j.proeng.2017.02.226

[8] M. Kekez, L. Radziszewski, Modelling of pressure in the injection pipe of a diesel engine by computational intelligence. Proc. Inst. Mech. Eng. Part D-J. Automob. Eng. 225, 12 (2011) 1660-1670.

[9] A. Bąkowski, L. Radziszewski, Z. Skrobacki, Assessment of uncertainty in urban traffic noise measurement, Procedia Engineering 177 (2017) 281-288.

[10] A. Bąkowski, L. Radziszewski, M. Žmindak, Analysis of the coefficient of variation for injection pressure in a compression ignition engine, Procedia Engineering 177 (2017) 297-302. https://doi.org/10.1016/j.proeng.2017.02.228

[11] R. Ulewicz, F. Novy, J. Selejdak, Fatigue strength of ductile iron in ultra-high cycle regime, Adv. Mater. Res.-Switz. 874 (2014) 43-48. https://doi.org/10.4028/www.scientific.net/AMR.874.43

[12] R. Ulewicz, Hardening of steel X155CrVMo12-1 surface layer, J. Balk. Tribol. Assoc. 21 (2015) 166-172.

[13] T. Chang-bin, L. Dao-xin, W. Zhan, G. Yang, Electro-spark alloying using graphite electrode on titanium alloy surface for biomedical applications, Appl. Surf. Sci. 257 (2011) 6364-6371. https://doi.org/10.1016/j.apsusc.2011.01.120

[14] A.J. Izenman, Modern Multivariate Statistical Techniques. Regression, Classification, and Manifold Learning. Springer, New York, 2008.

[15] E. Skrzypczak-Pietraszek, J. Pietraszek, Chemical profile and seasonal variation of phenolic acid content in bastard balm (Melittis melissophyllum L., Lamiaceae), J. Pharmaceut. Biomed. 66 (2012) 154-161. https://doi.org/10.1016/j.jpba.2012.03.037

[16] J. Pietraszek, Response surface methodology at irregular grids based on Voronoi scheme with neural network approximator, Adv. Soft Comp. (2003) 250-255.

[17] J. Pietraszek, A. Gadek-Moszczak, T. Torunski, Modeling of Errors Counting System for PCB Soldered in the Wave Soldering Technology, Adv. Mater. Res.-Switz. 874 (2014) 139-143. https://doi.org/10.4028/www.scientific.net/AMR.874.139

[18] B.S. Everitt, S. Landau, M. Leese, D. Stahl, Cluster Analysis, Wiley, Hoboken, 2012.

[19] E. Skrzypczak-Pietraszek, J. Pietraszek, Seasonal Changes of Flavonoid Content in Melittis melissophyllum L. (Lamiaceae), Chem. Biodivers. 11 (2014) (4) 562-570.

[20] T. Styrylska, J. Pietraszek, Numerical Modeling of Non-Steady-State Temperature-Fields with Supplementary Data, Z. Angew. Math. Mech. 72 (1992) (6) T537-T539.

[21] A. Gadek-Moszczak, N. Radek, S. Wronski, J. Tarasiuk, Application the 3D Image Analysis Techniques for Assessment the Quality of Material Surface Layer Before and After Laser Treatment, Adv Mater Res-Switz 874 (2014) 133-138.
https://doi.org/10.4028/www.scientific.net/AMR.874.133

[22] J. Korzekwa, A. Gadek-Moszczak, M. Bara, The Influence of Sample Preparation on SEM Measurements of Anodic Oxide Layers, Prakt. Metallogr.-Pr. M. 53 (2016) (1) 36-49.

[23] A. Gadek-Moszczak, History of Stereology, Image Anal. Stereol. 36 (2017) (3) 151-152.

[24] A. Gadek-Moszczak, P. Matusiewicz, Polish Stereology – a Historical Review, Image Anal. Stereol. 36 (2017) (3) 207-221.

Mechanical Properties of Anti-Graffiti Coating Systems used in the Railway Industry

RADEK Norbert [1, a*], PASIECZYŃSKI Łukasz [2,b], MAKRENEK Medard [3,c] and DUDEK Agata [4,d]

[1]Kielce University of Technology, Faculty of Mechatronics and Mechanical Engineering, Al. 1000-lecia P.P. 7, 25-314 Kielce, Poland

[2]Firma Handlowa Barwa, ul. Warkocz 3-5, 25-253 Kielce, Poland

[3]Kielce University of Technology, Faculty of Management and Computer Modelling, Al. 1000-lecia P.P. 7, 25-314 Kielce, Poland

[4]Czestochowa University of Technology, Institute of Materials Engineering, Al. Armii Krajowej 19, 42-200 Częstochowa, Poland

[a]norrad@tu.kielce.pl, [b]lpasieczynski@barwa.kielce.pl, [c]fizmm@tu.kielce.pl, [d]dudek@wip.pcz.pl

Keywords: Anti-Graffiti Coatings, Mechanical Properties, Rail

Abstract. The paper presents comparative tests results of selected properties of anti-graffiti paint systems for rolling stock industry. The system consists of a high solid corrosion protection primer, putty, filler, basecoat and anti-graffiti clearcoat. The research was based on an anticorrosion coatings system with a transparent layer with reduced adhesion of each subsequent types of contamination. The results of the research focus on the analysis of microstructure and measurements of nanohardness and adhesion.

Introduction

By applying new engineering materials or protective coatings, it is possible to improve the functional properties of machine parts so that they are resistant to corrosion, abrasion and erosion, and possess high fatigue strength [1-10].

The new materials, for instance alloy steels, are usually costly, which is undesirable, because the higher the cost of the material, the higher the price of a finished product. However, if an element is to be subjected to high loads, then strength rather than cost is a primary factor.

Applying protective coatings to machine parts is economically justifiable if the wear is local or if the coating material is expected to display properties different from those of the substrate.

Paint systems for rolling stock must fulfill mechanical and qualitative properties to protective and decorative properties maintenance longer on the vehicle. These requirements include adhesion, resistance to weather conditions (humidity, UV, corrosion) as well as hardness and specialized properties such as anti-graffiti. In addition to the above-mentioned requirements, as well as ease of application and operation, coating systems intended for rolling stock must also have adequate fire performance [11].

The paper presents comparative tests results of selected mechanical properties of anti-graffiti coating systems for rolling stock industry.

Materials

Coatings were applied with a SATA spray gun on the S355 carbon steel; before the application the surface of the steel was polished with 80-grit sandpaper. The coating systems consisted of the following layers: anti-corrosion epoxy primer, repair filler, primer filler, basecoat and anti-graffiti XPC60011, XPC60012, XPC60036, BO100-AGR clearcoats. Each layer was applied and dried in accordance with the requirements of technological cards. The prepared samples were conditioned at 23 °C and 50 % humidity for minimum 7 days in order to perform tests on dry coating.

Microstructure analysis

A microstructure analysis was conducted for anti-graffiti coating systems using the *JEOL* JSM-*7100F* scanning electron microscope with field emission and the Hirox KH-8700 light microscope.

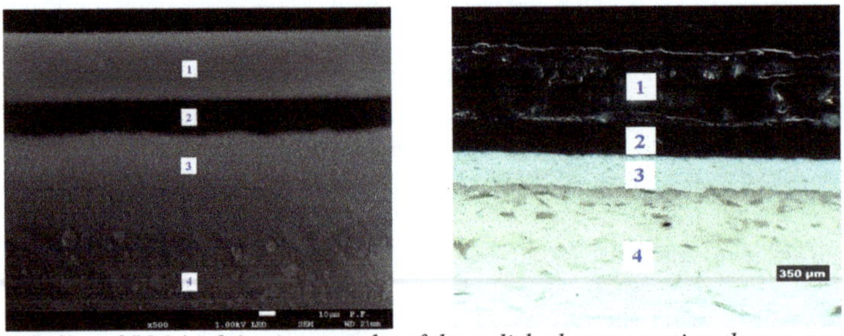

Fig. 1. SEM (left) and LM (right) micrographs of the polished cross-section throug a anti-graffiti XPC 60012 coating system on S355 carbon steel substrate: 1- anti-graffiti layer, 2 - base layer, 3 - undercoat layer, 4 - putty

The thickness of the obtained coating systems was from approx. 2300 to approx. 2500 µm. There are clear boundaries between the individual layers (Fig. 1). Fig. 1 shows a clear boundary between the varnish layers and the putty. Also, the varnish layers are free of pores and microcracks.

Adhesion tests

A scratch test was conducted to test the adhesion of the anti-graffiti coating systems without a putty. The adhesion tests were conducted using a Revetest Scratch Xpress instrument (CSM Instruments, Switzerland). The measurements were performed at a load increase rate of 11 N/min, a table feed rate of 6, 73 mm/min and a scratch length of 30 mm.

A special Rockwell diamond cone indenter, with a corner radius of 200 µm, was used to scratch the samples at a gradually increasing normal load. The information about the cracking or peeling of the layers was obtained based on the measurements of the material resistance (tangential force) and the registration of acoustic emission signals. The lowest normal force causing a loss of adhesion of the coating to the substrate is called a critical force and is assumed to be the measure of adhesion. The results are presented in Table 1.

Table 1. Results of scratch adhesion tests

System	Critical stylus load [N]			Mean value† [N]
	Measurement			
	1	2	3	
XPC 60011	45.10	47.05	46.90	46.35 ± 1.09
XPC 60012	47.45	46.95	42.55	45.65 ± 2.70
XPC 60036	45.15	46.05	43.85	45.02 ± 1.11
BO100-AGR	41.95	47.35	47.15	45.48 ± 3.06

†scatter intervals estimated at 90% confidence level

It becomes evident from the obtained data that anti-graffiti coating systems have good adhesion with the substrate material. The mean value of the critical force calculated from three measurements performed on the individual anti-graffiti coating systems was from 45.02 to 46.35 N. In addition, a low scatter of critical stylus loads indicates that the varnish layers are homogeneous and very tight.

Nanoindentation of anti-graffiti coating systems

Production of coatings with the required properties is a major research challenge. In the manufacturing process it is necessary to indicate controllable parameters and properties expected from the resulting coating. A multitude of controlled parameters and their values make large number of combinations effects costs high. In our case, it is the hardness and modulus of elasticity of paint coatings.

The hardness and elastic modulus were investigated by the nanoitender technique. This measurement technology was possible due to the development of instruments that continuously measure force and displacement. In the measurement, the load force equalled 3 mN and unload rating was 40 mN/min. Due to the type of material tested used creep of about 3 seconds. The hardness is determined by the penetration depth of the indenter and the modulus of elasticity determined by the slope of the unload curve. Hardness measurements were carried out in several selected places on the surface of the paint coating. Fig. 1 shows measurement curves that reflect the displacement of the bar as a function of force for an anti-graffiti XPC 60012 coating system.

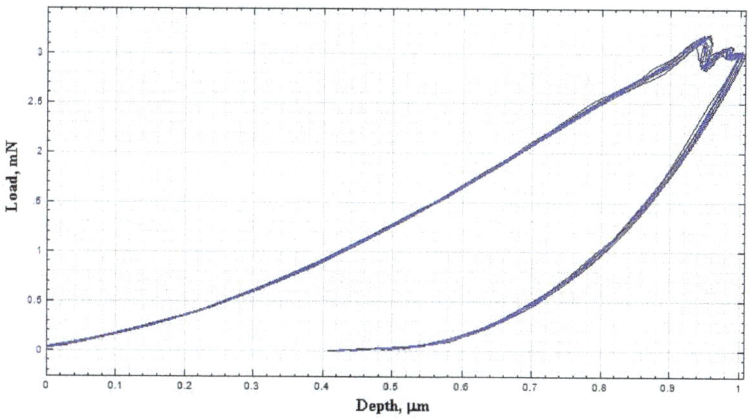

Fig. 1. Graphical interpretation of the indentation of the indenter in the material depending on the load of 3 mN and unload rate of 40 mN/min

Practically all measuring curves overlap. This proves the accuracy and repeatability of measurements. On the basis of 10 measurements, the values of average hardness and elasticity modulus were determined and placed in Table 2.

As shown in Table 2, four coating systems were tested. Table 2 contains the average values of hardness and elastic modulus together with the standard error.

Table 2. Value of hardness and modulus of elasticity with errors

System	Hardness [GPa]	Elastic modulus [GPa]
XPC 60012	0.229 ± 0.005	2.700 ± 0.050
BO100-AGR	0.259 ± 0.004	2.900 ± 0.016
XPC 60011	0.265 ± 0.004	2.843 ± 0.017
XPC 60036	0.258 ± 0.003	2.842 ± 0.022

Summary

In the present study, it was clearly presented that the anti-graffiti coating systems are characterized by good mechanical properties. They have a homogeneous structure and are free of defects.

Further research will be targeted at the determination of corrosion and erosion resistance. With deeper investigation on the impact of particular factors, it should include statistical multivariate analyzes [12, 13], including both, RSM (the response surface methodology) [14] as well as a factorial modeling [15] and also methods of the image analysis [16-20] to detailed investigate scratched surfaces.

References

[1] R. Ulewicz, F. Novy, J. Selejdak, Fatigue strength of ductile iron in ultra-high cycle regime, Advanced Materials Research 874 (2014) 43-48. https://doi.org/10.4028/www.scientific.net/AMR.874.43

[2] R. Ulewicz, Hardening of steel X155CrVMo12-1 surface layer, Journal of the Balkan Tribological Association 21 (2015) 166-172.

[3] M. Madej, The effect of TiN and CrN interlayers on the tribological behavior of DLC coatings, Wear 317 (2014) 179-187. https://doi.org/10.1016/j.wear.2014.05.008

[4] M. Madej, D. Ozimina, Electroless Ni-P-Al_2O_3 composite coatings, Kovove Materialy-Metallic Materials 44 (2006) 291-296.

[5] N. Radek, E. Wajs, M. Luchka, The WC-Co electrospark alloying coatings modified by laser treatment, Powder Metallurgy and Metal Ceramics 47 (2008) 197-201. https://doi.org/10.1007/s11106-008-9005-7

[6] N. Radek, J. Konstanty, Cermet ESD coatings modified by laser treatment, Archives of Metallurgy and Materials 57 (2012) 665-670. https://doi.org/10.2478/v10172-012-0071-y

[7] J. Pietraszek, N. Radek, K. Bartkowiak, Advanced statistical refinement of surface layer's discretization in the case of electro-spark deposited carbide-ceramic coatings modified by a laser beam, Solid State Phenomena 197 (2013) 198-202. https://doi.org/10.4028/www.scientific.net/SSP.197.198

[8] N. Radek, J. Pietraszek, B. Antoszewski, The average friction coefficient of laser textured surfaces of silicon carbide identified by RSM methodology, Advanced Materials Research 874 (2014) 29-34. https://doi.org/10.4028/www.scientific.net/AMR.874.29

[9] M. Scendo, N. Radek, J. Trela, Influence of laser treatment on the corrosive resistance of WC-Cu electrospark coatings, International Journal of Electrochemical Science 8 (2013) 9264-9277.

[10] A. Góral, W. Żórawski, M. Makrenek, Sz. Kowalski, Microstructure and properties of cold sprayed composite coatings, Journal of Achievements in Materials and Manufacturing Engineering 81 (2017) 49-55. https://doi.org/10.5604/01.3001.0010.2037

[11] J. Radziszewska-Wolińska, Fire properties of anticorrosion coatings to rolling stock, Technical Transactions 3-M (2016) 79-86.

[12] A.J. Izenman, Modern Multivariate Statistical Techniques. Regression, Classification, and Manifold Learning. Springer, New York, 2008.

[13] E. Skrzypczak-Pietraszek, J. Slota, J. Pietraszek, The influence of L-phenylalanine, methyl jasmonate and sucrose concentration on the accumulation of phenolic acids in Exacum affine Balf. f. ex Regel shoot culture, Acta Biochim. Pol. 61 (2014) (1) 47-53.

[14] J. Pietraszek, Response surface methodology at irregular grids based on Voronoi scheme with neural network approximator, Adv. Soft Comp. (2003) 250-255.

[15] J. Pietraszek, M. Kolomycki, A. Szczotok, R. Dwornicka, The Fuzzy Approach to Assessment of ANOVA Results, Lect. Notes Artif. Int. 9875 (2016) 260-268. https://doi.org/10.1007/978-3-319-45243-2_24

[16] A. Gadek-Moszczak, S. Zmudka, Description of 3D microstructure of the composites with polypropylene (PP) matrix and Tuf particles fillers, Solid State Phenom. 197 (2013) 186-191. https://doi.org/10.4028/www.scientific.net/SSP.197.186

[17] A. Gadek-Moszczak, J. Pietraszek, B. Jasiewicz, S. Sikorska, L. Wojnar, The Bootstrap Approach to the Comparison of Two Methods Applied to the Evaluation of the Growth Index in the Analysis of the Digital X-ray Image of a Bone Regenerate, Stud. Comput. Intell. 572 (2015) 127-136. https://doi.org/10.1007/978-3-319-10774-5_12

[18] A. Gadek-Moszczak, L. Wojnar, Objective, Quantitative and Automatic X-Ray Image. Analysis of the Bone Regenerate in the Ilizarov Method, ECS10: The10th European Congress of Stereology and Image Analysis (2009) 453-458.

[19] A. Szczotok, S. Roskosz, New possibilities of light microscopy research resulting from digital recording of images, Mater. Sci.-Poland 23 (2005) (2) 559-565.

[20] A. Szczotok, Metallographic Study of the Casting Made from Cmsx-6 Sc Nickel-Based Superalloy, Arch. Metall. Mater. 62 (2017) (2) 581-586.

Keyword Index

Author Index

About the Editors

Agnieszka SZCZOTOK, M.Eng., Ph.D.

is an assistant professor in the Institute of Materials Engineering, Faculty of Materials Engineering and Metallurgy, Silesian University of Technology, Katowice, Poland. She has scientific interests in: modern methods of materials characterization, an image analysis application, a quantitative description of the materials structures and statistics in materials science analysis.

Jacek PIETRASZEK, M.Eng., Ph.D., D.Sc.

is an associate professor in the Department of Software Engineering and Applied Statistics, Cracow University of Technology, Krakow, Poland. He has scientific interests in the experimental design methodology in a mechanical engineering, applied statistics and data mining with artificial intelligence support.

Norbert RADEK M.Eng., Ph.D., D.Sc.

is an associate professor in the Centre for Laser Technologies of Metals, Kielce University of Technology, Kielce, Poland. He has scientific interests in the materials science with a focus on a special coatings of a surface layer modified with a laser beam and their technological capabilities in a machine building.

www.ingramcontent.com/pod-product-compliance
Ingram Content Group UK Ltd.
Pitfield, Milton Keynes, MK11 3LW, UK
UKHW020657041026
15590UKWH00001B/328

9 781945 291807